At a Glance

General Anatomy

Trunk

Upper Limb

Lower Limb

Head and Neck

Topography of Peripheral Nerves and Vessels

Color Atlas
of Human Anatomy

in 3 volumes

Volume 2:   General Anatomy
            by Helga Fritsch and Wolfgang Kuehnel

Volume 3:   Nervous System and Sensory Organs
            by Werner Kahle and Michael Frotscher

Volume 1

Locomotor System

7th Edition

Werner Platzer, MD

Professor Emeritus
Former Chairman, Institute for Anatomy
University of Innsbruck
Innsbruck, Austria

Doctor of Science
Wake Forest University
Winston-Salem
North Carolina, USA

215 color plates

Thieme
Stuttgart · New York · Delhi · Rio

Library of Congress Cataloging-in-Publication Data is available from the publisher

This book is an authorized and revised translation of the 11ᵗʰ German edition published and copyrighted 2013 by Georg Thieme Verlag, Stuttgart. Title of the German edition: Taschenatlas Anatomie, Band 1: Bewegungsapparat

Translator: Terry Telger, Fort Worth, Texas, USA (parts from previous English edition)

Illustrator: Professor Gerhard Spitzer, Frankfurt, Germany, in cooperation with Stephanie Gay, Lothar Schnellbächer and Stefan Spitzer

11th German edition 2013	2nd Indonesian edition 2000
6th English edition 2008	4th Italian edition 2007
	6th Japanese edition 2011
1st Bulgarien edition 2005	1st Polish edition 1996
1st Chinese edition 2000	2nd Portuguese edition 2007
1st Czech edition 1996	1st Serbo-Croatian edition 1991
6th Dutch edition	3rd Croatian edition 2011
4th French edition 2007	4th Spanish edition 2008
1st Hungarian edition 1996	1st Turkish edition 1987
2nd Greek edition 2009	

© 2015 Georg Thieme Verlag KG
Thieme Publishers Stuttgart
Rüdigerstrasse 14, 70469 Stuttgart,
Germany, +49 [0] 7 11 8 93 14 21
customerservice@thieme.de

Thieme Publishers New York
333 Seventh Avenue, New York, NY
10001, USA, 1-800-782-3488
customerservice@thieme.com

Thieme Publishers Delhi
A-12, second floor, Sector-2, NOIDA-201301,
Uttar Pradesh,
India, +91 120 45 56 00
customerservice@thieme.in

Thieme Publishers Rio
Thieme Publicações Ltda.
Argentina Building, 16ᵗʰ floor, Ala A,
228 Praia do Botafogo,
Rio de Janeiro 22250-040 Brazil,
+55 21 37 36-36 31

Cover design: Thieme Publishing Group
Typesetting by Druckhaus Götz GmbH, Ludwigsburg, Germany
Printed in Germany by CPI Books GmbH, Leck

ISBN 978-3-13-533307-6

Also available as e-book:
eISBN 978-3-13-149487-0

Important note: Medicine is an ever-changing science undergoing continual development. Research and clinical experience are continually expanding our knowledge, in particular our knowledge of proper treatment and drug therapy. Insofar as this book mentions any dosage or application, readers may rest assured that the authors, editors, and publishers have made every effort to ensure that such references are in accordance with **the state of knowledge at the time of production of the book.**

Nevertheless, this does not involve, imply, or express any guarantee or responsibility on the part of the publishers in respect to any dosage instructions and forms of applications stated in the book. **Every user is requested to examine carefully** the manufacturers' leaflets accompanying each drug and to check, if necessary in consultation with a physician or specialist, whether the dosage schedules mentioned therein or the contraindications stated by the manufacturers differ from the statements made in the present book. Such examination is particularly important with drugs that are either rarely used or have been newly released on the market. Every dosage schedule or every form of application used is entirely at the user's own risk and responsibility. The authors and publishers request every user to report to the publishers any discrepancies or inaccuracies noticed. If errors in this work are found after publication, errata will be posted at www.thieme.com on the product description page.

Some of the product names, patents, and registered designs referred to in this book are in fact registered trademarks or proprietary names even though specific reference to this fact is not always made in the text. Therefore, the appearance of a name without designation as proprietary is not to be construed as a representation by the publisher that it is in the public domain.

MIX
Papier aus verantwortungsvollen Quellen
FSC® C083411

Contents

Head and Neck 281

Topography of Peripheral Nerves and Vessels 333

Preface

No text book would be able to reach such a large audience if it did not offer exactly what students are looking for. It is with great joy that I thank those who have purchased this book during the past almost forty years. and especially those students who have been able to convey to the author just what it is they need in a book. The idea originated more than 50 years ago in Vienna where I was a lecturer. In many conversations before and after these lectures, and later in Innsbruck (from 1968), I came to realize what the students needed, and with the the help of my assistants we were able to develop the book.

As in previous editions, we have again expanded the number of Clinical Tips, continuing to acknowledge the fact that morphology is essential in patient care, and that understanding diseases would not be possible without knowledge of the fundamentals of morphology.

The "Latin Equivalents" pages have also been expanded in each chapter for interested readers. I have also included more eponyms in this edition, in recognition of the growing but regrettable trend toward the more frequent use of proper names in anatomical terms. The Index of Proper Names has been updated and expanded appropriately.

I am grateful to Thieme Medical Publishers and especially to Ms Angelika Findgott and the production team in creating this new edition.

This edition is dedicated to Professor Beatrix Volc-Platzer, who was and is a model daughter. She is an excellent doctor and we hope that future generations of doctors will look up to her and profit from her knowledge and her humaneness.

Werner Platzer, MD

General Anatomy

The Body

Parts of the Body (A, B)

The body is divided into the main part of the body (*trunk in the broad sense*) and the upper and lower limbs, or *extremities*. The trunk is divided into the head, the neck, and the torso (*trunk in the narrow sense*). The torso consists of the *thorax, abdomen,* and *pelvis*.

The upper extremity is joined to the trunk by the shoulder girdle and the lower extremity by the pelvic girdle. The shoulder girdle consists of the clavicles (**1**) and the scapulas (**2**), which lie on the trunk and move upon it. The pelvic girdle, which consists of the two hip (coxal) bones (**3**) and the sacrum (**4**), forms an integral part of the trunk.

General Terms (A–G)

Principal Axes

The *longitudinal (vertical) axis*, or long axis (**5**) of the body, is vertical when the body is in an upright posture.

The *transverse (horizontal) axis* (**6**) is perpendicular to the long axis and runs from left to right.

The *sagittal axis* (**7**) runs from the back to the front surface of the body in the direction of the arrow (Gr: *sagitta*) and is perpendicular to the other two axes.

Principal Planes

Median plane, the plane through the longitudinal axis and the sagittal axis; it is also called the *midsagittal plane* (**8**). It divides the body into two almost equal halves, or *antimeres* (hence is also called *plane of symmetry*). It includes the longitudinal and sagittal axes.

Sagittal or *paramedian plane* (**9**), any plane that is parallel to the midsagittal plane.

Frontal or *coronal plane* (**10**), any plane that contains the transverse and longitudinal axes and is parallel to the forehead and perpendicular to the sagittal planes.

Transverse planes (**11**) lie perpendicular to the sagittal and coronal planes. They are horizontal in the upright posture and contain the sagittal and transverse axes.

Directions in Space

cranial = toward the head (**12**)
superior = upward with the body erect (**12**)
caudal = toward the buttocks (**13**)
inferior = downward with the body erect (**13**)
medial = toward the middle, toward the median plane (**14**)
lateral = away from the middle, away from the median plane (**15**)
medius = in the midline (**16**)
median = in the median plane
deep (profundus) = toward the inside of the body (**17**)
peripheral, superficial = toward the body surface (**18**)
rostral = toward the rostrum (beak), toward the oral and nasal region
anterior = toward the front (**19**)
ventral = toward the abdomen (**19**)
posterior = toward the back (**20**)
dorsal = toward the back (**20**)
proximal = toward the trunk or point of attachment (**21**)
distal = away from the trunk or point of attachment (**22**)
ulnar = toward the ulna (**23**)
radial = toward the radius (**24**)
tibial = toward the tibia (**25**)
fibular = toward the fibula (**26**)
palmar (volar) = on or toward the palm of the hand (**27**)
plantar = on or toward the sole of the foot (**28**)

Directions of Movement

flexion = the act of bending
extension = the act of straightening
abduction = movement away from the median plane
adduction = movement toward the median plane
rotation = movement around an axis
circumduction = circular (circumferential) movement

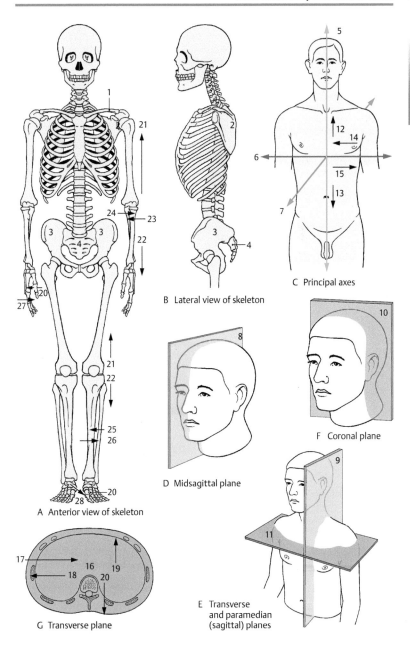

C Principal axes

B Lateral view of skeleton

F Coronal plane

D Midsagittal plane

A Anterior view of skeleton

E Transverse and paramedian (sagittal) planes

G Transverse plane

The Cell (A)

The smallest living entity is the *cell*. There are unicellular organisms, *protozoa*, and multicellular organisms, *metazoa*. Human cells range in size from 5 to 200 µm. They live for different lengths of time. Some cells survive for only a few days, for example granular leukocytes of the blood, and others survive the whole of the human life span, for example nerve cells.

Cells differ in shape depending on their function (e.g., muscle cells are elongated).

Each cell consists of the cell body, *cytoplasm* (**1**), and the nucleus, *karyoplasm* (**2**), which contains one or more *nucleoli* (**3**). The nucleus is separated from the cytoplasm by a double membrane, the *nuclear envelope* (**4**).

Cytoplasm

The cytoplasm is subdivided into **organelles**, **cytoskeleton**, and **cell inclusions**. These structures are contained in a fluid matrix, the **cytosol**.

The cell membrane, the *plasma membrane* or *plasmalemma* (**5**), appears as a trilamellar structure in electron micrographs. The cell surface is irregular and may exhibit fine processes, *microvilli*. The cell membrane is covered by a coating, the *glycocalyx*, which is approximately 20 nm thick. The glycocalyx is species-specific as well as cell-specific, thus facilitating cell–cell recognition.

Organelles

The *endoplasmic reticulum* (*ER*) (**6**) consists of a system of interconnected cisterns; it may be granular (rough ER) (**6**) or agranular (smooth ER). The rough ER has small granules, *ribosomes*, attached to the cytoplasmic side of its membrane. The ribosomes are approximately 15 to 25 nm in diameter and are made up of ribonucleic acid and protein molecules. The rough ER is in-

volved in protein synthesis, while the smooth ER fulfills various other functions (e.g., it plays a role in lipid metabolism in hepatocytes).

The *mitochondria* (**7**) are of special importance as they provide the cell with energy. They are long, flexible, rod-shaped organelles that move about in the cytoplasm. They vary in number and size depending on the type and functional state of the cell.

The *Golgi apparatus* (**8**) consists of several *dictyosomes*, or *Golgi stacks*. Each dictyosome consists of a stack of disc-shaped cisterns. The Golgi apparatus is responsible for formation and supplementation of the glycocalyx but is also involved in the synthesis and modification of carbohydrates and polypeptides produced in the ER.

Other organelles are the *lysosomes* (**9**) and *peroxisomes* (microbodies).

Cytoskeleton

The cytoskeleton consists of *microtubules* (including the *centrioles*, **10**, and *basal bodies*), *actin filaments* (microfilaments), and various cell-specific *intermediate filaments*. The two centrioles usually lie near the nucleus; together with the specialized cytoplasm surrounding them, the *centroplasm*, they form the *centrosome* (microtubule-organizing center). The cytoskeleton plays a major role in cell movement as well as intracellular movement (see p. 6).

Cell Inclusions

These include ribosomes, *lipids* (**11**), *glycogen* (**12**), *pigments* (**13**), *crystals*, and other insoluble components contained within a liquid matrix.

14 Vacuoles

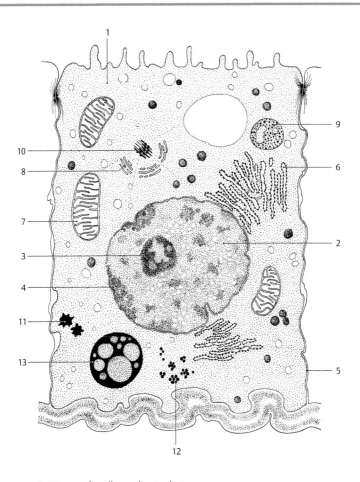

A Diagram of a cell according to electron-
   microscopic findings
   (from Faller, A.: Der Körper des Menschen,
   13th ed. Thieme, Stuttgart, 1999)

Cell Nucleus (A, B)

The **nucleus** (**A**), composed of karyoplasm, is essential for the life of the cell. Its size depends on the size of the cell. Normally cells possess one or more nuclei. The nucleus is usually visible in living cells because it is more refractive than the cytoplasm; it is separated from the cytoplasm by the delicate birefringent nuclear membrane (**1**). Upon fixation, a network-like structure, *chromatin* (**2**), becomes visible in the *interphase nucleus* (the resting nucleus between cell divisions). The chromatin carries the genetic material; it condenses in the *dividing nucleus* to form the *chromosomes*.

The micronucleus, or *nucleolus* (**3**), consists of proteins and is rich in ribonucleic acid (RNA). The number and size of the nucleoli vary a great deal among different cells. In the cells of females, each active nucleus contains a clump of chromatin, the *sex chromatin* (Barr body, **4**), which is attached to the nuclear membrane or the nucleolus. It can be used to determine the sex of a cell and thus of an individual. The sex chromatin is particularly easy to see in white blood cells (granulocytes) where it assumes the shape of a drumstick. In order to make the diagnosis of female sex, at least six drumsticks must be visible in 500 granulocytes.

Vital Cell Functions (C–H)

Every cell displays **metabolic activity**, which can be divided into *structural metabolism* and *functional metabolism*. Structural metabolism is the ability of a cell to assimilate ingested material to build up cellular structures, while functional metabolism is involved in cellular functions.

The uptake of particulate material is called *phagocytosis*, that of liquids *pinocytosis*. The release of substances by glandular cells is called *secretion*. The sum of oxidative processes within the cell is called *cell respiration*.

Among cellular **movements**, *cytoplasmic movement* is the most important one and includes movements of mitochondria, vesicles, and inclusions. More pronounced movements occur during each cell division. The cells themselves move by *ameboid movement* initiated by cyto-

plasmic processes called *pseudopodia*. Ameboid movement is especially pronounced in white blood cells (such as granulocytes and monocytes). Certain cells move by means of *cilia*, or *kinocilia*, which arise from basal bodies (*kinetosomes*). When joined together, ciliated cells form a **ciliated epithelium** and create *ciliary movement*. A cell with only one prominent cilium (*flagellum*) is called a *flagellated cell.*

Reproduction of cells takes place by cell division. We distinguish between *mitosis*, *meiosis*, and *amitosis*. Each cell division requires division of the nucleus. The interphase nucleus changes into the dividing nucleus, and the chromosomes become visible and perform characteristic movements (*karyokinesis*) toward the two poles of the *mitotic spindle.*

The process of **mitosis** is subdivided into different phases, called the *prophase* (**C**), *prometaphase* (**D**), *metaphase* (**E**), *anaphase* (**F, G**), and *telophase* (**H**). The nuclei of the two daughter cells are subsequently reorganized into interphase nuclei (*reconstruction phase*).

During **meiosis** (*reductional division*) the number of chromosomes per cell is reduced by half from the diploid to the haploid complement. The reduction takes place in both male and female germ cells during the first (or second) meiotic division and is required in preparation for fertilization.

During **amitosis** (*direct nuclear division*) the nucleus is divided by simple cleavage without chromosomal condensation and without the formation of a mitotic spindle. The distribution of chromosomes is therefore random. The nuclear division may or may not be followed by division of the cell.

For more details, see Histologie, Zytologie und Mikroanatomie des Menschen by Leonhardt, H., 8th ed. Thieme, Stuttgart, 1990; Taschenatlas der Zytologie, Histologie und mikroskopischen Anatomie by Kühnel, W., 11th ed. Thieme, Stuttgart, 2002, and 12th ed., Thieme, Stuttgart, 2008.

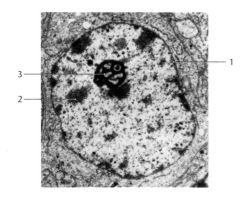

3 —
2 —
— 1

A Cell nucleus, x 12,000; electron micro-
graph

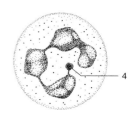

— 4

B White blood cells with sex chromatin attached to the segmented
nucleus, x 1,000
(panels A and B taken from Leonhardt, H.: Human Histology
and Cytology, 8 th ed. Thieme,
Stuttgart, 1990)

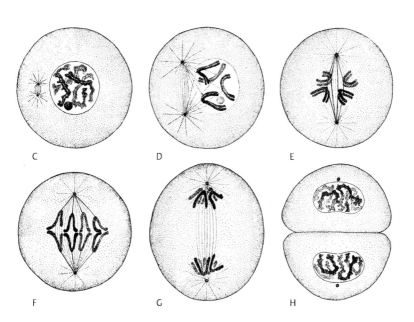

C

D

E

F

G

H

C–H Diagram of mitosis
(from Leonhardt, H.: Human Histology,
Cytology, and Microanatomy,
8 th ed. Thieme, Stuttgart, 1990)

Tissues

Tissues are aggregations of similarly differentiated cells and their derivatives. Multiple tissues may be associated to form an **organ**. The manner in which different cells are associated determines the different types of tissues. A more common system of classifying tissues is based not on the manner of association of cells but on their histologic structure and physiologic functions. **Epithelial**, **supportive**, and **muscular tissues** are described in this volume. Nervous tissue is discussed in Volume 3.

Epithelial Tissues (A–G)

Epithelial tissues are associations of closely adjoining cells. They can be classified according to **function**, as well as the **organization** and **shape** of their epithelial cells.

On the basis of their **functions**, superficial, glandular, and sensory epithelia can be distinguished. **Superficial epithelium** is, first of all, *a protective epithelium* that forms a covering for the external and internal body surfaces, prevents bacteria from entering the body, and keeps the body from drying up. Moreover, epithelia such as the *secretory and absorptive types* bring about the exchange of materials; that is, they can, on the one hand, take up substances (absorption) and, on the other hand, eliminate various substances (secretion). Epithelial tissue is also responsive to stimuli. This reception of stimuli takes place via the superficial epithelium through the induction of various specialized epithelial cells.

Glandular epithelium is a collective term for all epithelial cells that form a secretion and release it to an external or internal surface by an excretory duct (**exocrine glands**) or release it directly into the vascular system as a hormone (**endocrine glands**).

Exocrine glands can be classified as *endoepithelial* or *exoepithelial* depending upon their relationship to the superficial epithelium. Likewise, these glands can be divided into *eccrine*, *apocrine* and *holocrine* glands on the basis of the amount and manner of their secretions.

Eccrine cells are always ready to secrete and occur within the respiratory, digestive, and genital tracts (see Vol. 2). Apocrine glands are represented by the mammary and sweat glands; holocrine glands are represented by the sebaceous glands.

The **sensory epithelia** are specialized epithelia within the sensory organs and are discussed under that heading.

All epithelial cells rest upon on a basement membrane (*basal membrane*) which represents the boundary layer to the underlying connective tissue.

On the basis of their **organization**, epithelia can be divided into **simple** (single-layered, **A, B, C**), **stratified** (multilayered, **D**), or **pseudostratified** (**F**) epithelia. In the stratified epithelium only the deepest layer of cells makes contact with the basement membrane, whereas in the pseudostratified epithelium all cells contact the basement membrane, but not all the cells reach the surface.

Epithelial cells can be classified by their **shape** as **squamous** (**A**), **cuboidal** (**B**), or **columnar** (**C**).

Squamous epithelium, a markedly protective epithelium, may be *nonkeratinized* or *keratinized*. The epithelium of the skin is keratinized squamous epithelium, whereas nonkeratinized squamous epithelium (**E**) is found in parts of the inner surfaces of the body that are particularly vulnerable to mechanical stresses, such as the oral cavity. Simple nonkeratinized squamous epithelium consists of attenuated, pavement-like cells that include serous membranes (**mesothelium**) and the epithelial lining of blood and lymphatic vessels (**endothelium**). Columnar and cuboidal cells that have processes, or cilia, are classified as **ciliated epithelium** (**F**), which lines the respiratory tract, for example.

Cuboidal and columnar epithelia possess secretory and absorptive properties. They are found, for example, in the renal tubules (cuboidal) and in the intestinal tract (columnar). **Transitional epithelium** (**G**) is a special form of epithelium. Its cells can adapt themselves to different conditions of tension (distension and contraction) and make up the epithelium that lines the excretory portion of the urinary tract.

General Anatomy

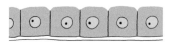

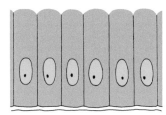

A Simple squamous epithelium
(pavement epithelium)

B Simple cuboidal epithelium

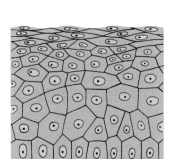

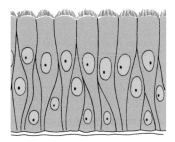

C Simple columnar epithelium

D Stratified columnar epithelium

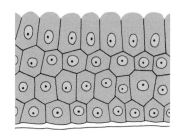

E Squamous stratified epithelium
(nonkeratinized)

F Pseudostratified ciliated epithelium

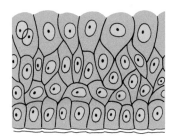

G Transitional epithelium

Connective Tissue and Supporting Tissues

These tissues consist of complex aggregations of cells, including **fixed** and **free cells**, and **intercellular substance**. The fixed cells are named according to the type of tissue, for example, connective tissue cells, cartilage cells, bone cells, etc. The intercellular substance in mature supporting tissue consists of *ground substance* and *differentiated fibers*.

The principal types are

Connective tissue: embryonic, reticular, interstitial, and rigid connective tissue and fatty (adipose) tissue

Cartilage tissue: hyaline cartilage, elastic cartilage, and fibrocartilage

Bone

Connective Tissue (A, B)

In addition to fixed and free cells, the intercellular substance contains reticular, collagenous, and elastic fibers, and ground substance, (proteoglycans and glycoproteins).

Fixed cells: **fibrocytes** (highly branched cells; their precursors, the fibroblasts, are able to produce intercellular substance and fibers), **mesenchymal cells**, **reticular cells**, **pigment cells**, and **fat cells**.

Free cells: **histiocytes** (polymorphic cells), **mast cells** (capable of ameboid movement) and, less commonly, **lymphocytes**, **plasma cells**, **monocytes**, and **granulocytes**.

The **intercellular substance** contains fibers—*reticular (lattice) fibers*—which resemble collagen in their structure (see below). They form fiber networks around capillaries, in basement membranes, around renal tubules, and elsewhere. The second group of *collagen* fibers consist of fibrils held together by an amorphous adhesive substance. They are found in all kinds of supporting tissues. They are wavy, almost unstretchable, and always occur grouped in bundles. This type is found particularly in tendons, the tympanic membrane, etc. Different types of collagen (I and III) are found in connective tissue, and these are dependent on the structure of the collagen molecules. Finally, there are the (yellowish) *elastic* fibers, which are also arranged in networks. They occur in arteries near the heart, certain

ligaments (ligamenta flava, see p. 56) and elsewhere. The intercellular substance also includes the **ground substance**, which is partly produced by the tissue cells. It is involved in the exchange of materials between tissue cells and the blood.

Embryonic connective tissue: contains mesenchymal cells and a mucinous, gelatinous ground substance. The most important type is mesenchyme.

Reticular connective tissue (**A**) contains reticular fibers and *reticular cells*, which are able to phagocytize and store material. They have a remarkably active metabolism. This type of connective tissue can be divided into *lymphoreticular* (in lymph nodes, etc.) and *myeloreticular* (bone marrow) connective tissue.

Interstitial connective tissue is a loose tissue with no particular shape. Its main purpose is to fill in the gaps between different structures (muscles, etc.) while also allowing for mobility between tissue layers. In addition to these functions, interstitial connective tissue takes part in general metabolism and regeneration. As well as cells (fibrocytes, fat cells) it contains collagen, elastic, and lattice fibers, and ground substance.

Rigid connective tissue (**B**) contains a high proportion of collagen fibers and fewer cells and less ground substance than interstitial connective tissue. It is found in the palmar and plantar aponeuroses, in tendons, etc.

Fatty tissue contains large cells with a flattened, eccentrically located nucleus. *Monovacuolar white fatty* (adipose) *tissue* should be distinguished from *plurivacuolar brown fat*. The latter is more abundant in infants and less so in adults (e.g., the renal fat capsule). In addition to fat cells, it contains interstitial connective tissue and shows some lobular structure. **Depot fat**, which depends on nutritional status, is distinguished from **structural fat**, which is independent of nutrition. The latter occurs in joints, bone marrow, the buccal fat pads, etc. Depot fat is most common in the subcutaneous fat layer. It is broken down according to requirements and the cells take on the form of reticular cells. After very marked weight loss (cachexia), these areas fill up with a collection of fluidserous fat cells.

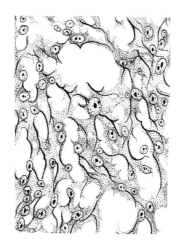

A Reticular connective tissue, x 300

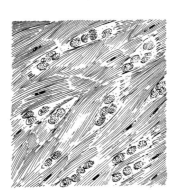

B Dense connective tissue in the corium, x 300
(panels A and B taken from Leonhardt, H.:
Human Histology, Cytology, and Microanatomy, 8th ed. Thieme, Stuttgart, 1990)

Cartilage (A–C)

Cartilage is compressible as well as flexible, yet **resistant to pressure and to bending**, and is soft enough to be cut. It consists of cells and intercellular substance, which is almost devoid of vessels and nerves. The nature of the intercellular substance determines the type of cartilage, which can be subdivided into **hyaline**, **elastic**, and **fibrous** forms.

Cartilage cells, or *chondrocytes*, are fixed cells rich in water, glycogen, and fat. They have a vesicular appearance, with a spherical cell shape and spherical nucleus. The *intercellular substance*, which is very rich in water (up to 70%), forms the basis of the protective function of cartilage. Cartilage is almost avascular and free of nerves; it is composed of fibrils or fibers and an amorphous ground substance containing proteoglycans, glycoproteins, lipids, and electrolytes.

Hyaline Cartilage (A)

Hyaline cartilage is slightly **bluish** and milky and contains abundant collagenous fibrils (converted to gelatin by boiling) and scattered elastic networks within its intercellular substance. In articular cartilage, the collagen fibrils are always aligned in the direction of the greatest stresses. The cells occupying the cartilaginous lacunae are surrounded by a capsule that is separated from the remaining intercellular substance by the *cellular halo*. The cells, which can be organized more or less into rows or columns (see p. 16), form, together with the cellular halo, a *chondrone* or *territory*. This grouping always consists of several daughter cells originating from one cell. Cartilage is surrounded externally by a connective tissue covering, the *perichondrium*, which is more or less continuous with the cartilage itself.

Hyaline cartilage exposed to pressure (articular surfaces at the lower limb) contains more glycosaminoglycans (chondroitin sulfate) than less stressed hyaline cartilage (e.g., articular surfaces of the upper limb).

The lack of sufficient blood vessels may favor degenerative processes inside the cartilage. These are initiated by the "unmasking" of collagenous fibers; that is, the collagenous fibrils become visible in the microscope. Since the content of water and chondroitin sulfate decreases with age, the stress capacity of hyaline (articular) cartilage decreases.

Calcification of hyaline cartilage occurs very early in life.

Hyaline cartilage is found in joint cartilage and rib cartilage, in respiratory tract cartilage, in epiphyseal disks and in the precursors of those parts of the skeleton that undergo chondral ossification. **Epiphyseal disk cartilage** contains columns or rows of cartilage cells, a structure that enables growth of cartilage (see p. 16) and subsequently of the bone that follows it.

Elastic Cartilage (B)

In contrast to the bluish hyaline cartilage, elastic cartilage is **yellowish** in color. Its intercellular substance is rich in elastic fibers and contains fewer collagen fibrils. The large proportion of elastic fibers makes this type of cartilage particularly pliable and elastic. It does not contain calcified deposits. It is found in the auricle, the epiglottis, etc.

Fibrocartilage (C)

Fibrocartilage, also known as connective tissue cartilage, contains fewer cells than the other types but has many *bundles of collagen fibers*. It is found particularly in parts of the intervertebral disks (see p. 54) and of the symphysis pubis (see p. 22).

A Hyaline cartilage (rib cartilage),
x 180

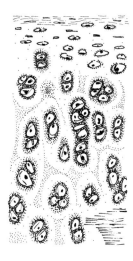

B Elastic cartilage (ear cartilage),
x 180

C Fibrocartilage (intervertebral
disk), x 180
(Figs. A–C taken from Leonhardt,
H.: Human Histology, Cytology,
and Microanatomy, 8th ed.
Thieme, Stuttgart, 1990)

Bone (A, B)

Bone tissue (osseous tissue) consists of bone cells (*osteocytes*), *ground substance*, *collagenous fibrils*, a *cement substance*, and *various salts*. The ground substance and collagenous fibrils form the intercellular substance, the *osteoid*. The fibrils belong to the organic part, the salts to the inorganic part. The most important salts are calcium phosphate, magnesium phosphate, and calcium carbonate. In addition, compounds of calcium, potassium, and sodium with chlorine and fluorine are found.

Clinical tip: The salts confer hardness and strength. A salt-free or "decalcified" bone is pliable. A deficiency in calcification may result from vitamin deficiency as well as from hormonal disturbances. A vitamin deficiency may arise, for example, when there is an absence of ultraviolet light exposure resulting in a failure to convert provitamins into vitamins. Inadequate calcification leads to a softening of the bone, for example in rickets.

The organic constituents, like the salts, are also responsible for the strength of a bone. When there is inadequate organic material, the elasticity of the bone is lost, and as a result the bone becomes brittle and can no longer withstand stress. The relationship between inorganic salts and collagenous fibrils becomes altered during life. In the newborn the content of inorganic salts amounts to about 50% and this rises to 70% in the elderly along with a loss of elasticity, as the bone becomes less flexible and shock-resistant. Destruction of the organic matter can also be induced artificially by exposure to heat.

Two types of bone can be distinguished on the basis of the arrangement of its fibrils: **woven bone** (reticulated) and **lamellar bone**. Nonlamellar, woven bone corresponds structurally to ossified connective tissue and in humans primarily occurs only during development. In the adult it is found only in the capsule of the inner ear and along the sutures of the cranial bones.

The substantially more common and more important **lamellar bones** (A, B) exhibit a distinct stratification produced by layers of parallel collagen fibrils that are called *lamellae* (**1**). These lamellae alternate with layers of *osteocytes* (**2**). The lamellar arrangement takes place around a vascular canal, the *central canal*, or *haversian canal* (**3**), which, together with its lamellae, constitutes an osteon or *haversian system* (**A**). The collagenous fibers are approximately 2 to 3 μm thick and are arranged spirally in such a way that a right (**4**) and a left spiral (**5**) lamella (5–10 μm thick) alternate with one another, producing an increase in stability.

Between the osteons are *interstitial lamellae* (**6**), which are the remnants of former osteons. The vascular canals in the osteons communicate with smaller *oblique canals*, which are called *Volkmann's canals* (**7**). The structure and organization of the osteons are dependent on the stresses in the bone. When there is a change in stress, the osteons become reconstructed, as evidenced by macroscopic observation. In this case, attention should be especially paid to the behavior, within the femur, of the *trajectories*, the lines of tension, which are developed in response to the stresses.

The nourishment of bone takes place from the periosteum (see p. 20). Bone marrow is nourished via the nutrient foramina (nutrient arteries).

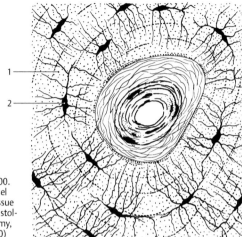

A Haversian system (osteon), x 400. In the center is a haversian vessel with perivascular connective tissue (from Leonhardt, H.: Human Histology, Cytology, and Microanatomy, 8th ed. Thieme, Stuttgart, 1990)

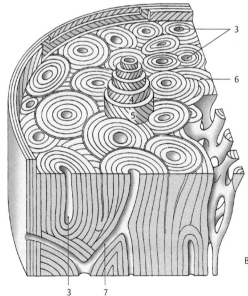

B Diagram of the compact part of the diaphysis of a long bone

Development of Bone (A–C)

Bone formation (*osteogenesis*) is based on the activity of *osteoblasts* (**1**), which are specialized mesenchymal cells. Osteoblasts secrete an intercellular substance, *osteoid*, which consists initially of soft ground substance and collagen fibers. Osteoblasts develop into *osteocytes*, the definitive bone cells. At the same time multinucleated *osteoclasts* (**2**) develop; these bone-degrading cells are associated with the absorption and remodeling of bone.

We distinguish *direct bone formation (intramembranous ossification)* (**A**) from *indirect bone formation (chondral ossification)* (**B, C**).

Intramembranous ossification, *osteogenesis membranacea* (**A**), is the development of bone from connective tissue. The latter contains many mesenchymal cells that develop via osteoblasts (**1**) into osteocytes. At the same time, osteoclasts (**2**) develop and collagen fibers also appear. The original bone is membrane bone and is later remodeled into lamellar bone. The skull cap, the facial bones, and the clavicles develop as intramembranous bones.

Chondral ossification, *osteogenesis cartilaginea* (**B, C**), requires preformed parts of skeletal cartilage (cartilage models), which will then become replaced with bone. Growth is possible only as long as cartilage still remains. The prerequisite for replacement bone formation is the presence of *chondroclasts*; these are differentiated connective tissue cells that degrade cartilage and thus enable the osteoblasts to form bone. Two types of replacement bone formation are recognized—*endochondral* (**C**) and *perichondral ossification*.

Endochondral ossification (**3**) begins inside the cartilage, and occurs predominantly in the epiphyses. The **epiphyses** are the ends of the long bones (see p. 20), while the shafts are called **diaphyses**. *Perichondral ossification* (**4**), which originates in the perichondrium (**5**), is confined to the diaphysis. The *epiphyseal disk* (growth plate) (**6**), which is necessary for growth in length, forms a layer between the epiphysis and the diaphysis. That part of the shaft adjacent to the epiphyseal disk is called the **metaphysis** and develops first on an endochondral basis (see below).

Clinical tip: An *apophysis* is a bony protuberance that does not arise from its own ossification center but develops purely in response to tendon traction. An example is the mastoid process (see pp. 288 and 290).

Within the epiphyseal cartilage, the processes of ossification occur in separate zones. In the epiphysis there is the *zone of reserve cartilage*, a capping of hyaline cartilage that is not affected by bone formation in the epiphyseal plate. Next to this inactive cartilage is the *zone of growth* (**7**), where the cartilage cells form columns. Here the cartilage cells divide, thus increasing in number. The next layer closer to the shaft is the *zone of maturation* (**8**); it contains vesicular cartilage, and calcification is already occurring. It is followed by the *zone of ossification*, where cartilage is degraded by chondroclasts and replaced with bone by osteoblasts. Some remnants of cartilage remain, so that the endochondral bone (**9**) of the diaphysis can be distinguished from the perichondral bone. It will later be replaced by perichondral bone. The endochondral bone is destroyed by the invading osteoclasts.

The increase in bone diameter in the region of the diaphysis is brought about by the deposition of new bony material on the outer surface beneath the cellular layer of the periosteum. The *bone marrow cavity* (**10**) becomes larger as a result of bone destruction. All growth processes are regulated by hormones.

The bony anlages in the epiphyses first appear after birth, except for those in the distal femoral epiphysis and the proximal tibial epiphysis. In both of these epiphyses, and in the cuboid bone, osteogenesis begins just before birth in the 10th intrauterine month (a sign of maturity).

Clinical tip: After closure of the epiphyseal disk X-rays show a fine line, later, in adolescence, known as the **epiphyseal disk scar**.

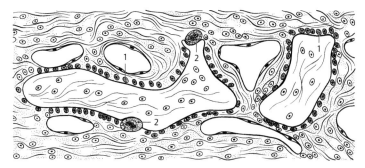

A Intramembranous ossification

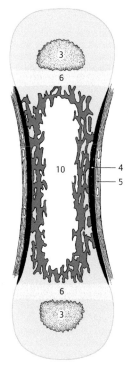

B Chondral ossification of a long bone
(diagram). Endochondral ossification
in the epiphyses and perichondral
ossification in the diaphysis

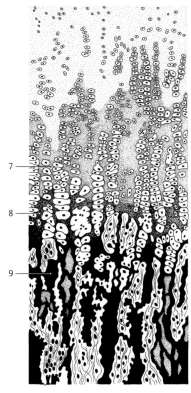

C Ossification in the region of
the epiphyseal disk cartilage

Muscular Tissue (A–D)

Muscular tissue is characterized by elongated cells containing myofibrils formed from myofilaments. These myofibrils are responsible for the contractility of the muscle cells. Three types of muscular tissue can be distinguished on the basis of fine structure and physiologic characteristics: smooth (**A**), striated (**B, D**), and cardiac muscle (**C**).

Smooth Muscle (A)

Smooth muscle consists of spindle-shaped cells, each being 40 to 200 μm long and 4 to 20 μm thick, with a central nucleus. These myofibrils are difficult to demonstrate and do not have transverse striations. Transverse reticular fibers join adjacent muscle cells and bind groups into functional units. Smooth muscle is not under voluntary control; axons synapse directly with the muscle cells (see Vol. 3).

Hormonal influences may cause smooth muscle to increase in length and to proliferate; that is, there may not only be an increase in the size of the cells but cells may also be newly formed. An example is the uterus, the muscle fibers of which may reach a length of 800 μm.

Striated Muscle (B, D)

Striated muscle consists of muscle cells (muscle fibers) which may be 10 to 100 μm thick and up to 15 cm long. The nuclei lie immediately beneath the surface of the cells in the direction of the long axis of the muscle fibers. The myofibrils are easily visible and are responsible for the longitudinal striations. The transverse striations are due to the periodic alternation of smaller, lighter, singly refractive (isotropic) zones (I bands) and wider, darker, double-refractive (anisotropic) zones (A bands). The A bands contain a light zone (H band) with a fine, dark middle line (M band), and the I bands show a delicate, anisotropic intermediate line (Z band). The myofibrillar segment that lies between two Z bands is called a **sarcomere**.

Each skeletal muscle cell contains several nuclei. The cytoplasm (*sarcoplasm*) contains a variable number of mitochondria (*sarcosomes*). According to their function, a distinction is made between *twitch* muscle fibers and *tonic* muscle fibers. The twitch muscle fibers include red (fast twitch) muscle fibers with high myoglobin and mitochondria content (for long-term stress performance) and white muscle fibers with high myofibril content (for short-term maximum stress performance).

The color of a muscle is due to its blood supply and the myoglobin in solution in the sarcoplasm. In addition, the color is determined also by the water content and the abundance of fibrils. This explains why different muscles differ in color. Thinner fibers with fewer fibrils and less water content are light in color, while thicker fibers appear darker.

The *sarcolemma* encloses individual muscle fibers as a connective tissue sheath. There is a delicate layer of connective tissue, the *endomysium*, between the fibers. Several muscle fibers are surrounded by the *internal perimysium*, and together they form the primary muscle bundle (fascicle).

The *external perimysium* is a connective tissue layer that combines several primary bundles to form a muscle.

Striated skeletal muscles are voluntary muscles, and they are innervated via motor end plates (neuromuscular junctions) (see Vol. 3).

Striated Cardiac Muscle (C)

The muscle fibers of the heart contain a large amount of sarcoplasm and form networks. Transverse striations are present, but the sarcomeres are shorter and the I band is narrower than in skeletal muscle. In cardiac muscle fibers the nuclei lie centrally. *Sarcosomes* are far more numerous than in skeletal muscle. In addition, cardiac muscle tissue contains highly refractile, transverse *intercalated disks*, which lie at the position of a Z band. Further details are given in Volume 2.

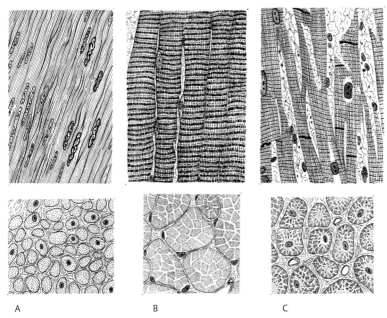

Longitudinal section (top row) and transverse section (bottom row) of smooth muscle (A), striated muscle (B), and cardiac muscle (C), x 400 (from Leonhardt, H.: Human Histology, Cytology, and Microanatomy, 8th ed. Thieme, Stuttgart, 1990)

A   B   C

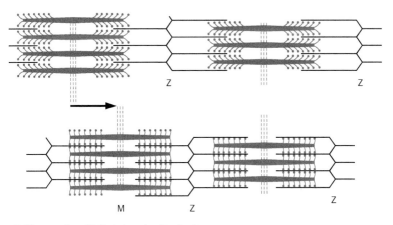

D Diagram of myofibrils during relaxation (top) and contraction (bottom)

General Features of the Skeleton

Classification of Bones (A–F)

The **bones** form the bony **skeleton** and, with the joints, they represent the passive locomotor system, which is controlled by the active locomotor apparatus, the musculature. The different shapes of bones are dependent on their function and their position in the body. Macroscopically, two differently constructed portions can be distinguished. A rather *dense compact* or *cortical bone* (**1**) is generally observed on the surface. Within the short and flat bones and in the epiphyses and metaphyses of the long bones, there is a spongelike mesh-work formed of individual bony trabeculae, *trabecular* or *spongy bone* (**2**). Between the meshes is the bone marrow or medulla. In the flat bones of the skull, the compact material is called the *external* (**3**) and *internal* (**4**) *laminae* and in between them is the *diploë* (**5**), corresponding to the spongy bone.

Long Bones (A–C)

A long bone as, for instance, the humerus (**A**), consists of a *body* (**6**) and two *ends* (**7**). In the center of the shaft (body) of a long bone (**B**, **C**) is the bone marrow or *medullary cavity* (**8**), which contains red or yellow bone marrow. This cavity is the reason for the name "tubular bones." Tubular bones grow mainly in *one* direction.

Flat Bones (D)

Flat bones consist of two layers of compact bone between which there may be found spongy material. Flat bones include the scapula and several bones of the skull, for example the parietal bone (**D**). Basically, growth in flat bones proceeds in *two main* directions.

Short Bones (E)

The short bones, which include, for instance, the small bones of the wrist (e.g.,

the capitate bone [**E**]), have a spongy core surrounded by compact bone.

Irregular Bones

These include all those bones, such as vertebrae, which do not belong to any of the preceding groups.

Pneumatized Bones (F)

These bones contain air-filled cavities lined by mucous membrane (**9**). They are found in the skull (ethmoid, maxilla [**F**], etc.).

Sesamoid Bones

These mostly occur in the skeleton of the hands and feet. They may also be found in tendons, for example the *patella*, the largest sesamoid bone in the body.

Periosteum

The **periosteum** covers all parts of the bone that are not joint surfaces. It consists of a *fibrous layer* and an *osteogenetic layer* forming the cambium layer. It contains many blood and lymph vessels and nerves. The latter account for the pain felt after a blow to a bone. Larger blood vessels in the outer layer send numerous capillaries to the inner, cell-rich layer. This is the site of the osteoblasts, which build up bone. After fractures, formation of new bone starts in the periosteum.

Blood vessels and nerves reach the bone through nutrient foramina. Some bones have canals that also serve for the passage of vessels, usually only veins, which are known as emissary veins. They are found, for example, in the vault of the skull.

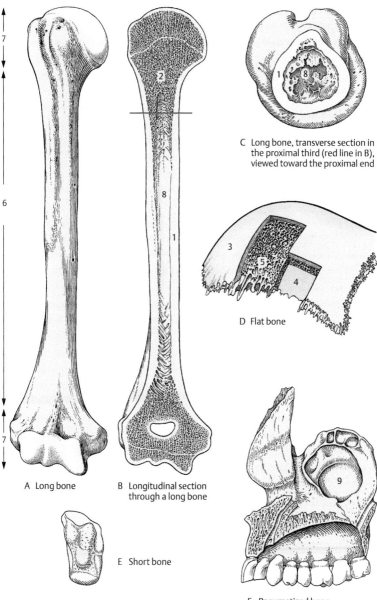

C Long bone, transverse section in the proximal third (red line in B), viewed toward the proximal end

D Flat bone

A Long bone

B Longitudinal section through a long bone

E Short bone

F Pneumatized bone

Joints between Bones

The individual bones of the skeleton are connected either *continuously* or *discontinuously*. Continuous bony joints comprise the large group of **synarthroses**, in which two bones are joined directly by various tissues.

Continuous Joints between Bones (A–H)

Fibrous Joint (A–E), Syndesmosis

In a syndesmosis two bones are joined by collagenous or elastic connective tissue. The union may be expansive or narrow. The *interosseous membrane* (1) in the forearm is a very taut syndesmosis consisting of collagenous connective tissue. More elastic syndesmoses are the *ligamenta flava* between the vertebral arches.

The **sutures of the skull** are a particular type of syndesmosis (**B–E**). These sutures retain connective tissue, which has persisted between the bones developing from connective tissue. Only when the connective tissue has completely disappeared does the growth of the skull cease and do the sutures fuse. The sutures of the skull are classified according to their shape: *serrate suture* (**B**) with sawlike edges, as in the sagittal suture; *squamous suture* (**C, D**) where one bone overlaps another, as between the parietal bone and the temporal bone; and last, *plane suture* (**E**) as between the nasal bones.

A specialized type of fibrous joint is the **gomphosis**, a peg-and-socket joint found in the fixation of the teeth in the alveoli of the jaw. Here, the tooth is joined to the jaw by connective tissue, which permits a slight degree of displacement.

Cartilaginous Joint (F), Synchondrosis

The second, large group of continuous bony joints is formed by the synchondroses (**2**), which are joints of hyaline cartilage between two bones. During adolescence, these are always found in the *epiphyseal disks*. Hyaline cartilage material is also present between the first, sixth, and seventh ribs and the sternum. The cartilaginous material disappears from those sites where it only permits growth. Epiphyseal disks or cartilage are subsequently completely replaced by bony material.

Symphysis (G)

Symphyses are also cartilaginous joints in which two bones are bound by fibrocartilage and connective tissue, for example between the two pubic bones (*pubic symphysis* **G**).

Bony Union (H), Synostosis

This is the firmest possible joint between two bones, for example between the parts of the hip bone, or between epiphyses and diaphyses after growth has ceased.

> **Clinical tip:** Synovial joints may sometimes become synostotic. However, they are then not called synostoses, but ankyloses (stiffened joint). An **ankylosis** presupposes that the joint was previously movable, and the alteration is usually the result of a disease process. Physiologic ankylosis is regarded as the fusion of the articular processes of the sacral vertebrae.

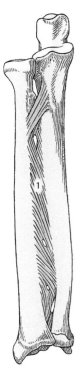

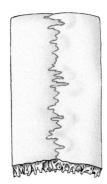

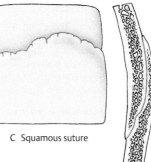

C Squamous suture

B Serrate suture

D Squamous suture
  in cross section

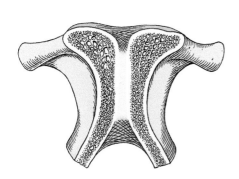

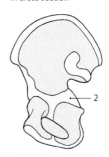

A Interosseous membrane

E Internasal suture

F Hip bone, medial view; cartilaginous interstices still present

G Symphysis

H Hip bone, lateral view;
  cartilaginous interstices closed

General Anatomy

Discontinuous Joints between Bones (A–C)

These joints, **diarthroses** or **synovial joints**, consist of *articular surfaces* (**1**), an *articular capsule* (**2**), a *joint cavity* (**3**) between the articular surfaces, and, according to need, some *additional features* (strengthening ligaments, intercalated disks, articular lips [labra], and bursae).

In a joint with two articular surfaces or bodies, that articular body which is moved is the *movable segment*; the one at comparative rest is the stationary or *fixed segment*.

To assess the degree of mobility of a joint, it is necessary to determine the *angle of excursion* (**4**), that is, the angle between its *initial* and *final* positions. The angle of excursion of a joint may be reduced by various factors. They include, in addition to the tension of the articular capsule, additional ligaments that restrict movement (*ligamentous limitation*, see p. 26), bony processes (*bony limitation*), and limiting surrounding soft tissues (*soft tissue limitation*). The *midposition* (**5**) is that position between the initial and final positions in which all parts of the joint capsule are under equal tension.

> **Clinical tip:** The range of movement of a joint is now stated in terms of the neutral-0 position based on the SFTR method of *Russe* and *Gerhardt* (**C**). The neutral-0 position of all joints is that occurring in an upright posture with the arms hanging at the sides and the palms facing forward. There is a distinction between anatomical and anthropological methods of measurement. Movements are measured in the **S**agittal plane, **F**rontal plane, and **T**ransverse plane and during **R**otation (SFTR). In the numbers given, it should be remembered that the first figure always refers to extension, retroversion, abduction, external rotation, supination, or a movement to the left corresponding to the function of the joint. The second number is the neutral-0 position and the third is the final position in opposition to that of the first movement.

Articular Surfaces

A joint possesses at least two articular surfaces. They are usually covered by hyaline cartilage (**6**) and occasionally by fibrocartilage or connective tissue interspersed with fibrocartilage.

The cartilage is tightly interlocked with the bone, and its surface is shiny and smooth. The thickness of the cartilage layer varies from 2 to 5 mm, although the patella has some very thick areas, up to 6 mm. The cartilage is nourished via the synovial fluid as well as by diffusion from the capillaries in the synovial membrane.

Joint Capsule

The joint capsule may be taut or loose and is attached to the bone near the cartilage-covered surfaces. It consists of two layers, the inner *synovial membrane* (**7**) and an *outer fibrous membrane* (**8**). The synovial membrane contains elastic fibers, blood vessels, and nerves. The amount of blood supply is directly related to the degree of activity so that very active joints are more richly vascularized than less active ones. The synovial membrane possesses inward-facing processes containing fat, the *plicae synoviales* (**9**), synovial folds, and *synovial villi*. The fibrous membrane is of variable thickness and contains a large quantity of collagen fibers and very few elastic ones. Irregularities in the thickness of the fibrous membrane may result in weak spots through which the synovial membrane may protrude; these cyst-like protrusions are called *ganglia* by the surgeon.

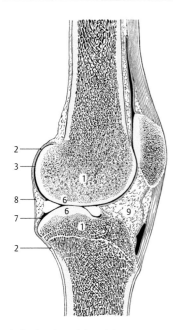

2
3
8
6
7
6
1
9
1
2

A Section through knee joint

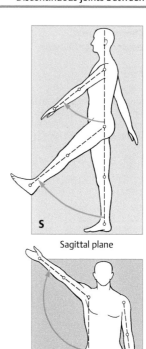

S

Sagittal plane

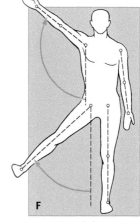

F

Frontal plane

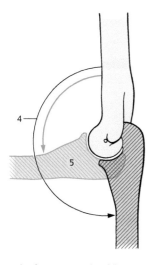

4
5

B Angle of excursion and middle position

T

R

Transverse plane and rotation

C Neutral-0 method and SFTR recording

Discontinuous Joints between Bones, continued

Joint Cavity (A, C)

A joint or articular cavity (**1**) is a cleftlike capillary space that contains *synovial fluid*. This is a clear, viscous, mucin-containing fluid resembling albumin. The fluid acts as a lubricant and aids nutrition of the articular cartilage. Its viscosity is determined by its content of hyaluronic acid and is temperature-dependent—the lower the temperature, the higher the viscosity of the synovial fluid. Since synovial fluid may also be regarded as a dialysate of blood plasma, its chemical and physical characteristics may be of diagnostic value in a variety of diseases.

Additional Features (A–D)

Ligaments (**2**). Ligaments are designated by their function as *reinforcing ligaments* (for the joint capsule), *guiding ligaments* (in movements), or *restrictive ligaments* (to constrain movements). According to their position there are *extracapsular*, *capsular*, and *intracapsular* ligaments.

Articular disks or **menisci** (**3**) consist of collagenous connective tissue containing fibrocartilage. A disk divides the joint cavity completely; a meniscus, only partly. They affect the direction of movement, ensure good contact between the moving parts, and may, in certain circumstances, produce two completely independent joint spaces, as, for instance, in the mandibular and sternoclavicular joints. Regeneration of disks after injury or removal is possible.

Articular labra (**4**) consist of collagenous connective tissue with scattered cartilage cells and serve to enlarge the joint surface.

Bursae and **synovial pouches** may communicate with the joint cavity (**5**). They form large or small, thin-walled sacs lined by synovial membrane (**6**). They create a weak point in a joint but also serve to enlarge the joint space.

Maintenance of Contact

There are various forces that act on the two articular surfaces and maintain contact between them. First, there are the muscles that span the joint and ensure a certain degree of contact between the articular surfaces. Next, there may be accessory capsular ligaments to increase the degree of contact. In addition, there is a certain degree of surface adhesion and, as another important factor, atmospheric pressure. Atmospheric pressure holds the articular surfaces together with a force equal to the product of the area of the smaller joint surface and the air pressure.

Clinical tip: Joints are subject to **age-related changes;** the avascular articular cartilage (**7**) loses its elasticity with aging.

Surfaces covered by cartilage undergo age-related alterations (**8**) and may degenerate. Outgrowths from the cartilage margins may occur, which are sometimes invaded by bone-forming cells. In such instances the cartilage becomes ossified and restricts joint mobility. Such processes may affect small joints such as intervertebral joints, and they may occur in young people if the joints in question are overstressed.

The "vacuum phenomenon," first described by Fick, refers to linear or crescent-shaped lucencies that appear in radiographs of joints and are caused by tissue gases entering the joint.

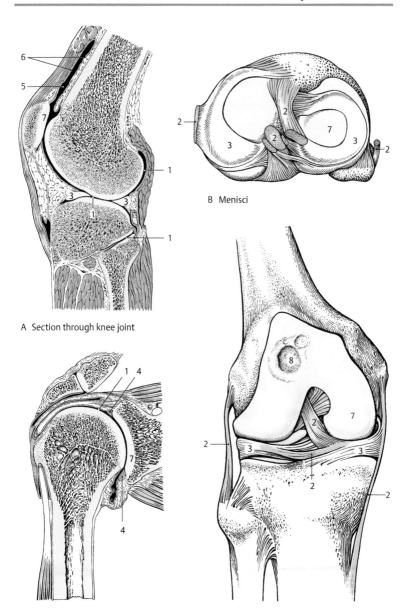

A Section through knee joint

B Menisci

C Section through shoulder joint

D Knee joint, anterior view

Classification of Joints (A–F)

Joints may be classified by various criteria. One classification is related to the **axes** and subdivides joints into monaxial, biaxial, and multiaxial articulations. A second classification divides the joints according to their **degrees of freedom**, which indicate the mobility of articular surfaces relative to each other. Joints are therefore divided into those with one, two, or three degrees of freedom. Another classification makes use of the **number of articular surfaces** and so separates simple from complex joints. A *simple joint* consists of only two surfaces contained in one capsule. If more than two surfaces are present in the capsule, the joint is called a *complex joint* (e.g., elbow joint, **B**).

Different types of joints may be combined. *Joints combined of necessity* are found at different points on two bones (e.g., proximal and distal radioulnar joints). *Forcibly combined joints* are activated by one or more muscles that span several joints, for example hand and finger joints by the flexors of the fingers (see p. 173).

Joints may also be classified according to the **shape of the articular surfaces**:

A *plane joint*, a joint with two flat surfaces, possesses two degrees of freedom, and gliding movements are possible (e.g., the small vertebral joints, zygapophyseal joints).

A *hinge joint* or *ginglymus* (**A**) consists of a convex and a concave articular surface. The concave articular surface often has a ledge-shaped elevation that fits into a groove of the convex one. Tense lateral ligaments (**1**) help to fix the joint more firmly. Hinge joints have one degree of freedom (e.g., the humeroulnar articulation, **B**). Ginglymus and trochoid articulations (below) are collectively known as *cylindrical joints*.

Trochoid joints include the pivot joints and the rotary joints. Both have one axis and one degree of freedom, and both have one convex cylindrical surface and a corresponding concave joint surface. The joint

axis runs through the cylindrical surface. In a pivot joint the convex (peglike) surface rotates within the concave surface, which is enlarged by ligaments (annular ligament, **2**; e.g., in the proximal radioulnar joint, **B**). In a rotary joint the concave articular surface rotates around the convex surface (e.g., the distal radioulnar joint).

Ellipsoidal or condylar joints have a convex and a concave elliptical joint surface. They have two degrees of freedom and are multiaxial, with two principal axes. When the movements are combined, a circumduction is possible, for example the radiocarpal joint.

A *saddle joint* (**C**) consists of two saddle-shaped articular surfaces each having a convex and a concave curvature. It has two degrees of freedom and two main axes, but is in fact multiaxial. Circumduction is possible (e.g., the carpometacarpal joint of the thumb, **D**).

Ball-and-socket or *spheroidal joints* (**E**) are multiaxial and consist of a globular bony head within a cup or socket. They have three degrees of freedom and three principal axes (e.g., shoulder joint, **F**). A special type of ball-and-socket joint is the *enarthrosis* in which the socket extends beyond the equator of the head. The hip joint is usually an enarthrosis in which the socket (acetabulum) is enlarged solely by the articular labrum.

A special type of joint is the fixed joint or *amphiarthrosis*. This type has very limited mobility since both the ligaments and the capsule are taut and the articular surfaces are rough, as in the sacroiliac joint.

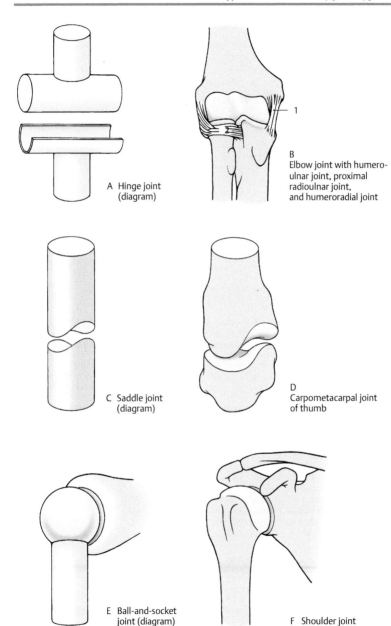

A Hinge joint (diagram)

B Elbow joint with humero-ulnar joint, proximal radioulnar joint, and humeroradial joint

C Saddle joint (diagram)

D Carpometacarpal joint of thumb

E Ball-and-socket joint (diagram)

F Shoulder joint

General Features of the Muscles

Classification of Skeletal Muscles (A–F)

All skeletal muscles have both an *origin* and an *insertion*. The origin is always on the less mobile bone (fixed end) while the insertion is on the more mobile bone (mobile end). In the limb the origin is always proximal and the insertion distal. At the point of origin there is often a *muscle head*, which merges into the *belly* (**1**) and ends in a *tendon* (**2**). Muscle power is dependent on physiologic cross section, which is the sum of the cross sections of all the fibers. From this the absolute muscular strength is calculated.

The location of the muscle belly depends on the space available. An important factor for the muscle's activity is its effective terminal part. The tendon of a muscle may, for example, be bent around a portion of the skeleton, a *muscular trochlea*, which functions as a fulcrum (*hypomochlion*). A long tendon may prove advantageous if there is a shortage of space. The best examples of this are the long finger muscles, which have their muscle bellies in the forearm but exert their actions more distally on the fingers.

Different muscle shapes are distinguished according to the relationship of the muscle fibers to the tendon. **Fusiform muscles** (**A**) have long fibers and produce large-amplitude movements that are not very forceful. Fusiform muscles have relatively short tendons. Another type is the **unipennate muscle** (**B**), which has a long, continuous tendon to which the short muscle fibers are attached. This provides a relatively large physiologic cross section, resulting in greater muscle power. A **bipennate muscle** (**C**) has the same structure as a unipennate muscle, but the fibers are attached to both sides of the tendon. There are also **multipennate muscles**.

There are several types of muscle origin, for example two-, three-, and four-headed muscles, in which the individual heads fuse into a single muscle belly and terminate in a common tendon. Examples of this type include the biceps (**D**) and the triceps brachii muscles.

If a muscle has only one head but one or more *intermediate tendons* (**3**), it is called a digastric or **multigastric muscle** (**E**). One such muscle with two bellies, the **digastric muscle**, has two successive, almost identical large muscle segments. A **flat muscle** (**F**) of a triangular shape, the *triangular muscle* with a flat tendon or *aponeurosis* (**4**), is distinguished from a quadrangular flat muscle, the *quadratus*.

Muscles that extend over one or more joints are called *uniarticular*, *biarticular*, or *multiarticular* muscles. They may produce different and in some cases even opposing movements at the various joints. Examples are the interossei muscles of the hand, which produce flexion of the proximal joint but extend the middle and distal joints of the fingers.

The muscles that work together to produce one movement are called **synergists**, and those that produce opposing movements are called **antagonists**. The combination of synergists and antagonists can vary in different movements. For example, several muscles that are synergists for flexion of the wrist become antagonists during radial abduction.

It is essential for their function that muscles have a *tone*, even at rest. One property of muscles is called *active* or *passive insufficiency*. In active insufficiency a muscle becomes exhausted when it has attained its maximal shortening. Passive insufficiency refers to a loss of muscular action in a terminal joint position, as illustrated by the inability to make a fist when the hand is flexed. In muscle action we distinguish an *active moving function* from a *passive restraining function*. A muscle may function both passively as a restraint and actively to produce movement.

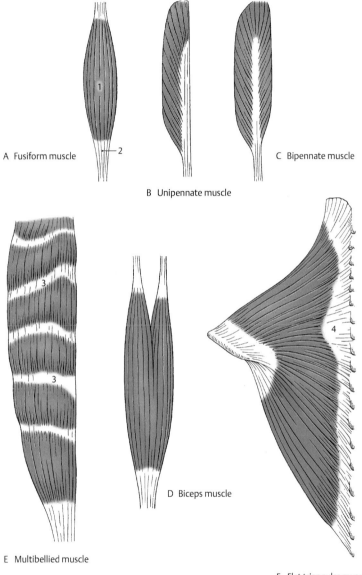

A Fusiform muscle

B Unipennate muscle

C Bipennate muscle

D Biceps muscle

E Multibellied muscle

F Flat triangular muscle

Auxiliary Features of Skeletal Muscles (A–D)

A number of auxiliary structures are essential for muscle function. They include:

(a) Connective tissue sheaths, or **fasciae**, which enclose individual muscles or muscle groups and allow them to move relative to one other.

(b) **Tendon sheaths** (**A**, **B**), which increase the gliding ability of tendons. The inner or *synovial layer* has an inner visceral layer (**1**), which lies in immediate contact with the tendon (**2**), and a parietal layer (**3**), which is connected to the *mesotendon* (**4**). The synovial fluid between the visceral and parietal layers acts as a lubricant to aid movement of the tendon. The outside of the synovial sheath is covered by a *fibrous layer* (**5**).

(c) A **synovial bursa** (**C**, **6**) protects a muscle that slides directly around a bone.

(d) **Sesamoid cartilages** and **sesamoid bones** (**D**) are found where tendons are subjected to pressure. The largest sesamoid bone is the patella (**7**), which is part of the knee joint and is also connected to the tibia via the patellar ligament (**8**) and the quadriceps tendon (**9**).

(e) **Fat pads** are placed between individual muscles and may reduce friction. Variable numbers of fat pads (e.g., the axillary fat pad) are located throughout the body.

Investigation of Muscle Function

Muscle function can be evaluated in a variety of ways. The simplest methods are *palpation* and *inspection*. The shape of a muscle may be demonstrated by particular movements.

Anatomical methods permit the demonstration of individual muscles in preparations. The origin, course, and insertion of a muscle can be determined, but an exact evaluation of its function cannot be obtained from a cadaver. Thus, dissection is an indirect method that only allows inferences and does not take into account the interaction of different muscles.

Electrical stimulation may be used to investigate muscle function, the stimuli being applied where the nerve enters the muscle ("motor point"). One disadvantage of this method is that it is useful only for superficial muscles. Another is that it produces maximal contraction without allowing for the fact that other muscles may hamper or reduce this maximal contraction.

Electromyography is another method in which the action potentials of fibers are recorded by an electrode inserted directly into the muscle. This method has shown that, with an increase in effort, more and more motor units (muscle fibers with their motor end plates and nerves, see Vol. 3) become activated. Electromyography has demonstrated that all fibers are never active at the same time. While some fibers are at rest, others contract, resulting in a uniform increase or decrease in tension.

As in other methods, the accuracy of electromyography is limited by the difficulty of determining the relative contributions of different muscles to any given movement.

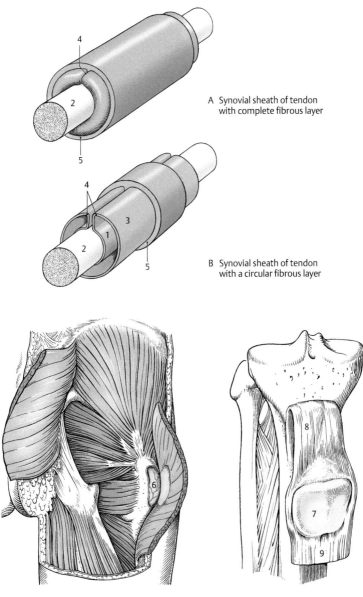

A Synovial sheath of tendon
with complete fibrous layer

B Synovial sheath of tendon
with a circular fibrous layer

C Synovial bursa

D Sesamoid bone (patella)

Anatomical Terms and their Latin Equivalents

General Anatomy	Anatomia generalis
Articular surface	Facies articularis
Ball-and-socket joint	Articulatio sphaeroidea
Bony union	Junctura ossea
Cartilage tissue	Textus cartilaginous
Cartilaginous joint	Junctura cartilaginea
Connective tissue	Textus connectivus
Ellipsoidal or condylar joint	Articulatio ellipsoidea
Fatty tissue	Textus adiposus
Fibrous joint	Junctura fibrosa
Flat bones	Ossa plana
Hinge joint	Ginglymus
Joint capsule (cavity)	Capsula (cavitas) articularis
Lamellar bone	Os compactum (lamellare)
Long bones	Ossa longa
Medullary cavity	Cavitas medullaris
Plane (pivot) joint	Articulatio plana (trochoidea)
Pneumatized (sesamoid) bones	Ossa pneumatica (sesamoidea)
Saddle joint	Articulatio sellaris
Short (irregular) bones	Ossa brevia (irregularia)
Simple (complex) joint	Articulatio simplex (composita)
Smooth muscle	Musculus nonstriatus
Striated muscle	Musculus striatus
Synovial fluid	Synovia
Woven bone	Os spongiosum (primitivum)

Systematic Anatomy of the Locomotor System

Trunk

Vertebral Column

The **vertebral column** is the foundation structure of the trunk. It consists of 33 or 34 *vertebrae* and *intervertebral disks*.

The vertebrae are divided into
- 7 cervical
- 12 thoracic
- 5 lumbar
- 5 sacral
- 4 or 5 coccygeal vertebrae

The sacral vertebrae fuse to form the *sacrum* and the coccygeal vertebrae fuse to form the *coccyx*. Thus the sacral and coccygeal vertebrae are false vertebrae, while the others are true vertebrae.

Cervical Vertebrae (A–G)

Of the seven vertebrae comprising the cervical spine, three can be readily distinguished: the first, or **atlas**, the second, or **axis**, and the seventh, the **vertebra prominens**. Only small differences characterize the third, fourth, fifth, and sixth cervical vertebrae. The *vertebral body* (**1**) is continuous posteriorly with the *vertebral arches* (**2**), each of which consists of an anterior *pedicle* (**3**) and a posterior *lamina* (**4**). At the junction of these two parts, a *superior articular process* (**5**) projects cranially and an *inferior articular process* (**6**) extends caudally. A recess, the *superior vertebral notch* (**7**), is present between the superior articular process and the vertebral body, whereas a larger *inferior vertebral notch* (**8**) is found between the inferior articular process and the body. The articular processes bear *articular surfaces* or *facets* (**9**); the superior articular facet is directed backward, the inferior articular facet forward. The vertebral arches terminate in a *spinous process* (**10**), which is directed posteriorly and which, in the third to sixth cervical vertebrae, is bifid at its tip. The cervical vertebral body and its arches enclose a relatively large *vertebral foramen* (**11**). The *transverse process* (**12**) extends laterally and includes a vertebral and costal element (see p. 52) that incompletely fuse

during development so that a *transverse foramen* (**13**) is preserved. The transverse process also has an *anterior tubercle* (**14**) and a *posterior tubercle* (**15**), between which runs the *groove for a spinal nerve* (**16**).

In the **third cervical vertebra**, the articular facets on the superior articular processes form an angle of 142° to each other open posteriorly (aperture angle, *Putz*), whereas in the fourth to seventh cervical vertebrae, this angle is approximately 180°.

The anterior tubercle of the **sixth cervical vertebra** may be especially prominent and is called the *carotid tubercle* (**17**). The upper end plates of the bodies of the third to seventh cervical vertebrae have raised lateral margins, the *uncal processes* or *unci* (**18**, see p. 58).

The **seventh cervical vertebra** has a large spinous process that is significant as the highest palpable spinous process of the vertebral column. It is therefore called the *vertebra prominens*. Its transverse process usually lacks an anterior tubercle (**E**).

▬ **Variants:** The transverse process of C7 (**G**) is incompletely developed and the costal element has incompletely fused (**19**) so that the part arising from this anlage can be distinctly differentiated from the vertebra. If the costal element is preserved independently, a **cervical rib** develops (**20**). Cervical ribs are usually present bilaterally. When they are present only on one side, they are more frequently found on the left side than on the right. The transverse foramen may be bipartite in different vertebrae.

> **Clinical tip:** The presence of a cervical rib may cause a triad of disorders, known also as **Naffziger syndrome:**
> Vascular complaints
> Complaints arising from the brachial plexus (sensory disturbances, especially of the ulnar nerve)
> Palpable mass in the greater supraclavicular fossa

Trunk

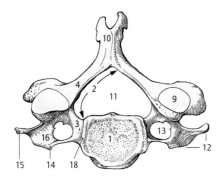

A Fourth and fifth cervical vertebrae superior view

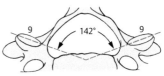

B Third cervical vertebra superior view (section)

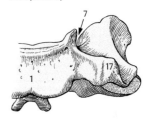

C Sixth cervical vertebra anterior view (section)

D Vertebra prominens superior view

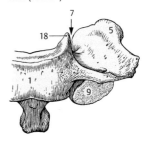

E Vertebra prominens anterior view (section)

F Cervical vertebra lateral view

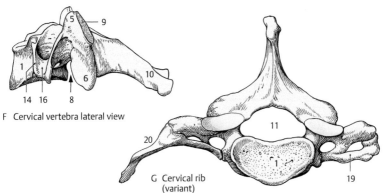

G Cervical rib (variant)

Trunk

Cervical Vertebrae, continued

First Cervical Vertebra (A–C)

The **atlas** differs basically from the other vertebrae in that it lacks a vertebral body. It consists of a smaller *anterior arch* (**1**) and a larger *posterior arch* (**2**). Both arches have small protuberances in the median plane, the *anterior* (**3**) *and posterior* (**4**) *tubercles*. The posterior tubercle may sometimes be xvery poorly developed. Lateral to the large *vertebral foramen* (**5**) of the atlas are the *lateral masses* (**6**), each of which has a *superior* (**7**) and an *inferior* (**8**) *articular facet*. The superior articular facet is concave and its medial margin is often indrawn. A bipartite superior articular facet is sometimes present. The inferior articular facet is flat or may be very slightly deepened and almost circular. On the inner side of the anterior arch is the articular facet for the dens, the *fovea dentis* (**9**). From the *transverse foramen* (**11**), which is located in the *transverse process* (**10**), the *groove for the vertebral artery* (**12**), runs across the posterior arch.

■■■ **Variants:** The groove for the vertebral artery may be replaced by a *canal* (**13**). Rarely, the atlas is divided into two halves joined by cartilage. Equally rarely, unilateral or bilateral assimilation of the atlas, that is, bony fusion with the skull, may be observed.

Second Cervical Vertebra (D–F)

The **axis** differs from C3 to C6 by the presence of the *dens* (**14**). The dens forms a toothlike process on the upper surface of the body of the axis and terminates in a rounded point called the *apex* (**15**). The anterior surface of the dens has a definite articular surface—the *anterior articular facet* (**16**). Its posterior surface may have a smaller articular facet—the *posterior articular facet* (**17**).

The lateral articular facets slope laterally. The poorly developed *transverse process* (**18**) contains the *transverse foramen*.

The shape of the lateral articular facets is somewhat complex. Although they may appear almost flat in a bony (macerated) preparation, they are more ridged when their cartilaginous covering is present. This covering is important in the joint between the atlas and the axis (see p. 60). The *spinous process* (**19**) is large and often, although not always, has a bifurcated tip. It develops from the joined parts of the *vertebral arch* (**20**), which in common with the *vertebral body* (**21**) encompass the *vertebral foramen* (**22**).

> **Clinical tip:** Isolated fractures of the arch of the atlas may occur, especially after motor vehicle accidents, and require differentiation from congenital variants of the atlas (see p. 44). A fracture of the dens is the typical axis fracture. Care is required because free proatlas segments (see p. 52) may rarely be found within the atlantooccipital membrane.
> The position of the dens axis relative to the body of C2 depends on the curvature of the cervical spine. In the absence of lordosis (see p. 62) it faces slightly backward. Its longitudinal axis then makes an angle with the vertical through the body of the second cervical vertebra.

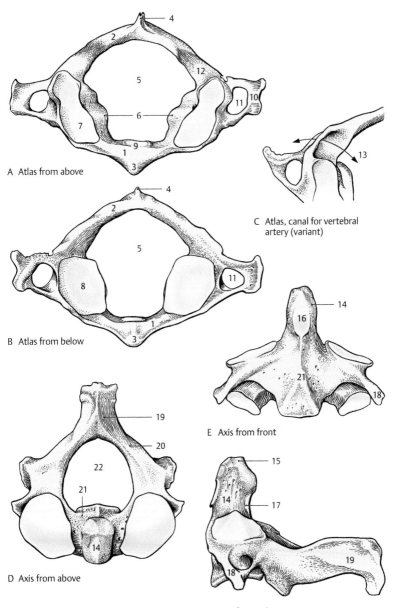

A Atlas from above

C Atlas, canal for vertebral artery (variant)

B Atlas from below

E Axis from front

D Axis from above

F Axis from side

Trunk

Thoracic Vertebrae (A–D)

The 12 **thoracic vertebrae** each have a *vertebral body* (**1**), which has incompletely ossified upper and lower end plates of compact bone and posterior openings for emergence of the basivertebral veins. Laterally the vertebral body usually has two *costal facets* (**2**), each of which forms half of an articular facet (**D**) for articulation with the head of a rib. The first, 10th, 11th, and 12th thoracic vertebrae are exceptions.

The first thoracic vertebra (**D**) has a complete articular facet (**3**) at the superior border of its body and a demifacet (**4**) at the inferior border. The 10th vertebra (**D**) has only a demifacet (**5**), while the 11th (**D**) has a complete articular facet (**6**) at its superior border. The 12th thoracic vertebra (**D**) bears the articular facet for the head of the rib in the midlateral surface of the vertebral body (**7**).

From the posterior surface of the body arises the *vertebral arch* with its *pedicles* (**8**), which are continued on each side into the *laminae of the vertebral arch* (**9**). The two laminae unite to form the *spinous process* (**10**). The spinous processes of the first through ninth thoracic vertebrae overlap each other like roof tiles, so that their tips lie one to one-and-a-half vertebrae lower than the corresponding vertebral bodies. They are triangular in cross section, in contrast to the spinous processes of the last three thoracic vertebrae, which are vertically oriented plates. These plates are not angled downward but just straight backward. On the upper margin of the pedicle of the arch is the poorly developed *superior vertebral notch* (**11**), and on the lower margin, the deeper *inferior vertebral notch* (**12**). The *vertebral foramen* (**13**) lies between the vertebral arch and the posterior surface of the body.

At the junction of the pedicle and lamina is the *superior articular process* (**14**) above and the *inferior articular process* (**15**) below. Laterally and a little posteriorly lie the *transverse processes* (**16**), which in the first to 10th thoracic vertebrae bear a *costal facet* (**17**) for articulation with the costal tubercle. The facets are concave only in the second through fifth thoracic vertebrae. On the first, sixth to ninth and 10th thoracic vertebrae, the facet is flattened. The shape of the facet imparts a differing mobility to the ribs (see p. 68).

Special Features: Like the cervical vertebrae, the first thoracic vertebra often has an *uncus corporis* (*Putz*; uncal process) on each side of its body. In the 11th and 12th thoracic vertebrae, the transverse processes may already be rudimentary. In this case, as occurs in the lumbar vertebrae (see p. 42), there may be an *accessory process* and a *mamillary process* on each side.

Clinical tip: The vertebral notches, one caudal and one cranial, together form the *intervertebral foramen* (**18**), which serves for the passage of the spinal nerves. Processes affecting the bones in this area may produce a narrowing which in turn may cause **nerve lesions**.

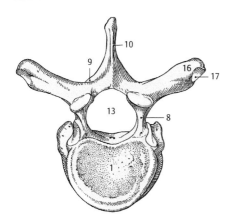

A Thoracic vertebra from above

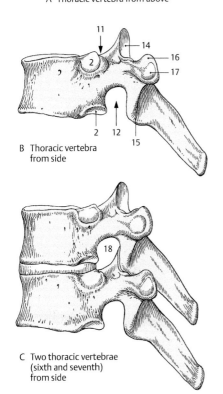

B Thoracic vertebra from side

C Two thoracic vertebrae
(sixth and seventh)
from side

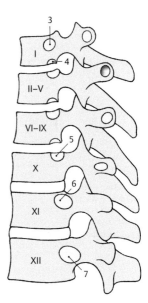

D Diagram of articular facets
of costovertebral joints

Lumbar Vertebrae (A–D)

The *bodies* (**1**) of the five **lumbar vertebrae** are much larger than those of the other vertebrae. The *spinous process* (**2**) is flat and is directed sagittally. The *lamina of the arch* (**3**) is short and sturdy, and the *pedicles of the vertebral arch* (**4**) are very thick, corresponding to those of the lumbar vertebra. The flattened lateral processes of the lumbar vertebrae may be called *costal processes* (**5**), and, since they originate from rib anlages, they are fused with the vertebrae. Behind the costal process is an *accessory process* (**6**) of variable size. Together with the *superior articular process* (**7**) and the *mamillary process* (**8**) resting on it, it represents the remnant of the *transverse process*. The *inferior articular process* (**9**) extends caudally. In essence, the articular facets face medially (**10**) on the superior articular processes and laterally (**11**) on the inferior articular processes. There is always a more or less marked angulation of these articular surfaces.

Between the superior and inferior articular processes is a region almost devoid of cancellous bone. Clinically it is known as the *pars interarticularis* (**12**).

As in all other vertebrae, there is a small *superior vertebral notch* (**13**) between the body of the vertebra and the superior articular process. The much larger *inferior vertebral notch* (**14**) extends from the posterior surface of the body as far as the root of the inferior articular process. The *intervertebral foramina* formed by the corresponding notches are relatively large in the lumbar vertebrae, whereas the *vertebral foramen* (**15**) is relatively small. At the posterior surface of the vertebral body inside the vertebral foramen, there is a large opening for the exit of a vein. The outer margins of the upper and lower surfaces (*intervertebral surfaces*) of the vertebral bodies of lumbar, as well as other, vertebrae exhibit a distinctly visible, ring-shaped, compact bony lamella, the marginal ridge or *ring epiphysis* (**16**). Can-

cellous bone occupies the central area of the vertebral body (**17**).

The compact ring corresponds to the ossified portion of the vertebral body epiphysis (see p. 52). Among the five lumbar vertebrae, the fifth lumbar vertebra can be distinguished from the others in that its vertebral body decreases in height from front to back.

■ **Variants:** Fairly often in the first lumbar vertebra and less commonly in the second lumbar vertebra, the costal process does not fuse with the bone and instead forms a **lumbar rib** (**18**). The last lumbar vertebra may fuse with the sacrum. This is called **sacralization** of the vertebra.

Clinical tip: Lumbar ribs may cause complaints because of their proximity to the kidney. Spondylolysis (see p. 44) may occur in the region of the pars.
An important diagnostic and therapeutic procedure is **lumbar puncture**, in which cerebrospinal fluid is withdrawn from the subarachnoid space with a spinal needle introduced in the midline, usually between the spinous processes of the third and fourth lumbar vertebrae.

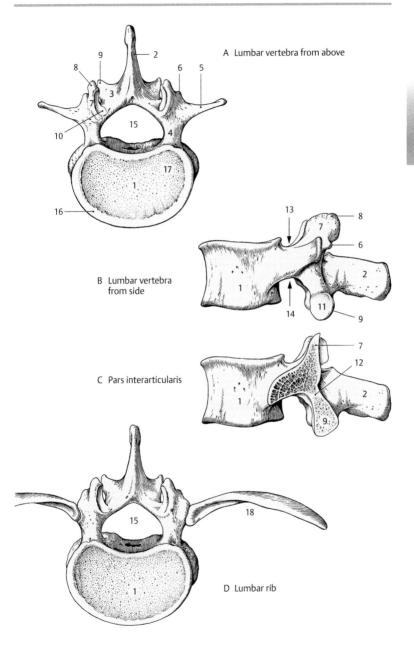

A Lumbar vertebra from above

B Lumbar vertebra from side

C Pars interarticularis

D Lumbar rib

Trunk

▇ Malformations and Variations of the Presacral Vertebrae (A–E)

Malformations of the vertebrae may be associated with more or less severe changes in the spinal cord. Various clefts or other abnormalities that may not have caused any symptoms can sometimes be detected incidentally on X-rays, ultrasound, CT, or MRI studies. Since these are developmental anomalies, some grouping will be done here. Moreover, only the free vertebrae will be considered—variations of the os sacrum are described on page 50. Likewise, cervical ribs (see p. 36) and lumbar ribs (see p. 42) will not be mentioned here.

Apart from such variations as the presence of a *vertebral artery canal* (see p. 38), or such malformations as *assimilation of the atlas* (unilateral or bilateral fusion with the skull base), the most common malformations are **clefts in the vertebral arches**. *Posterior clefts* must be distinguished from *lateral* ones and from *fissures at the root of the vertebral arches*, as well as from those *between the body and the arch*, as described by *Töndury*. In addition, there is the rare *anterior cleft of the anterior vertebral arch of the atlas*. Anterior and posterior vertebral clefts may be described as median clefts. Median posterior vertebral arch clefts may be associated with malformations of the spinal cord. According to *Töndury*, they arise during the mesenchymal phase of vertebral development.

Posterior clefts are common in the atlas (**A, B**) but they occur less often in the lower cervical vertebrae (**E**) and are very rare in the upper thoracic vertebrae. They are not uncommon in the lower thoracic and upper lumbar vertebrae and are most frequent in the sacrum (spina bifida, see p. 50).

Very infrequently the atlas has an **anterior median cleft** and in the example illustrated here there is also a posterior median cleft (**B**).

Lateral vertebral arch clefts (**C**) occur immediately posterior to the superior articular process (**1**), with the result that the inferior articular processes (**2**), together with the arch and spinous process, are separated from the other parts of the vertebra. This bony division is called *spondylolysis* and may lead to true slipping of the vertebra (*spondylolisthesis*).

Another malformation is the occurrence of **block vertebrae** (**D**), that is, the fusion of two or more vertebral bodies, as happens normally in the sacrum. Block vertebrae occur most commonly in the cervical, upper thoracic, and lumbar spine. The example shown here illustrates blocking of the second and third cervical vertebrae (**D**). Block vertebrae may have various causes, but the anomaly always arises during the mesenchymal phase of spinal development.

Clinical tip: Block vertebrae also develop during the course of various diseases, marked by the associated presence of exophytes or other definite pathologic changes. Block vertebrae may also result from motor vehicle accidents.

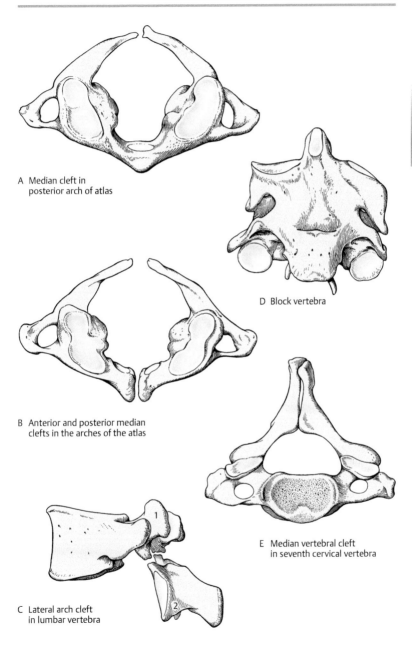

A Median cleft in
   posterior arch of atlas

B Anterior and posterior median
   clefts in the arches of the atlas

C Lateral arch cleft
   in lumbar vertebra

D Block vertebra

E Median vertebral cleft
   in seventh cervical vertebra

Trunk

Trunk

Sacrum (A, B)

The **sacrum** consists of the five sacral vertebrae and the intervertebral disks that lie between them. It has a concave anterior or **pelvic surface** (A) and a convex **posterior surface** (B). The *base of the sacrum* (**1**) has a surface that faces the last lumbar vertebra. The *apex of the sacrum* (**2**) faces downward and lies opposite to the adjoining coccyx.

Usually, the concave curvature of the **pelvic surface** (A) is not uniform but has its greatest depth approximately at the level of the third vertebra. Here the sacrum may even appear angulated. The pelvic surface has four paired pelvic *anterior sacral foramina* (**3**) as exits for the ventral rami of the spinal nerves (see Vol. 3). These foramina are not equivalent to the intervertebral foramina found in other vertebrae, which here lie directly next to the sacral canal, but are surrounded both by vertebral and rib anlages (see p. 52). They correspond to those foramina that are formed by vertebrae, ribs (or their anlages), and superior costotransverse ligaments. Between the right and left anterior sacral foramina lie the *transverse ridges* (**4**), which result from fusion of the adjacent surfaces of the vertebrae and intervertebral disks. That part of the sacral bone that lies lateral to the pelvic foramina is called the *pars lateralis* (**5**, p. 48).

The **posterior surface of the sacrum** (B) has a uniform convex curvature. Five longitudinal ridges, not always clearly developed, have their origin in fusion of the corresponding processes of the vertebrae. The *median sacral crest* (**6**) is formed in the midline by the fused spinous processes. Lateral to it, but medial to the *posterior sacral foramina* (**7**) is the *intermediate sacral crest* (**8**), which is usually the most poorly developed. It represents the fused remnants of the articular processes of the vertebrae. Lateral to the posterior foramina is the *lateral sacral crest* (**9**), which represents remnants of the transverse processes.

In the cranial prolongation at the upper end of the intermediate sacral crest, the *superior articular processes* (**10**) are found, which articulate with the last lumbar vertebra. Like the anterior sacral foramina, the eight posterior sacral foramina are not equivalent to the intervertebral foramina of other vertebrae. They correspond to those openings which are formed in common by the vertebra, rib (or rib anlages), and the costotransverse ligament. They are the exit points for the dorsal rami of the spinal nerves.

The median sacral crest terminates just above the *sacral hiatus* (**11**), which represents the inferior aperture of the vertebral canal at the level of the fourth sacral vertebra. It is bounded laterally by the *two sacral horns* (**12**).

Clinical tip: Local anesthetic can be injected into the sacral hiatus for the treatment of chronic low back pain. This therapy can anesthetize the pelvic region and lower limbs without affecting cardiac or respiratory function. Note that the needle must be angled on reaching the third sacral vertebra!

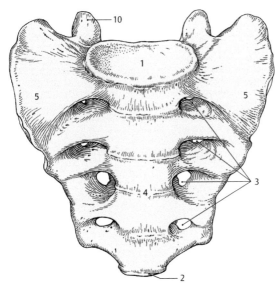

A Sacrum from front

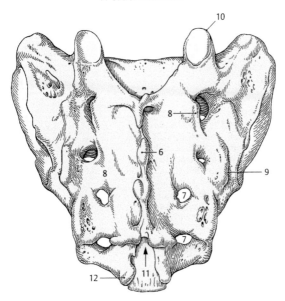

B Sacrum from back

Trunk

Sacrum, continued (A–D)

A view of the **sacrum from above** (**A**) shows in the middle the *base* (**1**), which forms the contact surface of the intervertebral disk with the last lumbar vertebra. Of all the intervertebral disks in the vertebral column, this one extends the farthest forward. It also projects farthest into the pelvis (see p. 62) and should by definition be called the promontory. However, in present-day usage, the most prominent point of the base of the sacral bone is called the **promontory**. On either side of the base are the *sacral alae* (**2**). They form the upper surface of the *pars lateralis*, which is formed on one side by the transverse processes and on the other by the rudiments of the ribs. Posterior to the base lies the entrance to the sacral canal and lateral to it are the two *superior articular processes* (**3**), which articulate with the last lumbar vertebra.

In a **lateral** view (**B**) of the sacrum, the *auricular surface* (**4**) for articulation with the hip bone can be seen. Posterior to it lies the *sacral tuberosity* (**5**), a roughened area for the attachment of ligaments.

The *sacral canal* lies within the sacrum and, corresponding in shape to the sacrum, is irregularly curved and of uneven width. At about the level of the third sacral vertebra the canal is narrowed. Channels that correspond to the intervertebral foramina and are formed from the fused superior and inferior vertebral notches open laterally from the sacral canal. The corresponding sacral foramina open anteriorly and posteriorly from these short channels (see p. 46).

Sex Differences: Males (**D**) have a longer and more curved sacrum. Females (**C**) have a shorter but broader sacrum, which is less curved than in the male.

Clinical tip: The promontory angle as described by Schmorl and Junghanns is normally in the range of 120 to 135°. It is measured at the most prominent point where lines tangent to the inferior border of L4 and the superior border of S2 intersect.

Coccyx (E, F)

The **coccyx**, which is usually formed from four or five vertebrae, is generally rudimentary. The surface that faces the sacrum has *cornua* (**6**) or *horns*, formed from the completely fused articular processes of the first coccygeal vertebra. The remainder of the coccygeal vertebrae consist only of small, round bones.

The coccygeal vertebrae decrease in size from above downward. Only the first coccygeal vertebra shows any resemblance to the structure of a typical vertebra. It has two lateral processes that represent remnants of the transverse processes.

Trunk

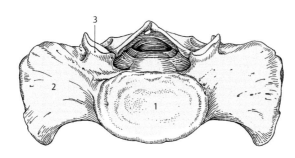

A Superior view of sacrum

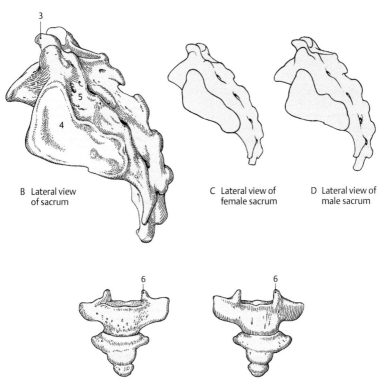

B Lateral view of sacrum

C Lateral view of female sacrum

D Lateral view of male sacrum

E Anterior view of coccyx

F Posterior view of coccyx

Trunk

■ Variations in the Sacral Region (A–D)

The vertebral column usually consists of **24 presacral vertebrae**, the remainder being arranged into five fused sacral vertebrae and three to four coccygeal vertebrae. Approximately one-third of individuals have an additional sacral vertebra, so that the sacrum consists of six vertebrae. Either one lumbar vertebra may be included in the sacrum (**A**), or the first coccygeal vertebra may be fused with it (**B**).

Situation (**A**) is called **sacralization** of a lumbar vertebra, and (**B**) is called sacralization of the coccyx or the first coccygeal vertebra. If either a lumbar or a coccygeal vertebra is fused with the sacrum, there are five sacral foramina on each side and the sacrum appears larger than in its typical form.

Fusion of the last lumbar vertebra may be unilateral, producing a **lumbosacral transitional vertebra**, which may lead to scoliosis of the spine (see p. 62). A lumbosacral transitional vertebra occurs also when there is **lumbarization** of the first sacral vertebra. In this case, posteriorly there is incomplete fusion of the first sacral vertebra with the rest of the vertebrae and there is no bony union in the region of the lateral parts, that is, in those areas that originated from remnants of ribs.

It should be noted that when lumbarization of a sacral vertebra occurs, there may nevertheless be five vertebrae if the first coccygeal vertebra is fused with the sacrum. An increased number of sacral vertebrae, that is, sacralization of a lumbar or coccygeal vertebra, is more common in males than in females.

Quite often an incomplete medial sacral crest is found (according to *Hintze* in 44% at 15 and 10% at 50 years of age). In these cases the posterior wall of the sacral canal appears to be defective (**C**). Apart from this, incomplete fusion of the spinous process of the first sacral vertebra with the spinous processes of the other sacral vertebrae produces a vertebral arch in the first sacral segment and so the medial sacral crest starts from the second vertebra.

In some cases none of the vertebral arches are fused, resulting in the absence of a posterior bony wall in the sacral canal. This anomaly is called **spina bifida** (**D**).

Clinical tip: When the spinal cord is intact and the skin of the area is undamaged, the condition is called **spina bifida "occulta."** It occurs in 2% of males and 0.3% of females. It is usually of no clinical importance.

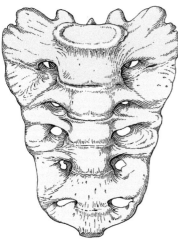

A   Sacralization of fifth lumbar vertebra

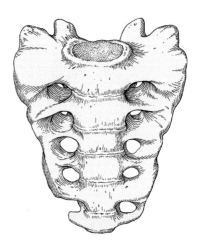

B   Sacralization of first coccygeal vertebra

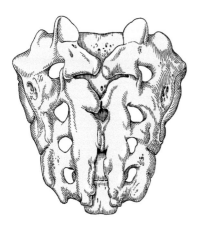

C   Incomplete medial sacral crest

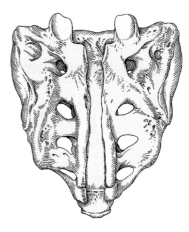

D   Spina bifida

Ossification of the Vertebrae (A–I)

Basically all vertebrae possess *three bony anlages*, two of which develop perichondrally and one endochrondrally. The perichondral cuffs (**1**) lie at the roots of the vertebral arches (pedicles), while the bony nucleus (**2**) is found in the body of the vertebra. Apart from these centers of ossification, individual vertebrae have *secondary epiphyseal bony anlages* that appear on the surface of the vertebral body, as well as in the transverse and spinous processes.

The **atlas** (**A**) develops from two lateral bony anlages (**1**), but in the 1st year of life the anterior arch may develop its own ossification center (hypochordal bar), which fuses with the other two between the ages of 5 and 9 years. The transverse processes of the atlas and axis contain rudimentary rib anlages (**3**).

In addition to the three bony anlages and the secondary epiphyses, the **axis** (**B, C**) has further ossification centers. The dens (**4**) is usually considered to arise from the bony anlage of the body of the atlas, although, according to another theory (*Ludwig*), it is formed from the dental processes. An ossification center (*ossiculum terminale*) develops in the *apex of the dens* (**5**), corresponding to the body of the proatlas, at a relatively late stage and does not fuse with the dens until the 25th year of life.

In the **other cervical vertebrae** (**D**) *three typical bony anlages* develop toward the end of the 2nd intrauterine month. Bony anlages appear in the transverse processes (**6**), which develop from the rib precursors (parietal bars), and from which the anterior tubercles and parts of the posterior tubercles are formed. The bony arches fuse in the 1st year. The fusion between the body and arches at the *neurocentral junction* occurs between the 3rd and 6th years of life. *Secondary epiphyseal anlages* appear at the ends of the transverse processes and the spinous processes between 12 and 14 years, and fuse with them at about 20 years. The *epiphyses of the vertebral bodies*, a cranial and a caudal cartilaginous plate, ossify from the 8th year onward in ring form (*annular epiphysis*) and fuse with the body from about the age of 18.

In the **thoracic region** (**E**) the bony anlages of the *pedicles* (**1**) develop first in the upper thoracic vertebrae. The endochondral center (**2**) of the vertebral body develops during the 10th week of intrauterine life, at first in the lower thoracic vertebrae. Fusion of the bony halves of the arches commences in the 1st year of life, and fusion between the arch and the body starts between the ages of 3 and 6. The *epiphyses of the vertebral bodies* ossify in a ringlike fashion.

The **lumbar vertebrae** (**F, G, K**) also ossify from *three bony anlages*; the bony centers (**2**) in the vertebral bodies appear first in the upper lumbar vertebrae (about the same time as in the bodies of the lower thoracic vertebrae) and the bony anlages in the vertebral arches (**1**) appear somewhat later. The costal processes (**7**) develop from the *rib anlages.*

The *secondary epiphyses* are represented by a bony anlage at the spinous process, as well as the ring ossified *annular epiphysis* (**8**) of the vertebral body, which is found both at its upper and lower surfaces.

In each of its segments the **sacrum** (**H, I**), develops, like the rest of the vertebrae, from three bony anlages, and, in addition, from a rib anlage (**9**) in the region of the lateral mass on each side. Thus, *each segment* of the sacrum has *five ossification anlages.* In the region of the transverse line there is an additional bony fusion of the margin with the intervertebral disks, the ossification of which begins at 15 to 16 years of age. The nuclei that arise from the rib rudiments appear in the 5th to 7th fetal months and fuse with the remaining ossification centers in the 2nd to 5th postnatal years. The sacral vertebrae fuse with one another in a caudocranial sequence up to the age of 25 to 35 years.

The **coccygeal vertebrae** develop from ossification centers that appear in the 1st year and fuse between the ages of 20 and 30 years.

A Atlas

B Axis from above

C Anterior view of axis

D Cervical vertebra

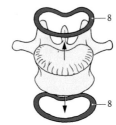

K Lumbar vertebra with anular epiphysis

E Thoracic vertebra

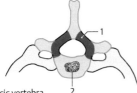

G Anterior view of lumbar vertebra

F Lumbar vertebra from above

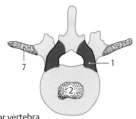

H Sacrum from above

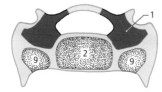

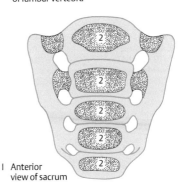

I Anterior view of sacrum

Trunk

Intervertebral Disks (A–D)

Each **intervertebral disk** consists of an outer tense *anulus fibrosus* (**1**) and a soft jellylike nucleus, the *nucleus pulposus* (**2**), which contains remnants of the embryonic notochord. The anulus fibrosus consists of concentrically arranged collagen fibers and fibrocartilage, which keep the nucleus pulposus under tension. The intervertebral disks are interposed between the bodies of the individual vertebrae. They appear conical when viewed in sagittal section. In the cervical and lumbar region they are higher in front and lower behind. The reverse is true in the thoracic region, where disks are lower in front and higher behind. Basically the thickness of the intervertebral disks increases in the craniocaudal direction.

The intervertebral disks include the *hyaline cartilage plates* (**3**) derived from the epiphyses of the vertebral bodies. This functional unit represents an important part of a motion segment (see p. 62). In addition, the intervertebral disks are also held in place by the longitudinal ligaments (**4**). The posterior longitudinal ligament is fused with the disks (see p. 56) over a broad surface, while the anterior longitudinal ligament is only loosely attached to them.

The intervertebral disks and the longitudinal ligaments form a functional entity and together are known as the **intervertebral joint**.

Function: The intervertebral disks act as shock absorbers. The nucleus pulposus distributes the pressure. Loading compresses them, and when it is released they regain their original shape after some time. In movements within the vertebral column (**C, D**) the intervertebral disks, as elastic elements, are compressed or stretched unilaterally.

Clinical tip: With increasing age, a reduction in internal pressure may result in shrinkage of the nucleus pulposus. This causes lessening of tension in the anulus fibrosus so it becomes torn more easily. Basically each tear begins in the region of the nucleus pulposus (*Schlüter*). Radially running tears (caused by excessive loads even in the young) should be distinguished from concentric tears. The latter are associated with degenerative processes. Finally parts of the intervertebral disk may be displaced.

Displacement with invasion of the adjacent vertebral body produces a **"Schmorl's nodule."** It is clearly visible in radiographs. **Herniated nucleus pulposus** occurs if the jellylike nucleus is pushed posterolaterally and laterally into the vertebral canal through a damaged anulus fibrosus. This may endanger the spinal cord, individual spinal roots, or spinal nerves.

Herniation of the nucleus pulposus is most common between the third and fourth lumbar vertebrae, as well as the fourth and fifth lumbar vertebrae. In addition, it often affects the lowest two cervical intervertebral disks between the fifth and sixth, or sixth and seventh cervical vertebrae.

Prolapse of a disk (i.e., of the nucleus) develops from a complete rupture of the anulus fibrosus. Reduction in the tension of the anulus fibrosus may lead to a loss of elasticity, followed by invasion of osteoblasts and ossification of parts of the disk.

A Intervertebral disk from above

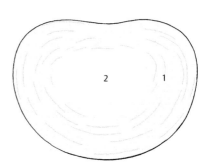

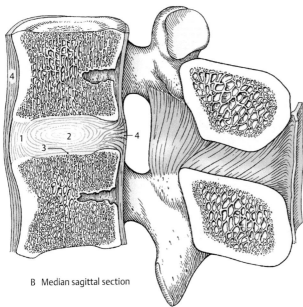

B Median sagittal section

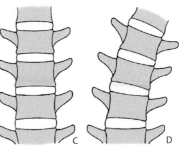

C Diagram of part of vertebral column in upright position

D Diagram of part of vertebral column, bent sideways

Trunk

Ligaments of the Vertebral Column (A–D)

The **anterior and posterior longitudinal ligaments:** these ligaments run anterior or posterior to the vertebral bodies.

The **anterior longitudinal ligament** (**1**) originates from the anterior tubercle of the atlas and descends along the anterior surface of the vertebral bodies as far as the sacrum. It widens inferiorly and is **always firmly bound to the vertebral bodies**, but not to the intervertebral disks.

The **posterior longitudinal ligament** (**2**) is divided into a superficial and deep layer and courses along the posterior surface of the vertebral body. The superficial layer arises as a continuation of the tectorial membrane (see p. 60) at the body of the axis and extends as far as the intervertebral disk between L3 and L4 (*Prestar* and *Putz*). The deep layer represents the continuation of the cruciform ligament of the atlas and extends into the sacral canal. In the cervical region the superficial layer is broad, whereas it becomes narrower in the thoracic and lumbar regions with the deep layer below L3/L4. The deep layer is very thin in the cervical region, whereas in the thoracic and lumbar segments it forms a rhombic expansion (**3**) at the intervertebral disks (**4**) and the upper marginal ridges of the vertebral bodies. In these regions a **firm union is established with the intervertebral disks**, thus affording them extensive protection. A narrow space is present between the vertebral body and the deep layer of the ligament for veins exiting from the vertebral body.

The longitudinal ligaments increase the stability of the vertebral column, particularly during flexion and extension movements. They have two functions, therefore: to restrict movement and to protect the intervertebral disks.

The **ligamenta flava** (**5**) extend segmentally between the vertebral arches (**6**). They border the medial and posterior sides of the intervertebral foramina. Their yellow color is due to an interrupted latticework arrangement of elastic fibers that form most of the bands. Even at rest these ligaments are under tension. During flexion of the spine they are stretched more tightly and **help the return of the vertebral column to the erect position**.

The **ligamentum nuchae** (not shown) extends from the external occipital crest to the spinous processes of the cervical vertebrae. The sagittal position provides attachment for muscles, and it continues beyond the neck as the interspinal and supraspinal ligaments.

The **intertransverse ligaments** (**7**) are short ligaments between the transverse processes.

The interspinous ligaments (**8**) are also short ligaments that stretch between the spinous processes (**9**).

The **supraspinous ligament** (**10**) begins on the spinous process of the seventh cervical vertebra and extends as far as the sacrum to provide a continuous connection between the vertebrae and the sacrum.

Long and short *perivertebral bands* occur lateral to the anterior longitudinal ligament, particularly in the lumbar and thoracic regions. These short bands (**11**), which extend parallel to the anterior longitudinal band, join adjacent intervertebral disks. Longer bands may arch over one disk.

12 Superior costotransverse ligament (p. 68)
13 Lateral costotransverse ligament (p. 68)
14 Radiate ligament of the head of rib (p. 68)

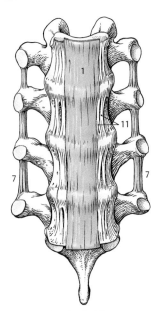

A Anterior longitudinal ligament

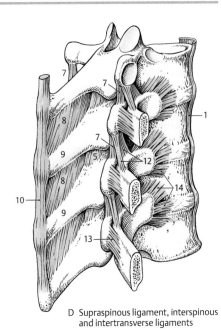

D Supraspinous ligament, interspinous and intertransverse ligaments

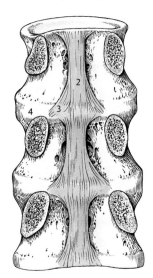

B Posterior longitudinal ligament

C Ligamenta flava

Trunk

Joints of the Vertebral Column (A–E)

Zygapophyseal Joints (A–B)

These are the vertebral synovial joints between the articular processes (**A**). Clinically they are also referred to as vertebral arch joints or "facet joints." The *articular capsules* become tenser in the craniocaudal direction. In the cervical region they are broad and lax with *meniscuslike infoldings*. These *synovial plicae* (**B**) enable the joints to bear a greater load. However, there is relatively little movement between any two adjacent vertebrae. It is only the combined action of all the participants (vertebrae and intervertebral disks) that results in corresponding movements. Movements in the **cervical** region consist of *lateral, forward*, and *backward flexion* and a limited degree of *rotation*. In the **thoracic** region mainly *rotation*, but to some extent also *flexion* and *extension*, is possible. Movements in the **lumbar** region are essentially limited to *flexion* and *extension*, although *slight rotation* is sometimes possible.

Movement in the individual segments of the vertebral column is dependent upon the position of the articular surfaces. With regard to the cervical vertebrae, the articular surfaces occupy an approximately coronal position. The articular surfaces of the third cervical vertebra exhibit a different position (see p. 37, B) in that they form an angle of 142° with one another (*Putz*). In the case of the thoracic vertebrae they describe sectors of a cylindrical shell and in the lumbar vertebrae most of the articular facets lie roughly parallel to the sagittal plane. The locations of these facets in the lumbar vertebrae, however, can exhibit great variation (*Putz*).

"Uncovertebral Joints" (C–E)

The uncovertebral joints are found in the **cervical** region. The *uncinate processes*, which are flat at first, begin to elevate in childhood. Between the ages of 5 and 10 years, fissures appear in the cartilage that

assume an articular character; thus "uncovertebral" joints are not present initially but develop *secondarily*. Approximately between the ages of 9 and 10 years, these structures extend as gaps into the disks. This initially confers functional advantages, but later in life the fissure may develop into a complete tear through the disk (**E**), with a risk of **nucleus pulposus herniation** (see p. 54). Although uncovertebral joints are initially physiologic structures, later they may become pathologic due to rupture of the disk.

> **Clinical tip:** Clinically the differential diagnosis between "uncovertebral joints" and traumatic or pathologic changes is very difficult. Damage to the disk is most common at C5, where it may be visible in a lateral radiograph as the **"lordotic crack."**

Lumbosacral Joint

The lumbosacral joint is the articulation of the last lumbar vertebra with the sacral bone. There is a highly variable relationship between the articular surfaces and the superior articular processes of the sacral bone. It is asymmetrical in 60% of people. The iliolumbar ligament (see p. 188) joins the costal process of L4 and L5 to the iliac crest and protects the lumbosacral joint from excessive loads during flexion and rotation (*Niethard*).

Sacrococcygeal Joint

The connection between the sacrum and the coccyx is often a *synovial joint*. It is strengthened by a superficial ligament and a deep posterior sacrococcygeal ligament, an anterior sacrococcygeal ligament, and a lateral sacrococcygeal ligament.

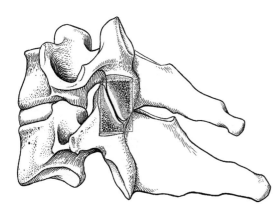

A Zygapophyseal joint (sagittal section)

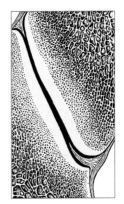

B Meniscoid folds in a facet joint (enlarged)

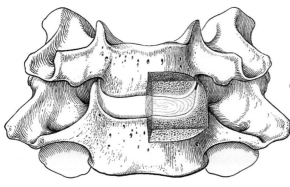

C Uncovertebral joint between C6 and C7 (frontal section)

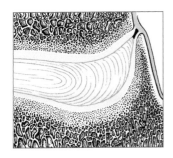

D Uncovertebral joint (enlarged)

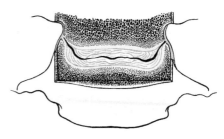

E Frontal section of split intervertebral disk in cervical spine region

Trunk

Joints of the Vertebral Column, continued

Atlantooccipital Joint (A, D, E)

The right and left **atlantooccipital articulation** is a combined joint between the atlas and the occipital bone, which in shape corresponds to an ellipsoid joint (**A, D**). The articular surfaces are the *superior articular facets* of the atlas and the *occipital condyles* (**1**). The joint capsules are lax and permit sideways bending and forward and backward movements. This **"upper craniovertebral joint"** is secured by ligaments, just like the **"lower craniovertebral joint"** (**atlantoaxial joint**).

Atlantoaxial Joints (B–E)

The atlantoaxial joint consists of the conjoined **median** and **lateral atlantoaxial joints**. Functionally it is a *rotary joint* in which movement of 26° to each side is possible from the midposition. In the lateral joints the articular facets are the *inferior articular facets of the atlas* (**2**) and the *superior articular facets of the axis* (**3**). The incongruity of the articular surfaces is reduced by the cartilaginous covering and *meniscoid synovial folds* (**4**). The folds appear triangular in sagittal section (**C**). The articular facets of the median atlantoaxial joints include the *anterior articular facet of the dens of the axis* (**5**), and the *facet for the dens on the posterior surface of the anterior arch of atlas* (**6**). In addition, in the region of the *transverse ligament of the atlas* (**7**), which extends behind the dens, there is another articular surface on the dens. The "lower craniovertebral joint," like the upper one, is secured by ligaments.

The **ligaments of both craniovertebral joints** are the *apical ligament of the dens* (**8**), which extends from the apex of the dens to the anterior margin of the foramen magnum. The *transverse ligament of the atlas* (**7**) connects the two lateral masses of the atlas. It passes posterior to the dens and stabilizes it. The transverse ligament is

strengthened by *longitudinal bands* (**9**) that run upward to the anterior margin of the foramen magnum and downward to the posterior surface of the body of the second cervical vertebra. The longitudinal bands and the transverse ligament of the atlas together form the *cruciate ligament of the atlas.*

The *alar ligaments* (**10**) are paired ligaments that arise on the dens and ascend to the lateral margin of the foramen magnum. They have a protective function, preventing excessive rotation between the atlas and the axis. The *tectorial membrane* (**11**) is a broad band that arises on the clivus and descends to join the posterior longitudinal ligament.

The *anterior* (**12**) *and posterior* (**13**) *atlantooccipital membranes* consist of broad fibrous bands extending between the anterior and the posterior arches of the atlas, respectively, and the occipital bone.

14 Ligamenta flava
15 Nuchal ligament
16 Zygapophyseal joint
17 Dura mater
18 Hypoglossal canal

I–III   Cervical vertebrae 1 to 3

Clinical tip: Nerve lesions are more common in the cervical spine (55%) than in any other vertebral region. Note, however, that lesions of the atlas and axis are fundamentally different from lesions of other cervical vertebrae.

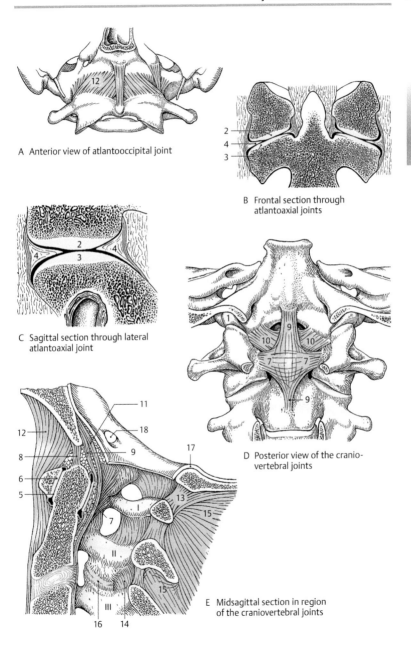

A Anterior view of atlantooccipital joint

B Frontal section through atlantoaxial joints

C Sagittal section through lateral atlantoaxial joint

D Posterior view of the craniovertebral joints

E Midsagittal section in region of the craniovertebral joints

Trunk

The Vertebral Column, Curvatures and Movements (A–H)

Curvatures of the Vertebral Column

In the sagittal plane the vertebral column of the adult shows two anteriorly convex secondary curvatures, **lordoses**, and two posteriorly convex primary curvatures, **kyphoses**.
The lordotic curves are in the cervical and lumbar regions (**1**) and the kyphotic curves are in the thoracic and sacral regions (**2**). The intervertebral disk between the fifth lumbar vertebra and the sacrum is sometimes called the promontory (see p. 48).

> **Clinical tip:** The curvature in the cervical region is quite variable. Three types occur between the ages of 20 and 30 years. "**True lordosis**" (**A**) is actually very uncommon. A double lordosis (**B**), also called a **lordotic bend** (see p. 38), is the most common and is typical of adults in the 3rd decade of life. In addition, there may be an almost complete absence of lordosis, the "**attenuated form**" (**C**). Investigation of differences between the sexes has shown that true lordosis is least common in females, that double lordosis occurs with equal frequency in both sexes, and that the attenuated type is more common in females than in males (*Drexler*).
> A lateral curvature is known as **scoliosis**. A slight degree of scoliosis is often present in radiographs, deviation to the right of the midsagittal plane being more common than to the left. The most common pathologic finding is increased kyphosis (adolescent kyphosis, kyphosis of old age).

The curvatures of the vertebral column develop as a result of the stresses of sitting and standing. Its load capacity is dependent on the degree of ossification of the vertebrae, so that the final posture (**D**) is not achieved until after puberty. The line of the center of gravity lies partly in front of and partly behind the vertebral column. The curvatures are already present by 10 months of age (**E**), but the line of the center of gravity (**3**) lies behind the vertebral column. In infants of 3 months (**F**), the curvatures are only indicated.
In adults the vertebral column is like an elastic rod, the mobility of which is restricted by ligaments. During the aging process the vertebral column undergoes various changes, so that in the elderly a reduction in the thickness of the disks produces a rather uniform kyphosis along the entire vertebral column, thereby reducing the mobility of the spine.

Movements of the Vertebral Column

Forward and backward bending (flexion and extension) occur primarily in the cervical and lumbar spine. Backward bending take place mainly between the lower cervical vertebrae, the 11th thoracic and second lumbar vertebrae and the lower lumbar vertebrae. Because of the greater mobility in this region, damage and injury to the spinal column due to overstrain is more frequent here than at other levels. In forward bending (blue) and backward bending (yellow) of the cervical (**G**) and lumbar (**H**) spine, changes are seen in the intervertebral disks that are subject to considerable stress. The degree of *lateral flexion* in the cervical and lumbar regions is approximately equal, but it is greatest in the thoracic region.
Rotation is possible in the thoracic and cervical region and particularly at the level of the atlantoaxial joint. Head rotation always goes hand in hand with movement of the atlantoaxial joint, movement of the cervical spine and slight movement of the thoracic spine. New research (*Putz*) has shown that rotation is also possible in the lumbar region. Movement of 3 to 7° may occur between two vertebrae.
Movements take place in "*motion segments*" (*Junghanns*), which are combined into "*motion zones*" (*Putz*). A motion segment is the range of movement that may occur between two vertebrae. This includes the intervertebral disks with the superior and inferior hyaline cartilage plates, vertebral joints, and ligaments, including all spaces.

Functional motion zones:
Craniovertebral joints–C3
C3–T1(T2)
T1 (T2)–(T11) T12
(T11) T12–Sacrum

> **Clinical tip:** Limitations of vertebral motion, called **restrictions**, most commonly affect the third cervical vertebra due to the position of the joint surfaces. They are also common in the lower cervical and lumbar spine. Restrictions cause significant pain due to nerve irritation, with neuralgia and muscle pain often radiating to the limbs. Spinal restrictions can usually be relieved by manual manipulation. Untreated restriction may eventually cause irreversible damage to the articular cartilage.

Trunk

A Typical cervical lordosis
(from a radiograph)

B Lordotic kink
(from a radiograph)

C Mild cervical lordosis
(from a radiograph)

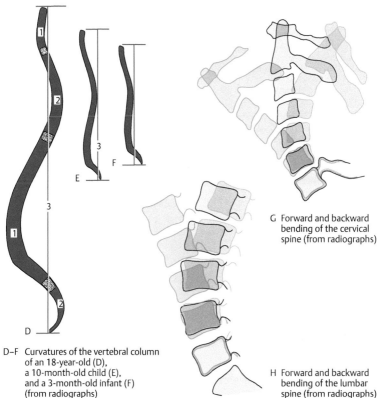

G Forward and backward
bending of the cervical
spine (from radiographs)

D–F Curvatures of the vertebral column
of an 18-year-old (D),
a 10-month-old child (E),
and a 3-month-old infant (F)
(from radiographs)

H Forward and backward
bending of the lumbar
spine (from radiographs)

Thoracic Cage

Ribs (A–F)

In each **rib** we distinguish a bony part, the **os costale**, and at the anterior end the **costal cartilage**.

There are 12 pairs of ribs, of which the upper seven are normally connected directly to the sternum and are called **true ribs**. The lower five ribs, **false ribs**, are joined indirectly (eighth to 10th) or not at all (11th–12th) to the sternum. The 11th and 12th ribs can be distinguished from the others as **floating ribs**.

Each **rib** has a *head* (**1**), *neck* (**2**), and *body* (**3**). The border between the neck and the body is defined by the *tubercle* (**4**). The head and the tubercle (*articular facet of the tubercle*, **5**) each have an articular surface. From the second to the 10th rib, the *articular facet of the head* (**6**) is divided in two by the *crest of the head of the rib* (**7**). On the upper margin of the neck of most ribs is the *crest of the neck of the rib* (**8**). Lateral and anterior to the tubercle is the *angle of the rib*. With the exception of the first, 11th and 12th, all ribs have a *costal sulcus* on their lower surface.

Curvatures. There are three rib curvatures—on edge, on the flat, and a torsion curvature. Although the *edge curvature*, which is the principal one in the first rib, is readily apparent, the *flat surface curvature* can only be seen on close inspection. It is present from the third rib on. If the upper surface of a rib is viewed near its anterior end and is followed toward the back, it will be seen that the surface slowly turns posteriorly. In addition to this curve, there is a longitudinal twist in the rib, which is most marked in the middle ribs and is called torsion. It is not present in the first, second or 12th ribs.

The **hyaline costal cartilage** begins to calcify with increasing age, more in males than in females. This reduces the mobility of the thorax (see p. 70).

Individual Features of Specific Ribs

The **first rib** (**A**) is small and flattened. On the inner circumference of its upper surface is an area of roughness, the *scalene tubercle* (**9**), to which the anterior scalenus is attached. Posterior to it lies the *groove for the subclavian artery* (**10**), and in front of it is the *groove for the subclavian vein* (**11**), which is not always clearly visible.

The **second rib** (**B**) has a rough area on its upper surface, the *tuberosity for the serratus anterior muscle* (**12**), from which one part of the serratus anterior originates.

The costal tubercle and costal sulcus are absent from **ribs 11 and 12** (**D**), and the costal angle is barely perceptible.

In two-thirds of cases the 10th rib ends freely; that is, it is not connected to the ninth rib or sternum. The first seven ribs are usually directly connected to the sternum, although sometimes the first eight may be so associated, and less commonly only the first six.

▨ **Variants:** The number of pairs of ribs is variable. There are usually 12 pairs but sometimes 11 or 13 are found. When there are 13 pairs, cervical (see p. 36) or lumbar ribs (see p. 42) may be present.

Malformations may lead to **fenestrated** or **bifid (forked) ribs** (**E**). The fourth rib is most commonly affected.

Ossification (F)

The cartilage anlages begin to ossify, progressing from dorsal to ventral by the end of the 2nd intrauterine month. By the end of the 4th intrauterine month, ossification ceases and the ventral part is preserved as the rib cartilage.

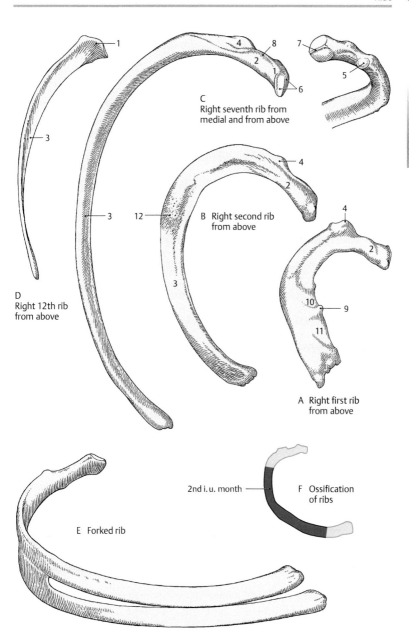

C
Right seventh rib from
medial and from above

B Right second rib
from above

D
Right 12th rib
from above

A Right first rib
from above

E Forked rib

2nd i. u. month

F Ossification
of ribs

Trunk

Trunk

Sternum (A–F)

The **sternum** consists of a *manubrium* (**1**), *body* (**2**) and *xiphoid process* (**3**). Between the manubrium and body lies the *sternal angle* (**4**), which is open toward the back. The xiphoid process is cartilaginous until maturity; with advancing age it may become ossified completely or remain partially cartilaginous. At the upper end of the manubrium sterni is the *jugular notch* (**5**) and lateral to it on either side the *clavicular notches* (**6**). The latter articulate with the clavicle. Just below the clavicular notch, the manubrium again has an additional paired *costal notch* (**7**) for a continuous cartilaginous joint with the first rib. At the sternal angle is a *notch* (**8**) for articulation between the sternum and second rib. The lateral borders of the body have costal notches for continuous connections with ribs 3 to 7. The costal notch for the seventh rib lies just at the junction of the body and xiphoid process. The manubrium and body of the sternum are usually joined by the *manubriosternal joint* (*synchondrosis*, see p. 68). A *xiphosternal joint* (*synchondrosis*) between the body and the xiphoid process is much less common.

The xiphoid process varies in shape. It may consist of one piece or it may be forked. Sometimes it contains a foramen and it may be bent forward or backward.

Sex Differences: The body of the sternum is longer in males than in females, and, for sterna of the same length, that of the male is narrower and slimmer than that of the female.

▇▇ **Variants:** Very rarely there are **suprasternal bones** (**C, 9**), also called the episternum, at the cranial end of the manubrium near the jugular notch. Sometimes there is an opening within the sternum, a **congenital sternal fissure** (**D, 10**), which arises during development.

Ossification (E, F)

The sternum develops from *paired sternal bands* that are formed by longitudinal fusion of individual rib anlages, followed by fusion of the sternal bands. In the region of the jugular notch a *paired suprasternal body* forms and subsequently regresses.

In the preformed cartilaginous part of the sternum, ossification starts from several bony centers. The first center usually appears in the manubrium between the 3rd and 6th intrauterine months. The remaining centers, usually paired, but partly unpaired, five to seven in number, then arise in the body of the sternum, the most caudal appearing in the 1st year. Fusion of the centers occurs between the ages of 6 to 20 (25) years. Secondary epiphyseal anlages may appear in the region of the clavicular notch and fuse with the manubrium between the ages of 25 and 30 years. Between the ages of 5 and 10 years, two ossification centers may develop in the region of the xiphoid process.

Clinical tip: Sternal puncture is performed by introducing a sternal puncture needle through the body of the sternum in the midline between the attachments of the second and third ribs. It must never be made at the level of the costosternal connections since synchondroses can be present here. Likewise, the lower two-thirds of the body of the sternum should **never** be punctured since a congenital sternal fissure (see above) conditioned by the paired ossific centers may be present.

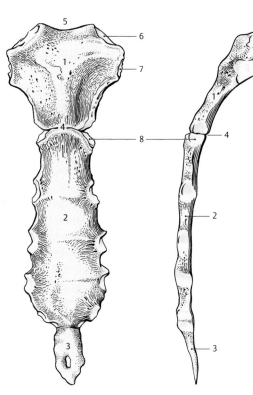

A Sternum from front

B Sternum from side

E Sternal ossification before birth

3rd to 6th i.u.m.

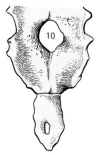

C Suprasternal bones

D Congenital sternal fissure

3rd to 6th g.m.

5th to 10th year

F Sternal ossification between 5 and 10 years

Joints of the Ribs (A–C)

Mobility of the ribs is a precondition for respiration. Connections exist between the ribs and vertebral column (joints) and also between the ribs and the sternum (diarthroses and synchondroses).

Costovertebral Joints (A, B)

Joints of the heads of ribs (**1**). Apart from the first, 11th, and 12th ribs, the joints of the heads of the ribs with the vertebral column represent double-chambered joints. Each rib articulates with the upper or lower borders of two neighboring vertebrae, and the intervertebral disk is connected by an *intra-articular ligament of head of rib* to the crest of the head of the rib. The capsule is strengthened by the *radiate ligament of head of rib* (**2**).

Costotransverse joints (**3**). With the exception of ribs 11 and 12, all ribs also articulate with the transverse processes of the vertebrae, so that here the two joints, head of rib and *costotransverse joints*, are necessarily combined. The articular surfaces of the costotransverse joints are the *articular facet of the costal tubercle* and the *costal fovea of the transverse process*. The capsules of these joints are delicate and are strengthened by ligaments, the *costotransverse ligament* (**4**), including the *lateral costotransverse ligament* (**5**) and the *superior costotransverse ligament* (**6**).

In the region of the 12th rib there is, in addition, the *lumbocostal ligament*, which extends from the costal process of the first lumbar vertebra to the 12th rib.

Movements. Sliding movements are possible for the first rib and ribs 6 to 9, and rotary motion about the neck is possible for ribs 2 to 5.

Sternocostal Joints (C)

Only some of the connections between the ribs and sternum are synovial joints. They are always present between the sternum and ribs 2 to 5, but ribs 1, 6, and 7 are joined to the sternum by *cartilaginous joints*, or *synchondroses* (**7**). The sternocostal joints are strengthened by ligaments that continue into the *sternal membrane* (**8**). An *intra-articular sternocostal ligament* (**9**) is always present at the second sternocostal joint. The other strengthening ligaments are the *radiate sternocostal ligaments* (**10**). In the sternocostal articulations one must keep in mind that the ribs (see p. 64) consist of bone and cartilage. The joints between the sternum and the ribs are formed by the cartilaginous part of the rib. This costal cartilage loses its elasticity at an early age due to calcium deposition.

The **interchondral joints** are a special type of articulation that occurs between the cartilages of the sixth to ninth ribs.

11 Manubriosternal symphysis (joint)
12 Clavicle
13 Xiphoid process

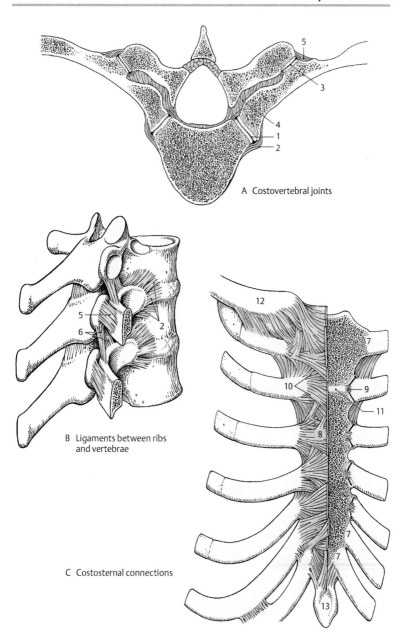

A Costovertebral joints

B Ligaments between ribs and vertebrae

C Costosternal connections

Trunk

Trunk

Boundaries of the Thoracic Cage (A–D)

The **thorax** consists of *12 thoracic vertebrae and their intervertebral disks, 12 pairs of ribs, and the sternum.* The thorax encloses the **thoracic cavity**, which has a *superior* (**1**) and an *inferior* (**2**) *aperture.* While the superior aperture is relatively narrow, the inferior one is very wide. The inferior thoracic aperture is limited by the *costal arch* (**3**) and *xiphoid process* (**4**) and the superior aperture by the two first ribs. The angle between the right and left costal arches is called the *infrasternal angle* (**5**).

The marked posterior curvature of the ribs and their posteriorly directed course between the transverse processes of the thoracic vertebrae and the costal angle causes the back of the chest wall to project posteriorly. The space located lateral to and behind the vertebral column is called the *pulmonary groove* of the thorax.

Movements of the Thoracic Cage (A–D)

Its elasticity makes for great resistance to stress. Movements of the thorax result from a summation of individual movements. As limiting positions we distinguish **maximal expiration** (**A, B**) on the one hand and **maximal inspiration** (**C, D**) on the other. During inspiration there is a widening of the thorax both in the anteroposterior and lateral directions. The expansion is made possible (1) by the mobility in the costovertebral joints, (2) by elasticity of the costal cartilages, which permit twisting, and (3) to a slight extent by increased kyphosis of the thoracic column.

During expiration the ribs are depressed, thus diminishing the size of the thorax in the anteroposterior and lateral dimensions. At the same time there is some decrease in the thoracic kyphosis. The infrasternal angle increases, becoming less acute during inspiration, while during expiration it becomes more acute.

The mobility of the thorax is reduced by calcification of the costal cartilages. The shape of the thorax is not the main determinant of respiratory capacity. The essential factor is its mobility, that is, the difference in volume between maximal expiration and maximal inspiration. Disorders not only of the cartilage but also of the joints cause reduction of total thoracic function.

The **forces that move the thorax** are generated by the *intercostal* (see p. 82) and *scalenus* (see p. 80) *muscles.* The intercostal muscles occupy the intercostal spaces. They are primitive metameric muscles, which are included among the intrinsic thoracic muscles. The latter also include the transversus thoracis and subcostal muscles. The musculature is innervated by ventral rami of the spinal nerves, the intercostal nerves.

Clinical tip: Erb's point 1 is a left parasternal auscultation point located in the third intercostal space in the plane of the cardiac valves (*red dot* in the diagrams).

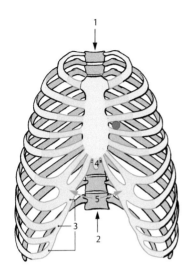

A Thorax-expiratory position from front

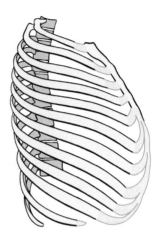

B Thorax-expiratory position from side

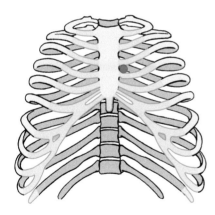

C Thorax-inspiratory position from front

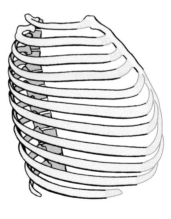

D Thorax-inspiratory position from side

Intrinsic Back Muscles (A, B)

This group includes *all the muscles inner-vated by the dorsal rami of the spinal nerves.* In the living body they form two longitudinal bulges that are lateral to the spinous processes and are most marked in the lumbar region. The muscles lie in a fibro-osseous canal formed by the bones of the vertebral arches, the costal processes, and the spinous processes. Posteriorly and laterally this canal is bounded by the thoracolumbar fascia (see p. 78). Because these muscles are difficult to demonstrate and are subject to considerable variation, currently they are classified without re-gard for their embryonic origins. Thus we no longer speak in terms of a "lateral tract" and "medial tract" but divide the muscles into three parts:

Erector spinae
  – iliocostalis
  – longissimus
  – spinalis

At the origin of all three muscles is the erector spinae aponeurosis (erroneously called the "sacrospinalis" in older publica-tions).

Spinotransverse
  – splenii
Interspinales (p. 74)
Intertransversarii (p. 74) and the
Transversospinales (p. 74)
  – rotatores
  – multifidi
  – semispinalis

Erector Spinae Muscles

The **iliocostalis** (**1**, **2**, **3**) consists of the ilio-costalis lumborum, iliocostalis thoracis and iliocostalis cervicis. The **iliocostalis lum-borum** (**1**) *extends from the sacrum, the ex-ternal lip of the iliac crest and the thora-columbar fascia to the costal processes of the upper lumbar vertebrae and the lower six to nine ribs.* The **iliocostalis thoracis** (**2**) *stretches from the lower six to the upper six ribs,* and the **iliocostalis cervicis** (**3**) *arises from the sixth to third ribs and inserts on the transverse processes of the sixth to fourth cervical vertebrae.*
Nerve supply: dorsal rami (C4–L3).

The **longissimus** (**4, 5, 6**) is subdivided into the longissimus thoracis (**4**) and cervicis (**5**) and the longissimus capitis (**6**). The **longissimus thoracis** *arises from the sacrum, the spinous processes of the lumbar verte-brae, and the transverse processes of the lower thoracic vertebrae and extends to the first or second ribs.* It is attached medially and laterally–medially to the accessory processes (**7**) of the lumbar vertebrae and to the transverse processes (**8**) of the thoracic vertebrae, and laterally to the ribs, the costal processes (**9**) of the lumbar vertebrae and the deep lamina of the thoracolumbar fascia. The **longissimus cervi-cis** *arises from the transverse processes of the upper six thoracic vertebrae and extends to the posterior tubercles of the transverse processes of the second to fifth cervical vertebrae.* The **longissimus capitis** *originates from the transverse processes of the three to five upper thoracic and the three lower cer-vical vertebrae and ends on the mastoid process* (**10**).
Nerve supply: dorsal rami (C2–L5).

Spinotransverse Muscles

The **splenius cervicis** (**11**) *extends from the spinous processes of the (third) fourth to (fifth) sixth thoracic vertebrae to the trans-verse processes of the first and second cervi-cal vertebrae.*

The **splenius capitis** (**12**) *arises from the spinous processes of the upper three thoracic and the lower four cervical verte-brae and ends in the region of the mastoid process* (**10**).
Nerve supply: dorsal rami (C1–C8).

The actions of all these muscles supplement one other. The first two are largely responsible for the erect posture of the body and then the two splenii, when contracted on one side, produce rotation of the head to the same side. They have an additional supporting function for the other intrinsic muscles of the back. In the thoracic and lumbar regions the intrinsic back muscles are held in place by the thoracolumbar fascia.

▬ **Variants:** Variations in the number of muscle slips is common.

I – XII: 1st to 12th ribs
13 Aponeurosis of the erector spinae muscle

The **levatores costarum** are on page 78.

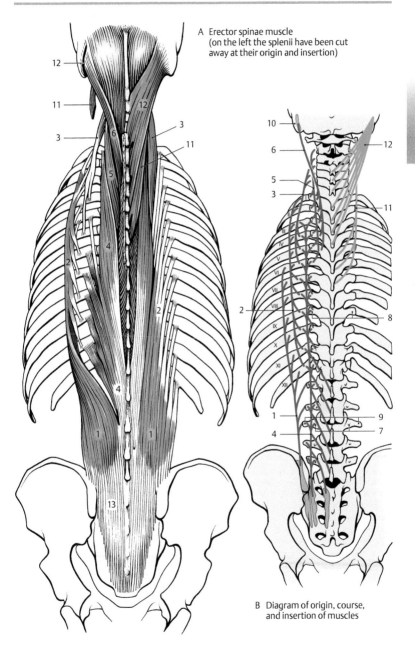

A Erector spinae muscle
(on the left the splenii have been cut
away at their origin and insertion)

B Diagram of origin, course,
and insertion of muscles

Trunk

Trunk

Intrinsic Back Muscles, continued (A–C)

Interspinales

The **interspinales** are arranged segmentally and are present in the cervical and lumbar regions. They are absent from the thoracic region, except between the first and second, second and third, and 11th and 12th thoracic vertebrae, and between the 12th thoracic and first lumbar vertebrae. *They link adjacent spinous processes.* On either side there are **six interspinales cervicis** (**1**), **four interspinales thoracis** (**2**), and **five interspinales lumborum** (**3**).
Nerve supply: dorsal rami (C1–T3 and T11–L5).

Intertransversarii

The **intertransversarii** lie lateral to the interspinales. The **six posterior intertransversarii cervicis** (**4**) connect the *adjacent posterior tubercles of the transverse processes of cervical vertebrae 2 to 7.*
Nerve supply: dorsal rami (C1–C6).
The **four medial intertransversarii lumborum** (**5**) connect the *mamillary* and *accessory processes of adjacent lumbar vertebrae.*
Nerve supply: dorsal rami (L1–L4).

Erector Spinae

The **spinalis** is made up of the spinalis thoracis, cervicis, and capitis. The latter is only occasionally present. The fibers of the **spinalis thoracis** (**6**) *arise from the spinous processes of the third lumbar through 10th thoracic vertebrae. They are inserted on the spinous processes of thoracic vertebrae 8 to 2*; the innermost fibers (from the 10th to eighth thoracic vertebrae) are the shortest. The fibers of the **spinalis cervicis** (**7**) *arise from the spinous processes of the second thoracic through the sixth cervical vertebrae and insert on the spinous processes of the fourth to second cervical vertebrae.*
Nerve supply: dorsal rami (C2–T10).

Transversospinales

The **rotatores breves** (**8**) and **longi** (**cervicis**), **thoracis** (**9**) (**et lumborum**) are most prominent in the thoracic region. *Each arises from a transverse process and runs to the next higher spinous process, or the one after, where it is inserted into the base.*
Nerve supply: dorsal rami (T1–T11).

The **multifidus** (**10**) consists of a number of small fasciculi (**M. multifidus, lumborum, thoracis**, and **cervicis**), which extend from the sacrum to the second cervical vertebra. It is best developed in the lumbar region. The individual fascicles *arise from the superficial aponeurosis of the longissimus muscle, the posterior surface of the sacrum, the mamillary processes of the lumbar vertebrae, the transverse processes of the thoracic vertebrae, and the articular processes of the seventh to fourth cervical vertebrae. The muscle bundles cross two to four vertebrae and are inserted in the spinous processes of the appropriate higher vertebrae.*
Nerve supply: dorsal rami (C3–S4).

The **semispinalis**, which overlies the multifidus laterally, is divided into thoracic, cervical, and cephalic (*capitis*) parts. Individual muscle bundles cross five or more vertebrae. The fibers of the **semispinalis thoracis** and **cervicis** (**11**) *arise from the transverse processes of all thoracic vertebrae. They are inserted into the spinous processes of the upper six thoracic and lower four cervical vertebrae.* The **semispinalis capitis** (**12**), which is one of the strongest muscles of the neck, *arises from the transverse processes of the upper four to seven thoracic vertebrae and the articular processes of the five lower cervical vertebrae. It is inserted between the superior and inferior nuchal lines of the skull.*
Nerve supply: dorsal rami (T4–T6, C3–C6, and C1–C5).

The straight muscles function as extensors when both sides are innervated and unilaterally as lateral flexors when only one side is innervated. Oblique muscles function when unilaterally innervated as rotators and bilaterally innervated as extensors.

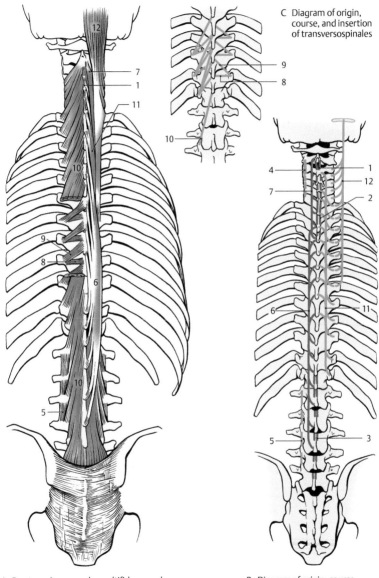

C Diagram of origin, course, and insertion of transversospinales

A Erector spinae muscle, multifidus muscle partially removed to make the rotator muscles visible)

B Diagram of origin, course, and insertion of straight muscle system

Trunk

Short Nuchal Muscles (A, B)

The paired short nuchal muscles, the rectus capitis posterior minor and major and the obliquus capitis superior and inferior, are part of the intrinsic back muscles and, except for the inferior obliquus capitis, they, too, belong to the straight system of the medial tract. Both recti originate from interspinal muscles, and the obliquus capitis superior originates from an intertransverse muscle.

Two other short neck muscles, the rectus capitis lateralis and the rectus capitis anterior, do not belong to the intrinsic muscles of the back. The former is one of the muscles that have migrated from the anterolateral body wall; it is described on page 78. The anterior rectus capitis, a prevertebral muscle, is described on page 80.

The **rectus capitis posterior minor** (1) *arises* from the *posterior tubercle of the atlas* and ascends upward in a fan shape. It *inserts* in the medial region of the *inferior nuchal line*. It is covered at the lateral aspect of its insertion by the rectus capitis posterior major muscle.

The **rectus capitis posterior major** (2) takes its *origin* from the *spinous process of the second cervical vertebra* and *inserts* on the *inferior nuchal line* lateral to the rectus capitis posterior minor muscle. It also widens out in the direction of its insertion in a similar fashion to the rectus capitis posterior minor.

The **obliquus capitis superior** (3) *originates from the transverse process of the atlas*. It is inserted on the occipital bone somewhat above and lateral to the rectus capitis posterior major.

The **obliquus capitis inferior** (4) *runs from the spinous process of the second cervical vertebra to the transverse process of the atlas*.

All the short nuchal muscles act on the craniovertebral joints. Bilateral contraction causes the straight and oblique muscles to tilt the head backward while unilateral contraction of the obliquus capitis superior turns the head sideways. Lateral rotation of the head is caused by synergistic contraction of the rectus capitis posterior major and obliquus capitis inferior.

Nerve supply: suboccipital nerve (C1).

Clinical tip: The rectus capitis posterior major and the obliquus capitis superior and inferior form the **suboccipital triangle** (trigonum a. vertebralis). Here the vertebral artery (see p. 346) can be located, lying on the posterior arch of the atlas. Between the artery and the posterior arch of the atlas lies the first cervical nerve, whose dorsal ramus, the suboccipital nerve (see p. 346 and Vol. 3), innervates these muscles. Suboccipital puncture is described on page 346.

▬ **Variants:** The rectus capitis posterior minor may be absent or very small on one side. The rectus capitis posterior major is rarely absent. Sometimes it may be divided into two muscles.

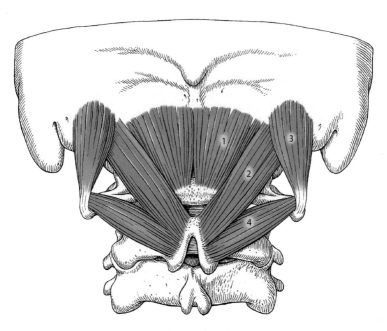

A Suboccipital muscles

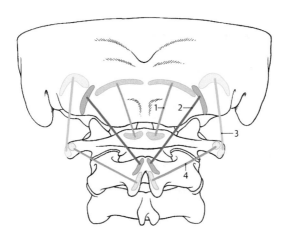

B Diagram of origin, course, and insertion of muscles

Body Wall

Thoracolumbar Fascia (A, B)

The **thoracolumbar fascia** (**1**) completes the fibro-osseous canal formed by the vertebral column and the posterior surfaces of the ribs. *It invests all intrinsic muscles of the back* (**2**) *and consists of three layers.* The **superficial = posterior layer** (**3**) is firmly bound to the erector spinae aponeurosis in the sacral region. Ascending in the body it becomes somewhat thinner and serves as an origin for the latissimus dorsi (**4**) and posterior inferior serratus (**5**). In the cervical region, where it has become very thin, it separates the splenius capitis and splenius cervicis from the trapezius (**6**) and becomes the nuchal fascia (**7**).

The **deep = anterior layer** (**8**) arises from the costal processes (**9**) of the lumbar vertebrae and separates the intrinsic back muscles (**2**) from those of the anterolateral body wall.

The internal abdominal oblique (**10**) and the transversus abdominis (**11**) arise from the deep layer, which extends as far as the iliac crest. The **middle layer** lies within the intrinsic back muscles.

The **nuchal fascia** (**7**) is continuous anterolaterally with the superficial cervical fascia (see p. 331). The nuchal ligament lies in the middle of the nuchal fascia.

Extrinsic Anterolateral Muscles (A)

The muscles described are innervated by the ventral rami of the spinal nerves, and in the course of development have migrated into the posterior body wall.

The **rectus capitis lateralis** *runs from the transverse process of the atlas to the jugular process of the occipital bone* and corresponds developmentally to an anterior intertransverse muscle. Its action produces lateral head flexion.
Nerve supply: C1.

The **anterior intertransversarii cervicis** are six small bundles running *between the anterior protuberances on the transverse processes of the cervical vertebrae.*
Nerve supply: C2–C6.

The **lateral intertransversarii lumborum** consist of five or six muscle bundles *between the costal processes of the lumbar vertebrae.*
Nerve supply: L1–L4.

The **levatores costarum** *arise from the transverse processes of the seventh cervical and the first to 11th thoracic vertebrae. They reach the costal angles* of the next rib below as the **short levatores costarum**, or the second rib below as the **long levatores costarum**. They are involved in spinal rotation.
According to *Steubl* these muscles are innervated by the dorsal rami of the spinal nerves and thus belong to the lateral tract of the intrinsic back muscles.
Nerve supply: dorsal rami of the spinal nerves.

The **posterior superior serratus** (**12**) *originates from the spinous processes of the last two cervical and the first two thoracic vertebrae and is inserted on ribs 2 to 5*, which it elevates.
Nerve supply: intercostal nerves (T1–T4).

The **posterior inferior serratus** (**5**) *arises from the thoracolumbar fascia* in the region of the 12th thoracic and first to third lumbar vertebrae *and usually extends* with four digitations *to the 12th to ninth ribs.* It lowers the ribs.
Nerve supply: intercostal nerves (T9–T12).

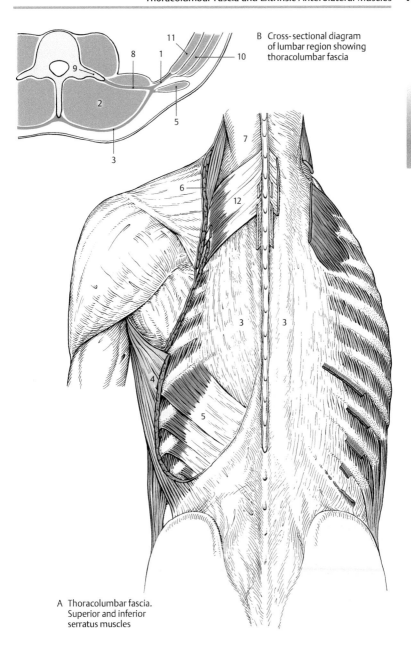

B Cross-sectional diagram of lumbar region showing thoracolumbar fascia

A Thoracolumbar fascia. Superior and inferior serratus muscles

Prevertebral Muscles (A, B)

The prevertebral muscles include the rectus capitis anterior, longus capitis, and longus colli.

The **rectus capitis anterior** (**1**) *extends from the lateral mass of the atlas* (**2**) *to the basal part of the occipital bone* (**3**). It helps to flex the head.
Nerve supply: cervical plexus (C1).

The **longus capitis** (**4**) *arises from the anterior tubercles of the transverse processes of the third to sixth cervical vertebrae* (**5**). It runs upward and is *attached to the basal part of the occipital bone* (**6**). The two longi capitis muscles bend the head forward. Unilateral action of the muscle helps to tilt the head sideways.
Nerve supply: cervical plexus (C1–C4).

The **longus colli** (**7**) is roughly triangular in shape because it consists of three groups of fibers. The **superior oblique fibers** (**8**) *arise from the anterior tubercles on the transverse processes of the fifth to second cervical vertebrae* (**9**) *and are inserted on the anterior tubercle of the atlas* (**10**). The **inferior oblique fibers** (**11**) *run from the bodies of the first to third thoracic vertebrae* (**12**) *to the anterior tubercle on the transverse process of the sixth cervical vertebrae* (**13**). The **medial fibers** (**14**) *extend from the bodies of the upper thoracic and lower cervical vertebrae* (**15**) *to the bodies of the upper cervical vertebrae* (**16**). Unilateral contraction of the muscle bends and turns the cervical vertebral column to the side. Together, both longi colli muscles flex the cervical spine forward. Electromyographic studies have shown that the homolateral muscle is also involved in lateral flexion and rotation of the cervical spine.
Nerve supply: cervical and brachial plexus (C2–C8).

Scalene Muscles (A, B)

The **scalene muscles** represent the cranial continuation of the intercostal muscles. They arise from the vestigial ribs of the cervical vertebrae. They are the most important muscles for quiet inspiration, as they lift the first two pairs of ribs and thus the superior part of the thorax. Their action is increased when the head is tilted backward. Unilateral contraction bends the cervical spine to one side. Occasionally there is a scalenus minimus, which arises from the seventh cervical vertebra and joins the scalenus medius. It is attached to the apex of the pleura.

The **scalenus anterior** (**17**) *arises from the anterior tubercles of the transverse processes of the (third) fourth to sixth cervical vertebrae* (**18**) *and is inserted on the anterior scalene tubercle* (**19**) *of the first rib.*
Nerve supply: brachial plexus (C5–C7).

The **scalenus medius** (**20**) *arises from the posterior tubercles of the transverse processes of the (first) second to seventh cervical vertebrae* (**21**). It is *inserted into the first rib* behind the subclavian artery groove *and into the external intercostal membrane of the first intercostal space* (**22**). In this way it can reach the second rib. The attachment at the first rib is located behind the groove for the subclavian artery.
Nerve supply: cervical and brachial plexus (C4–C8).

The **scalenus posterior** (**23**) *runs from the posterior tubercles on the transverse processes of the fifth to seventh cervical vertebrae* (**24**) *to the second (third) rib* (**25**). It can be absent.
Nerve supply: brachial plexus (C7–C8).

A **scalenus minimus muscle** may be present in approximately one-third of cases. It arises from the *anterior tubercle of the transverse process of the seventh cervical vertebra* and *reaches the fibrous vault of the pleura and the first rib.* If the muscle is absent, a *transverse cupular ligament (Hayek Ligament)* replaces it.
Nerve supply: brachial plexus (C8).

Clinical tip: Between the scalenus anterior and scalenus medius lies the **scalene interval** (**26**), through which pass the brachial plexus (see p. 360 and Vol. 3) and the subclavian artery. Retroversion of the arm may occlude the subclavian artery between the rib and the clavicle. Together with the longus colli, the scalenus anterior forms the medial wall of the **scalenovertebral triangle** (**27**; see p. 366).

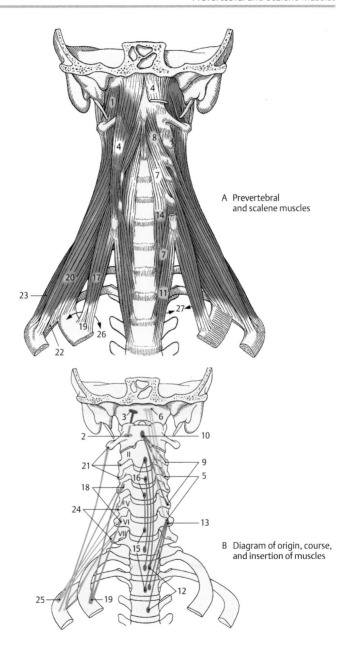

Trunk

A Prevertebral and scalene muscles

B Diagram of origin, course, and insertion of muscles

Trunk

Muscles of the Thoracic Cage

Intercostal Muscles (A–D)

In addition to the scalene muscles, the intercostals are necessary for movements of the chest wall. These are divided into

- **external intercostal**
- **internal intercostal**
- **subcostal and**
- **transverse thoracic muscles**

The outermost intercostal muscles, the **external intercostals** (**1**), *extend from the costal tubercle to the beginning of the rib cartilage* and continue in every intercostal space into the **external intercostal membrane** where the rib bone merges with the costal cartilage. Each of these muscles *originates from the inferior margin of a rib* and *inserts on the superior margin of a rib*. The external intercostals run from superoposterior to inferoanterior. According to their function they are known as inspiratory muscles (*Fick*). Electromyographic studies between 1950 and 1990 have shown, however, that the external intercostals are active only during forced inspiration and that quiet breathing depends on the action of the scalene muscles alone (see p. 80).
Nerve supply: intercostal nerves (T1–T11).

The **internal intercostals** (**2**) *run from the costal angle to the sternum* in every intercostal space. *They arise from the superior margin of the inner surface of the rib and are inserted in the region of the costal groove*. From the costal angle medially toward the vertebrae, the internal intercostals are replaced by ligamentous fibers, which are known as the **internal intercostal membrane**.

In the region of the costal cartilages they may be referred to as **intercartilaginous muscles** (**3**).

A portion of each inner intercostal muscle is separated to form the intercostales intimi, also called the **innermost intercostal muscles**. Between them and the internal intercostals lie the intercostal nerve and vessels.

The direction of the internal intercostals is opposite to that of the external muscles; that is, they run from inferoposterior to superoanterior.

According to *Fick* they are expiratory muscles; that is, they are activated only when the ribs are lowered. The intercartilaginous muscles; particularly those of the fourth to sixth intercostal spaces, act as inspiratory muscles by virtue of their position in relation to the sternum.
Nerve supply: intercostal nerves (T1–T11).

The **subcostals** (**4**), which lie in the region of the costal angles, consist mainly of fibers of the internal intercostal muscles that extend over several segments. They have the same function as the internal intercostals.
Nerve supply: intercostal nerves (T4–T11).

The **transversus thoracis** (**5**) *arises from the internal surface of the xiphoid process and the body of the sternum*. Its fibers run in a laterocranial direction and *are attached to the lower border of the second to sixth costal cartilages*.

The direction of the muscle slips fans out; that is, the uppermost slip ascends steeply upward, whereas the lowermost slip courses parallel to the transversus abdominis muscle. A sharp boundary between the transversus thoracis and the transversus abdominis is only then attained when the origin of the costal part of the diaphragm (see p. 102) is well developed from the seventh rib. The transversus thoracis functions in expiration.
Nerve supply: intercostal nerves (T2–T6).

■■■ **Variants:** Numerous variations are known. Right and left muscles are frequently formed asymmetrically. Sometimes it may be absent. The number of slips can vary.

Clinical tip: The internal thoracic artery and vein run anterior to the transversus thoracis. When the muscle is strongly developed, the artery is difficult to expose during coronary bypass operations.

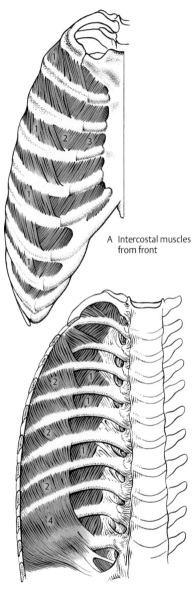

A Intercostal muscles from front

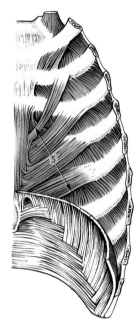

B Transversus thoracis, viewed from inside anterior thoracic wall

C View from inside posterior thoracic wall

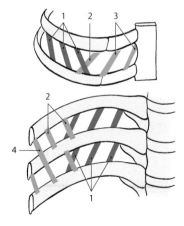

D Diagram of origin, course, and insertion of muscles

Abdominal Wall

The abdominal wall is bounded superiorly by the infrasternal angle and inferiorly by the iliac crest, the inguinal sulcus, and the pubic sulcus. Under the abdominal skin lies the more or less extensive subcutaneous fatty tissue, which is separated from the muscles by the superficial abdominal fascia. The framework of the abdominal wall is provided by the abdominal muscles. The superficial abdominal muscles are so arranged as to produce the greatest possible degree of effectiveness. Individual abdominal muscles develop from several myotomes and are therefore innervated by several segmental nerves. This makes possible regional contraction of the anterior muscles.

Superficial Abdominal Muscles:

Lateral group:
- **External abdominal oblique**
- **Internal abdominal oblique**
- **Transversus abdominis**

Medial group:
- **Rectus abdominis**
- **Pyramidalis**

Deep Abdominal Muscles:

- **Quadratus lumborum**
- **Psoas major**

Flattened ligaments, the aponeuroses of the lateral abdominal muscles, enclose the rectus abdominis to form the **rectus sheath** (see p. 88).

Superficial Abdominal Muscles

Lateral Group (A–C)

The **external abdominal oblique** (**1**) *arises by eight slips from the outer surface of the fifth to 12th ribs* (**2**). Between the fifth and (eighth) ninth ribs it interdigitates with the slips of the serratus anterior (**3**) and between the 10th and 12th ribs with those of the latissimus dorsi (**4**).

Fundamentally the direction of its fibers is from superolaterally and posterior toward inferomedial and anterior. The fibers that come from the three lowest ribs extend almost vertically down to the iliac crest and its labium externum (**5**), and the remainder run obliquely downward and medially, where they blend into a flat aponeurosis (**6**). The transition of the muscle fibers into the aponeurosis follows an almost vertical line that is covered by the margin between the cartilage and bone of the sixth rib. Just above the anterior superior iliac spine, the transition of the muscle fibers into the aponeurosis takes place in a transverse plane. One speaks here of a "muscle edge." The lowermost portion of this aponeurosis is continuous with the inguinal ligament (**Lig. of Vesalius**).

The **superficial inguinal ring** lies in the medial region directly above the **inguinal ligament** and is bordered by the *medial* (**7**) and *lateral* (**8**) *crus* as well as by *intercrural fibers* (**9**; see p. 96). The attachment of the external abdominal oblique is located in the midline. Here, the aponeuroses of the right and left muscles are interwoven with one another and with those of the other lateral abdominal muscles to form a fibrous raphe, the **linea alba** (**10**).
Nerve supply: intercostal nerves (T5–T12).

■■■ **Variants:** The muscle may have more or fewer slips of origin. Tendinous intersections may be present. There may also be connections with the nearby latissimus dorsi and serratus anterior.

Clinical tip: Up to 15 different terms, most of them antiquated, have been applied to the inguinal ligament over the years, ranging from crural arch and superficial crural arch to Poupart's ligament. Several of the terms have nothing to do with the actual inguinal ligament (see also Kremer et al, Chirurgische Operationslehre, Vol. 7/1, pp. 62–63).

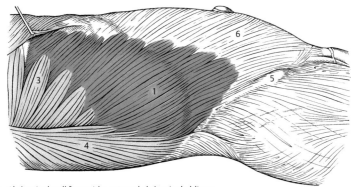

A Abdominal wall from side: external abdominal oblique

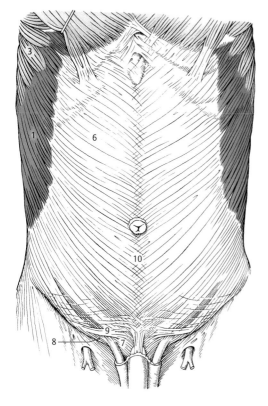

B Abdominal wall from front: external abdominal oblique

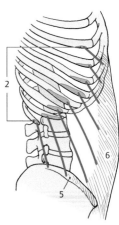

C Diagram of origin, course, and insertion of muscles

Superficial Abdominal Muscles

Lateral Group, continued (A, B)

The **internal oblique** (**1**) *originates* at the *intermediate line of the iliac crest* (**2**), at the *deep layer of the thoracolumbar fascia*, and at the *anterior superior iliac spine* (**3**). Individual fibers can also arise from the *inguinal ligament* (**4**).

The muscle takes a fan-shaped, predominantly ascending course, and thus **three parts** can be distinguished on the basis of their attachments.

Its **cranial portion** *inserts* at the *lower margins of the last three ribs* (**5**).

The **middle part** (**6**) *continues medially into the aponeurosis, which is divided into anterior and posterior layers*. These layers form the framework of the *rectus sheath* (see p. 88), and they reunite in the linea alba. The anterior layer completely covers the rectus abdominis, but the posterior layer ends about 5 cm below the navel as a cranially convex line, the arcuate line. As this margin is not always sharply defined, it is more correct to speak of an *area arcuata* (*Lanz*).

Its **caudal part** is continued in the male into the spermatic cord as the *cremaster muscle* (**7**). The development of the cremaster muscle is subjected to great variation. In the female the muscle bundles that reach the round ligament of the uterus are distinctly weaker and are designated as the round ligament part of the internal abdominal oblique.

Nerve supply:
Internal oblique: intercostal nerves (T10–T12 and L1).
Cremaster muscle: genital ramus of the genitofemoral nerve (L1–L2).

■■■ **Variants:** Reduction or increase in the number of slips inserting on the ribs as well as of tendinous intersections may occur.

The **transversus abdominis** (**8**) *arises by six slips from the inner surface of the cartilage of ribs 7 to 12* (**9**); its slips interdigitate with those of the costal part of the diaphragm. They are attached directly to the origins of the transversus thoracis muscle. *It also takes its origin from the deep layer of the thoracolumbar fascia, the inner lip of the iliac crest* (**10**), *the anterior superior iliac spine* (**11**), *and the inguinal ligament* (**12**). Its fibers run transversely to a medially concave line that is known as the *semilunar line*. The aponeurosis begins at this line. It is cranial to the line or area arcuata and participates in the formation of the posterior layer of the rectus sheath. Caudal to the area arcuata (see above), the aponeurosis only forms the anterior layer of the rectus sheath. The transversus abdominis participates via its aponeurosis in the linea alba. The **inguinal falx**, also called the conjoined tendon or ligament of Henle (see p. 92), a band that is concave laterally, runs from the aponeurosis to the lateral margin of the attachment of the rectus abdominis muscle.

Nerve supply: intercostal nerves (T7–T12) and L1.

■■■ **Variants:** The transversus abdominis may fuse completely in its lower region with the internal abdominal oblique, and because of this it is sometimes called the *complex muscle*. There are reports in the literature of its complete absence. The number of bands of origin may be increased or decreased.

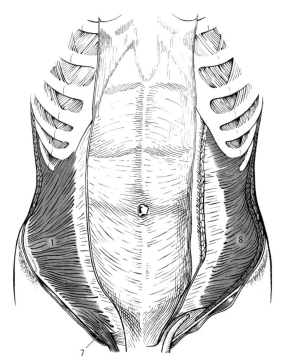

A Abdominal wall from front:
internal abdominal oblique muscle
and transversus abdominis

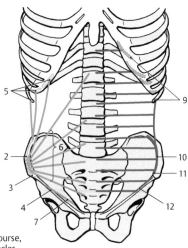

B Diagram of origin, course,
and insertion of muscles

Trunk

Superficial Abdominal Muscles, continued

Medial Group (A–D)

The **rectus abdominis** (1) arises by three slips from the *outer surface of the cartilages of the fifth to seventh ribs* (2), the *xiphoid process* (3), and the *intervening ligaments. It descends to the pubic crest* (see p. 186). In its course down near the level of the umbilicus there are three tendinous intersections; sometimes there are another one or two below it.
Nerve supply: intercostal nerves (T5–T12).

■■■ **Variants:** The muscle may arise from more ribs or, rarely, may be entirely absent.

The rectus abdominis lies within the **rectus sheath**. This is formed by the aponeuroses of the three lateral abdominal muscles coming together in such a way that above the *arcuate line* (4) the aponeurosis of the internal abdominal oblique (5) divides into an *anterior* (6) and *a posterior lamina* (7). The aponeurosis of the external abdominal oblique (8) strengthens the anterior lamina and that of the transversus abdominis (9) strengthens the posterior lamina of the sheath. In the region of the **linea alba** (10) there is partial intertwining of the fibers (B).

Between the individual aponeurotic fibers there is a fatty infiltrate. The linea alba extends as far as the symphysis and is strengthened at the superior margin of the pelvis (11). Below the arcuate line the rectus sheath is incomplete, since the aponeuroses of all the lateral abdominal muscles run in front of both rectus muscles, and the inner side of these muscles is covered only by the transversalis fascia (12; see p. 92) and the peritoneum (C). In the region of the origin of the rectus abdominis, the rectus sheath is a thin fascial structure representing a continuation of the pectoral fascia.

> **Clinical tip:** Separation of the rectus muscles and an abnormal increase in the width of the linea alba is of clinical importance (**rectus diastasis**; see p. 96).
> Only the anterior surface of the rectus abdominis muscle is fused to the rectus sheath in the region of the intersecting tendons. Therefore abscesses or collections of pus can only form between two intersections on the anterior surface, while on the posterior surface they may spread along the entire rectus muscle.

The small, triangular **pyramidalis** (13) *arises from the pubis, radiates into the linea alba* and lies within the aponeurosis of the three lateral abdominal muscles. It is supposed to be absent in 16 to 25% of cases.

Careful examination reveals that the pyramidalis is present in most cases, although variable in its development. We have found it in 90% of cases, so that in only 10% of cases no muscle fibers were seen. The sole function of the pyramidalis is to tense the linea alba.
Nerve supply: T12–L1.

Trunk

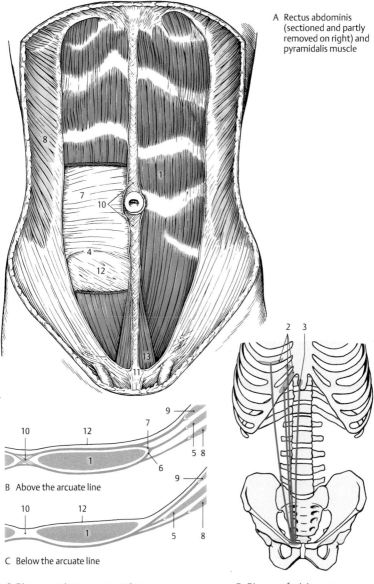

A Rectus abdominis (sectioned and partly removed on right) and pyramidalis muscle

B Above the arcuate line

C Below the arcuate line

B, C Diagrammatic transverse sections through anterior abdominal wall

D Diagram of origin, course, and insertion of muscles

Function of the Superficial Abdominal Muscles (A–D)

The superficial abdominal muscles with their aponeuroses form the basis of the anterior and lateral abdominal wall.

Together with the deep muscles, the psoas major and quadratus lumborum, they are necessary for movement of the trunk. In addition, the anterior and lateral abdominal muscles act on the intra-abdominal space. Their contraction causes a rise of intra-abdominal pressure. The diaphragm and pelvic floor are also involved. This is necessary, for example, during evacuation of the bowel. Finally they may be important during respiration, when the rectus abdominis contracts in forced expiration.

Basically all the superficial muscles act together to produce the different movements conditional on the tension of the aponeuroses within the linea alba. The direction of tension (**A**) in the individual muscle fibers is such that they supplement one another.

The rectus abdominis (green) runs craniocaudally and is subdivided into several segments. Most of the fibers of the external oblique (red) run obliquely from superolaterally to inferomedially, while those of the internal oblique (blue) extend inferolaterally to superomedially. The transverse abdominal muscle (violet) runs transversely from lateral to medial.

The function of each muscle may vary in individual movements.

Flexion (**B**) of the trunk is essentially a movement of the rectus muscles (green). They are assisted by the oblique muscles (not shown).

Lateral flexion (**C**) is achieved by contraction of the external oblique (red), the ipsi-lateral internal oblique (blue), the quadratus lumborum muscle (not shown), and the ipsilateral intrinsic back muscles (not shown).

Rotation (**D**) follows contraction of the internal oblique (blue) on the same side (i.e., the side toward which the body is rotated) and the external oblique on the opposite side.

It should be understood that the external oblique (red) and internal oblique (blue) of the same side sometimes act synergistically (in lateral flexion) (**C**) and sometimes are antagonists (**D**).

The transverse abdominal (violet) is mainly active in abdominal pressure, so that both transverse muscles may constrict the abdominal cavity. In addition, during expiration, their contraction may pull the diaphragm upward.

Clinical tip: During contraction of the abdominal muscles, particularly in reaching an upright posture from the supine position, it should be noted that the iliopsoas muscle (see p.94) plays an essential part. In a thin person, the tendinous intersections (see p.88) of the rectus muscles and the strands of origin of the external oblique muscles may be clearly seen. Any damage to the rectus muscles, such as a **rectus diastasis** (see p.96), can be detected. In addition, reflex contractions of the superficial abdominal muscles in intraperitoneal inflammations (reflex contraction of the abdominal muscles) may be observed.

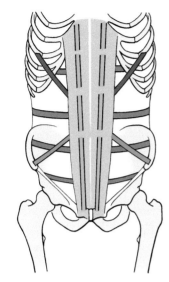

A Direction of tension of the muscle fibers

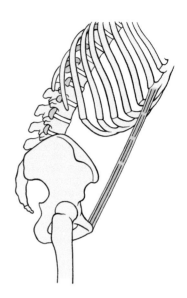

B Anterior flexion

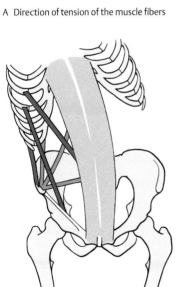

C Lateral flexion

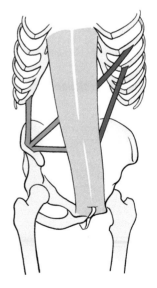

D Lateral rotation

Trunk

Fasciae of the Abdominal Wall (A, B)

The abdominal wall can be divided into
- The **skin**
- The **subcutaneous fatty tissue**
- The **connective tissue lamellae**
- The **superficial abdominal fascia**
- The **muscles and their fasciae**
- The **transverse fasciae**
- The **peritoneum**

The connective tissue lamellae permeating the subcutaneous fatty tissue form the **membranous layer of the subcutaneous tissue of the abdomen** or **Scarpa's fascia** (**1**), which is arranged in the caudal region of the abdominal wall in the inguinal regions and pubic region. The fatty tissue in this region is called the **fatty layer of the abdomen**, **the abdominal fat pad**, or **Camper fascia**. Both structures together form the **subcutaneous tissue of the abdomen**. The membranous abdominal layer, which is continuous onto the thigh, is of significance to the surgeon because the larger subcutaneous vascular trunks are situated between it and the true superficial abdominal fascia. A portion of the connective tissue lamellae that is continued in the direction of the sexual organs is also designated as the **fundiform ligament of the penis** (**2**) **or clitoris**.

The **abdominal fascia** (**3**) represents a thin plate that is strengthened only in the region of the linea alba (see p. 96) and covers the entire anterior abdominal musculature and its aponeuroses. The portion of the fascia situated in the midline continues into the elastic-rich fibers of the **suspensory ligament of the penis** (**4**) **or clitoris**. This ligament embraces the corpus cavernosum of the penis or clitoris with two crura.

In the region of the superficial inguinal ring the fascia fuses with the extension of the aponeurosis of the external abdominal oblique to form the **external spermatic fascia** (**5**), which provides the outer covering of the spermatic cord. With the aponeurosis of the external abdominal oblique it is more firmly bound also in the region of the inguinal ligament and then continues in the fascia of the thigh (**6**).

The inner loose abdominal wall fascia, the **transversalis fascia** (**7**), covers the inner surface of the abdominal muscles. It is taut in the umbilical region, where it may be called the **umbilical fascia** (**8**).

This fascia is also reached by connective tissue lamellae with embedded fat cells that pass upward from the apex of the urinary bladder. They contain the urachal cord and the cords of the umbilical arteries and can be designated as the vesicoumbilical fibrous septum. This septum strengthens the transversalis fascia.

The transversalis fascia blends inferiorly with the inguinal ligament to form the **iliopubic tract** (**9**), thus constituting the posterior wall of the inguinal canal (see p. 96). It extends from the inguinal ligament into the **iliac part** of the iliopsoas fascia, which covers the iliac muscle (**10**). Superiorly it covers the diaphragm and posteriorly the quadratus lumborum and psoas major as the **iliopsoas fascia**.

In the region of the inguinal canal the transversalis fascia, strengthened by aponeurotic fibers of the transversus abdominis, thickens to form the **interfoveolar ligament** (**11**; see p. 98). Attached medially to the rectus abdominis (**12**), the transversalis fascia extends as a band covering a radiation of the aponeurosis of the transverse abdominal muscle and is firmly attached to it. This band, which is laterally concave, extends behind the reflex ligament (see p. 96) to the lacunar ligament (see p. 100), where it is in close contact with the inguinal ligament, and is called the **inguinal falx** (**13**) or conjunctival tendon. Lateral to the interfoveolar ligament the transversalis fascia evaginates at the deep inguinal ring (**14**) to form the **internal spermatic fascia**. Below the inguinal ligament lies the femoral canal (**15**).

16 Cord of umbilical artery
17 Urachal cord

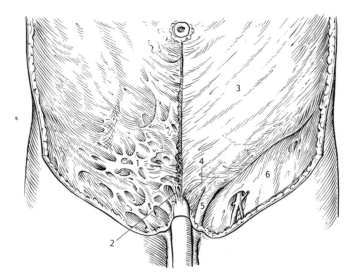

A Right: subcutaneous tissue of abdomen;
left: (external) superficial abdominal fascia

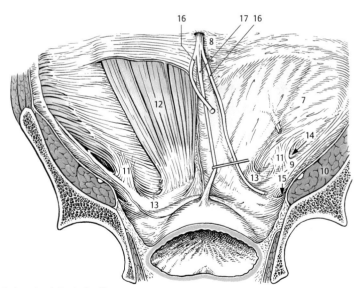

B Anterior abdominal wall
from inside with transverse fascia on the right

Trunk

Deep Abdominal Muscles (A, B)

The **psoas major** (**1**) is subdivided into a **superficial** and a **deep part**. *The superficial part arises from the lateral surfaces of the 12th thoracic and the first to fourth lumbar vertebrae* (**2**) and their *intervertebral disks. The deep part arises from the costal processes of the first to fifth lumbar vertebrae* (**3**).

The psoas major joins the iliacus and, surrounded by the iliac fascia, extends as the **iliopsoas** (**4**) through the lacuna musculorum to the *trochanter minor* (**5**). The lumbar plexus runs between the two layers of the psoas major (see also p. 234).
Nerve supply: direct branches from the lumbar plexus and the femoral nerve (L1–L3).

The psoas major extends over several joints and is capable of considerable elevation of the leg. The iliacus muscle (see p. 234), with which it joins to form the iliopsoas muscle, is a powerful flexor and thus supplements the action of the psoas major. In the recumbent position both psoas muscles help to lift the upper or lower half of the body. In addition the psoas major can give slight assistance in lateral flexion of the vertebral column.

Sometimes a **psoas minor** is found, *split off from the psoas major*, which enters into the iliac fascia and *inserts on the iliopubic eminence*. It acts as a tensor of the fascia (see p. 234).
Nerve supply: direct branch from the lumbar plexus (L1–L3).

Clinical tip: The fascia surrounds the psoas major as a tube, *psoas fascia*, stretching from the medial lumbocostal arch (see p. 102) to the thigh. Thus, any inflammatory processes in the thoracic region can extend within the fascial tube to appear as a **psoas (migratory) abscesses** as far down as the thigh.

The **quadratus lumborum** (**6**) *extends to the 12th rib* (**7**) *and to the costal processes of the first to third (fourth) lumbar vertebrae* (**8**). *It arises from the inner lip of the iliac crest* (**9**). This muscle consists of two incompletely separated layers. The anterior layer extends to the 12th rib and the posterior layer is attached to the costal processes.

The quadratus lumborum muscle lowers the 12th rib and aids lateral flexion of the body.
Nerve supply: T12 and L1–L3.

10 Median arcuate ligament
11 Medial arcuate ligament
12 Lateral arcuate ligament
13 Diaphragm (costal part)
14 External abdominal oblique
15 Pectineus

Trunk

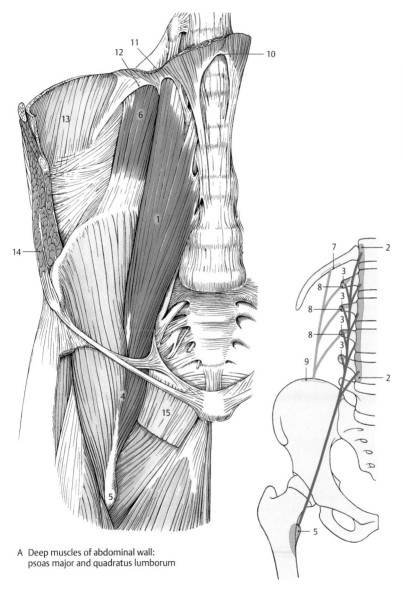

A Deep muscles of abdominal wall:
psoas major and quadratus lumborum

B Diagram of origin, course,
and insertion of muscles

Sites of Weakness in the Abdominal Wall (A–D)

Sites of weakness in the musculoaponeurotic abdominal wall are the sites at which **hernias** tend to develop. A hernia is the escape of abdominal contents from the original body cavity. These contents lie in a *hernial* sac, a secondary protrusion of the peritoneum that passes through the *hernial orifice* in the abdominal wall. **Sites of weakness in the abdominal wall** are the *linea alba, umbilicus, inguinal region, femoral canal, lumbar triangle,* and *surgical scars.*

Linea alba

The linea alba (**1**) is formed by the interlacing aponeuroses of the lateral abdominal muscles and is a tendinous raphe lying between the rectus sheaths. It ends at the upper margin of the symphysis. On the posterior surface it widens near its attachment and ends as a triangular plate, the **posterior attachment** (**adminiculum**) **of the linea alba.** Above the umbilicus (**2**) it is 1 to 2 cm wide, while below it the recti muscles (**3**) lie closer to each other and the linea alba is narrower. Under pathologic conditions when there is a fat pendulous abdomen, or during pregnancy, the two recti may separate, producing **rectus diastasis** (**A**). A relatively small **epigastric hernia** (**4**) may develop in the linea alba. It develops from an enlargement of a small hole within the linea alba. An epigastric hernia may expand into an anterior abdominal wall hernia.

Umbilicus

The umbilicus (**2**) is produced by fusion of the structures that originally protruded from the umbilicus with the adjacent tissues, and is reinforced by connective tissue. If the umbilical ring is stretched, as during pregnancy, an **umbilical hernia** (**5**) may occur.

Scars

Incisional hernias (**6**) may develop at the site of surgical scars.

Inguinal canal

The inguinal canal is produced by apposition of the lateral abdominal wall muscles and extends obliquely through the abdominal wall. The **anterior wall** of the canal is formed by the *aponeurosis of the external abdominal oblique* (**7**) and the **floor** by the *inguinal ligament.* The **posterior wall** consists of the *transversalis fascia,* while its **roof** is formed by the inferior border of the *transversus abdominis.* The **deep inguinal ring** (see p. 98) is the internal opening and the **superficial inguinal ring** (**8**) is a slit-like opening in the aponeurosis of the external abdominal oblique. The superficial inguinal ring (**8**) is only visible after dissecting off the external spermatic fascia (**9**) away from the external abdominal oblique. It is bounded by concentrated fiber bundles of the aponeurosis, the *medial crus* (**10**), the *lateral crus* (**11**), and the *intercrural fibers* (**12**). Posteriorly the superficial inguinal ring is reinforced by the *reflex inguinal ligament* (**13**), which represents a division of the inguinal ligament.

In the male, the spermatic cord, which is enclosed by the *cremasteric fascia* and *cremaster muscle* (**14**), runs through the inguinal canal. In the female, the *round ligament of the uterus* and *lymphatics* run through the inguinal canal (see Vol. 2). These lymphatic vessels arise from the uterine fundus and drain into the superior superficial inguinal lymph nodes (see p. 414).

15 Femoral hernia (p. 100)
16 Indirect inguinal hernia (p. 100)

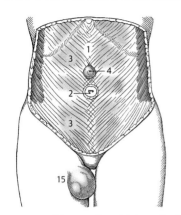

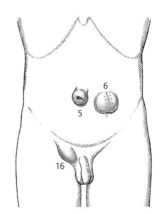

A Hernias of the anterolateral abdominal wall
   and the femoral region

B Hernias of the anterolateral
   abdominal wall

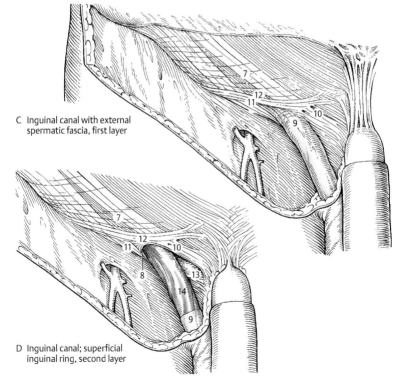

C Inguinal canal with external
   spermatic fascia, first layer

D Inguinal canal; superficial
   inguinal ring, second layer

Sites of Weakness in the Abdominal Wall, continued

Inguinal Canal, continued (A, B)

Incising the *aponeurosis* (**1**) of the external abdominal oblique reveals the *internal abdominal oblique* (**2**). In the male several of its fibers continue into the spermatic cord as the *cremaster muscle* (**3**). Another portion (**4**) of the fibers of the cremaster muscle originates from the inguinal ligament. Since the muscle fibers have developed very differently, the entire middle covering of the spermatic cord has been designated as the *cremasteric fascia with its accompanying cremaster muscle* (**5**), also called the Cooper fascia. In the female, these few muscle fibers are referred to as the round ligament portion of the internal oblique.

The *transversus abdominis* (**6**) forms the roof of the inguinal canal and is rendered visible only after cutting through the internal abdominal oblique (**2**) and the cremasteric fascia (**5**). The **deep inguinal ring** (**7**) is formed by the evagination of the *transversalis fascia* (**8**) which continues as the *internal spermatic fascia* (**9**), the innermost covering of the spermatic cord.

Abdominal Wall from Inside (C)

Both openings of the inguinal canal, the deep and superficial inguinal rings, represent sites of weakness in the abdominal wall. By examination of the abdominal wall from the inside (C), where the innermost layer, the peritoneum, is preserved, we see that it is depressed in two places, described as the **lateral inguinal fossa** (**10**), corresponding to the deep inguinal ring that lies beneath it, and the **medial inguinal fossa** (**11**), corresponding to the superficial inguinal ring.

Removal of the perineum reveals the *transversalis fascia* (**8**), which has various reinforcing tracts. Along the inguinal ligament is the *iliopubic tract* (**12**), and between the medial inguinal fossa and lateral inguinal fossa is the *interfoveolar ligament* (**13**). This band, called also the **ligament of Hesselbach**, is highly variable in its development. Caudally it is interwoven with the iliopubic tract. Cranially, it may radiate over a wide area and may participate as the *semilunar fold* in forming the medial boundary of the deep inguinal ring (**7**). The interfoveolar ligament may sometimes contain muscle fibers and is then known as the *interfoveolar muscle*. In this region the *inferior epigastric artery* and *vein* (**14**) are found, which create a peritoneal fold that is called the *epigastric fold* (**15**). Erroneously it is also known as the lateral umbilical fold, although it does not reach the umbilicus.

When examining the abdominal wall from the inside, we find the **supravesical fossa** (**16**) in addition to the lateral and medial inguinal fossae; it is medial to the latter and separated from it only by the *cord of the umbilical artery* (**17**). Hernias may develop at any of these three sites (see p. 100).

Clinical tip: The *inguinal triangle*, or **Hesselbach's triangle**, is the region delimited *medially* by the lateral margin of the rectus abdominis muscle, *caudally* by the pectineal ligament (see p. 100), and *laterally* by the external iliac artery and vein and inferior epigastric artery and vein. The triangle carries three weak sites of the abdominal wall, namely, the *medial inguinal fossa* (**11**), the *supravesical fossa* (**16**), and the *femoral canal* (**18**; see p. 100). It has recently regained importance in connection with minimally invasive surgery.

19 Reflex (inguinal) ligament
20 External spermatic fascia
21 Cut edge of the peritoneum
22 Medial umbilical fold (cord of umbilical artery)

Trunk

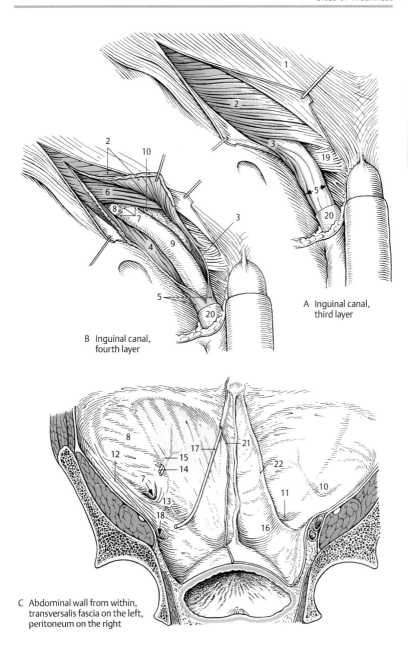

A Inguinal canal, third layer

B Inguinal canal, fourth layer

C Abdominal wall from within, transversalis fascia on the left, peritoneum on the right

Trunk

Sites of Weakness in the Abdominal Wall, continued

Hernias in the inguinal region (A)

The lateral, medial inguinal, and supravesicular fossae are regions of minimal resistance. In certain circumstances they become stretched, bulge out and **inguinal hernias** may occur. Two types of inguinal hernias are distinguished—direct and indirect—and *both traverse the superficial inguinal ring*. The **direct inguinal hernia** (**1**) *has its hernial orifice in the medial inguinal fossa*. An **indirect inguinal hernia** (**2**) *passes through the inguinal canal* (and is therefore also known clinically as *hernia of the canal*). It uses as *points of exit* the *lateral inguinal fossa* and the *deep inguinal ring*. Another type of hernia, the **supravesical hernia** (**3**), leaves the abdomen through the *supravesical fossa*; its hernial orifice, therefore, lies medial to the obliterated umbilical artery (**4**). The point of passage of this hernia through the abdominal wall is also the superficial inguinal ring.

A direct inguinal hernia and a supravesical hernia are difficult to distinguish externally. They are always **acquired hernias**, while indirect inguinal hernias may be acquired or **congenital**. During the descent of the testis in males, *the processus vaginalis*, an evagination of the serosa, is carried along into the scrotum. It later becomes obliterated and loses all previous connection with the peritoneal cavity, so that only a closed serous sac, the cavum serosum scroti, remains. In some cases, however, a connection persists and there may then be a congenital inguinal hernia with a patent processus vaginalis.

Femoral canal (B)

The femoral canal (**5**) represents an additional possible site for herniation. The femoral canal *lies behind the inguinal ligament* (**6**), within the vascular space (**7**), the medial femoral aperture. Laterally this is separated from the *muscular space* (**8**) by the *iliopectineal arch* (**9**). In the medial part

of the vascular space, medial to the large femoral vessels, lies the femoral canal (**5**). It is **bordered medially** by the **lacunar ligament** (**10**), which merges with the **posterior border** of the **pectineal ligament** (i.e., Cooper's ligament) across a ligamentous arch, the processus falciformis lacunaris. The canal is closed by loose connective tissues, the **femoral septum** (**11**).

The lymphatics pass through this femoral canal. It also contains the *deep inguinal lymph node* (**12**), also known as Cloquet's or Rosenmüller's node. In cases of excessive intra-abdominal pressure combined with weak connective tissue, a femoral hernia may result.

> **Clinical tip:** A **femoral hernia** can be differentiated from an inguinal hernia by its position in relation to the inguinal ligament and to the scrotum or the labium majus. Only an inguinal hernia can reach the scrotum or labia majora, while a femoral hernia appears in the thigh. Femoral hernias occur three times more often in women than in men.

Lumbar triangle

Between the iliac crest, the posterior border of the external oblique muscle, and the lateral border of the latissimus dorsi (see p. 140) there is often a triangular interval, the lumbar triangle. It contains fatty tissue and the internal oblique muscle. It is uncommon for **lumbar hernias** to occur through the triangle but it happens more often in men than in women.

13 Femoral vein
14 Femoral artery
15 Femoral nerve
16 Iliopsoas
17 Iliopectineal bursa
18 Lateral femoral cutaneous nerve
19 Pectineal muscle

Trunk

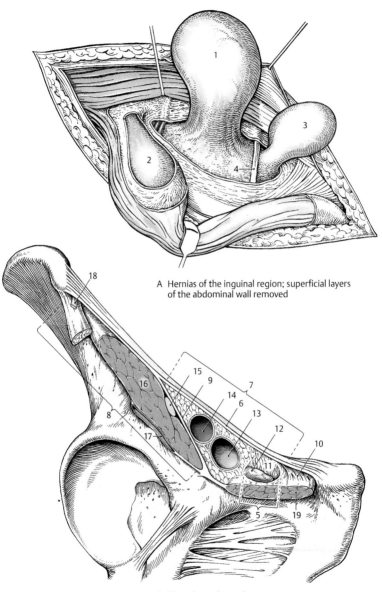

A **Hernias of the inguinal region; superficial layers of the abdominal wall removed**

B **Muscular and vascular spaces with femoral canal**

Diaphragm (A, B)

The **diaphragm** separates the thoracic and abdominal cavities. It consists of a **central tendon** (**1**) and a muscular portion, which can be divided into **sternal** (**2**), **costal** (**3**), and **lumbar** (**4**) **parts**.

Current nomenclature describes the lumbar part of the diaphragm as a uniform structure. Nevertheless, it is subdivided into a **left crus** and a **right crus** with three origins each, namely, at the lumbar vertebrae, the medial arcuate ligament, and the lateral arcuate ligament.

The **sternal part** (**2**), which *arises from the inner surface of the xiphoid process* (**5**), consists of muscle that is somewhat lighter in color than the rest and that radiates into the central tendon.

The **costal part** (**3**) *arises from the inner surfaces of the cartilage of ribs 7 to 12 by means of individual slips* that alternate with the slips of origin of the transversus abdominis.

The **lumbar part** (**4**) has a **medial** and a **lateral crus** and occasionally an **intermediate crus** splits off from the medial crus. The **right medial crus** (**6**) *arises from the bodies of the first to fourth lumbar vertebrae, and the* **left medial crus** (**7**) *from the bodies of the first to third lumbar vertebrae. The* **lateral crus** (**8**) *originates from two arches, formed by the medial arcuate ligament* (**9**)*, the psoas arcade or medial lumbocostal arch, and the lateral arcuate ligament* (**10**)*, quadratus arcade or lateral lumbocostal arch. The psoas arcade extends from the lateral surface of first (second) lumbar vertebral bodies to the costal process* (**11**) *of the first lumbar vertebra. The lateral arcuate ligament extends from this process to the apex of the 12th rib.*

Below these tendinous arches the psoas major (**12**) and quadratus lumborum (**13**) are visible. There are gaps between the lumbar, costal, and sternal parts of the diaphragm, which are points of minimal resistance. Between the lumbar and costal components lies the **lumbocostal triangle** (**14**), and between the sternal and costal parts is the **sternocostal triangle** (**15**).

The double-domed diaphragm, which is slightly depressed in the middle by the heart, is pierced by openings for the passage of various structures. Between the medial crura lies the **aortic hiatus** (**16**), which is bounded by tendons (median arcuate ligament). Through it passes the aorta and posteriorly to it the thoracic duct. The right medial crus (**6**) consists of three muscle bundles, of which that arising from the lumbar vertebrae is the largest and it reaches the central tendon (**1**) directly. A *second* bundle (**17**) arises from the *median arcuate ligament* (**18**), the tendinous border of the aortic hiatus (**16**), and forms the right border of the **esophageal hiatus** (**19**). The third bundle (**20**) also arises from the median arcuate ligament, but posteriorly, and forms the left border of the esophageal opening as the **"hiatus sling."** Only in exceptional cases does the left medial crus (**7**) participate in forming the border of the esophageal opening. The esophageal hiatus is bordered by muscle, and through it pass the esophagus and the anterior and posterior vagal trunks.

The **caval opening** (**21**) lies in the central tendon, and transmits the inferior vena cava and a branch of the right phrenic nerve. The greater and lesser splanchnic nerves, on the right the azygos vein and on the left the hemiazygos vein, pass through unnamed openings in the medial crus, or between it and the intermediate crus if present. The sympathetic trunk runs between the intermediate and lateral crura. Nerve supply: phrenic nerves ([C3] C4 [C5]).

Trunk

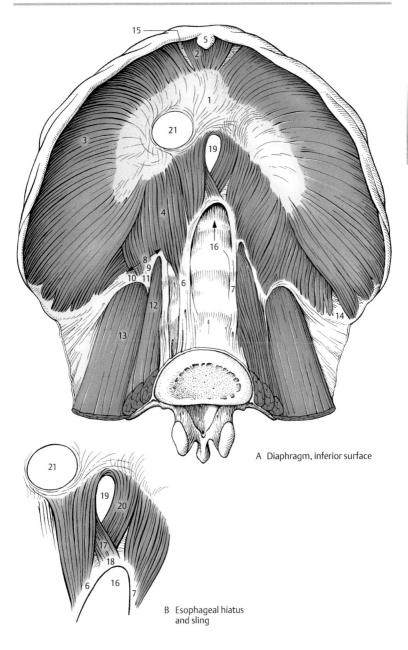

A Diaphragm, inferior surface

B Esophageal hiatus and sling

Position and Function of the Diaphragm (A)

In life the position and shape of the diaphragm depend on the phases of respiration, the position of the body and the degree of distension of the viscera.

As the principal respiratory muscle, the shape of the diaphragm changes greatly during the various phases of respiration. In the midposition between maximal expiration and inspiration, in the upright posture, the right dome of the diaphragm reaches the fourth intercostal space, and the left dome the fifth intercostal space. In *maximal expiration* (blue) *the projection on the anterior chest wall on the right lies at the upper margin of the fourth rib, and on the left in the fourth intercostal space. During maximal inspiration* (red), *the diaphragm sinks to about the first to second intercostal space*. The sternal part and its origin act as a fixed point. During expiration the muscle fibers rise and during maximal inspiration they descend toward the center of the tendon.

The **costodiaphragmatic recess** between the upper surface of the diaphragm and the ribs is flattened during maximal inspiration.

In the recumbent position convolutions of the abdominal viscera push the diaphragm upward and backward.

Clinical tip: Dyspneic patients prefer to sit rather than to lie and so relieve the thorax of the pressure of the abdominal contents.

Sites of Diaphragmatic Hernias (B)

Diaphragmatic hernias occur when the contents of the abdominal cavity enter the thorax. They may be congenital or acquired. True diaphragmatic defects (blue) must be distinguished from enlargement of pre-existing weak spots (red) such as the *esophageal hiatus* (**1**), the *lumbocostal triangle* (**2**) and *sternocostal triangle* (**3**). True diaphragmatic hernias usually occur in the *central tendon* (**4**) or the *costal part* (**5**). The majority of diaphragmatic hernias are prolapses, as they lack a hernial sac. They are known as **false diaphragmatic hernias**. **True hernias** with a sac are uncommon and occur only as para-esophageal hernias.

The most common congenital hernia is due to enlargement of the lumbocostal triangle (**2**). Another type of congenital hernia is **para-esophageal** in position and always occurs on the right side of the esophagus. It is a type of **hiatus hernia**, which, however, in the great majority of cases is an acquired sliding hernia. Sliding hernias have no hernial sac and develop through enlargement of the esophageal hiatus (**1**).

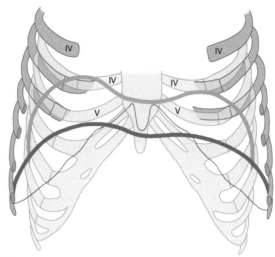

A Position of diaphragm
during maximal inspiration (red)
and maximal expiration (blue)

Trunk

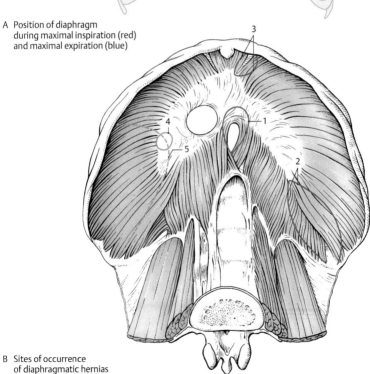

B Sites of occurrence
of diaphragmatic hernias

Pelvic Floor (A, B)

The pelvic floor forms the posterior and inferior boundaries of the trunk. It is formed by the **pelvic diaphragm** and **urogenital diaphragm**.

Pelvic Diaphragm

This consists of the **levator ani** and **coccygeus muscles**.

The **levator ani** (**1**) *arises from the pubic bone* (**2**), *the tendinous arch of the levator ani muscle* (**3**), *and the ischial spine* (**4**). Its fibers are divisible into the **puborectalis muscle** (**5**), the **puboperinealis** = prerectal fibers (**6**), the **pubococcygeal** (**7**) and the **iliococcygeal muscles** (**8**). The medial fibers of the puboperinealis form the *crura of the levator*, between which is enclosed the *urogenital hiatus*. The puboperinealis muscles extend into the perineum and thereby separate the urogenital tract from the anal tract. The urogenital hiatus is bounded laterally by the levator crura and posteriorly by the puboperineal muscles. The urethra and the genital canal pass through the urogenital hiatus, while only the rectum (anal canal) passes behind the prerectal fibers. Some of the fibers of the puborectalis end pararectally in the *external anal sphincter* (**9**), and some continue on to form a retrorectal sling behind the rectum. The fibers of the pubococcygeal and the iliococcygeal muscles extend laterally onto the *anococcygeal ligament* (**10**) and insert on this or directly onto the coccyx (**11**).

The genital hiatus is narrower in the male and broader in the female. Due to the width of the aperture of the genital hiatus a second closure mechanism—the urogenital diaphragm—is essential.

The **coccygeus** (**12**) *arises by means of a tendon from the ischial spine and ends on the coccyx*. It may be absent.

Function: The levator ani is concerned with intra-abdominal pressure. It bears the weight of the pelvic contents and thus has a supporting function. In its dynamic function it participates in closure of the rectum.

Urogenital Diaphragm

This consists mainly of the **deep transverse perineal muscle** (**13**). *It arises from the ramus of the ischium and from the inferior pubic ramus and extends to the urogenital hiatus*. The posterior part of the diaphragm is reinforced by the **superficial transverse perineal muscle** (**14**). *This arises from the ischial tuberosity* (**15**) *and radiates into the perineal body*. Anteriorly the urogenital diaphragm is completed by the **transverse perineal ligament** (**16**).

Both the urogenital diaphragm and the pelvic diaphragm are covered on their upper and lower surfaces by fascia, appropriately termed the *superior* and *inferior urogenital diaphragmatic fascia (perineal membrane)* and the *superior* and *inferior pelvic diaphragmatic fascia*, respectively. The ischiorectal (ischioanal) fossa lies between the pelvic and urogenital diaphragm and is open posteriorly.

Nerve supply: The pelvic diaphragm is innervated, as a rule, by a long branch from the sacral plexus, and the urogenital diaphragm by twigs from the pudendal nerve.

The term "urogenital diaphragm," which is a meaningful term, has been mostly discarded from the anatomical nomenclature (but not from clinical parlance) and has been replaced by the terms *perineal membrane with transverse perineal ligament* and *deep transverse perineal muscle*.

Clinical tip: Overstretching of the pelvic diaphragm in women leads to a prolapse of their internal genital organs, which can occur especially after childbirth. It is important to keep in mind that childbirth can also result in a laceration of the levator ani with a concomitant traumatic injury to the pelvic diaphragm. **Perineal hernias** rarely emerge through sites of muscular weakness in the pelvic floor, although they are substantially more frequent in women.

For further details of the pelvic floor, see Volume 2.

17 Sacrospinal ligament
18 Sacrotuberal ligament

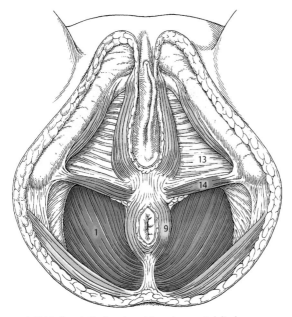

A Pelvic floor in the female: pelvic and urogenital diaphragm

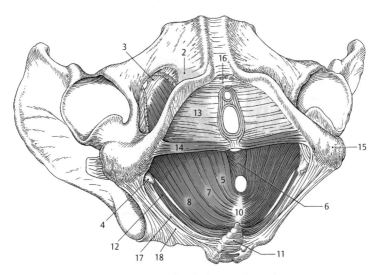

B Pelvic floor in the female: diagram of musculature

Trunk

Anatomical Terms and their Latin Equivalents

Trunk	Truncus
Alar ligaments	Ligamenta alaria
Caval opening	Foramen venae cavae
Conjoined tendon	Tendo conjunctivus
Costal notch	Incisura costalis
Esophageal (aortic) hiatus	Hiatus oesophageus (aorticus)
False (true) ribs	Costae spuriae (verae)
Floating ribs	Costae fluctantes
Iliopectineal arch	Arcus iliopectineus
Innermost intercostal muscles	Musculi intercostales intimi
Intervertebral joint	Symphysis intervertebralis
Joint of head of rib	Articulatio capitis costae
Nuchal fascia	Fascia nuchae
Pelvic (urogenital) diaphragm	Diaphragma pelvis (urogenitalis)
Rectus sheath	Vagina musculi recti abdominis
Scalene tubercle	Tuberculum musculi scaleni anterioris
Superficial (deep) inguinal ring	Anulus inguinalis superficialis (profundus)

Systematic Anatomy of the Locomotor System

Upper Limb

Upper Limb

Bones, Ligaments, and Joints

In the upper limb we distinguish the **shoulder girdle** and the **free upper limb**. The shoulder girdle is formed by the scapulae and the clavicles.

Shoulder Girdle

Scapula (A–E)

The shoulder blade or **scapula** (A–E) is a flat, triangular bone. It has a *medial border* (**1**), a *lateral border* (**2**) and a *superior border* (**3**), which are separated from one other by the *superior* (**4**) and *inferior* (**5**) *angles* and the truncate *lateral angle* (**6**). The anterior or *costal surface* is flat and slightly concave (*subscapular fossa*). It sometimes shows clear lines of muscle attachments. The *posterior surface* is divided by the *spine of the scapula* (**7**) into a smaller *supraspinous fossa* (**8**) and a larger *infraspinous fossa* (**9**). The scapular spine has a triangular base medially, which rises laterally to terminate in a flattened process, the *acromion* (**10**). Near the lateral end lies an oval *articular facet* (**11**) for articulation with the clavicle, the *clavicular facet*.

The *acromial angle* (**12**) is a palpable bony prominence that marks the site where the lateral acromial margin continues into the spine of scapula. The lateral angle bears the *glenoid cavity* (**13**). At its upper border is a small projection, the *supraglenoid tubercle* (**14**). Below the glenoid cavity lies the *infraglenoid tubercle* (**15**). The *neck of the scapula* (**16**) is adjacent to the glenoid cavity.

The *coracoid process* (**17**) lies above the glenoid cavity. It is bent at a right angle anterolaterally and its tip is flattened. Together with the acromion it protects the joint that lies beneath it. Medial to the base of the coracoid process, on the upper margin of the scapula, lies the *suprascapular notch* (**18**).

The scapula lies upon the thorax with the base of its spine at the level of the third thoracic vertebra. The inferior angle of the scapula should lie between ribs 7 and 8 and, when the arm hangs down, its medial margin should be parallel to the row of spinous processes. The **scapular plane** is the plane in which the scapular plate lies. It forms an angle of 60° with the plane of symmetry (midsagittal). The glenoid cavity faces laterally and anteriorly.

▬ **Variants:** The scapular notch may be transformed into a *scapular foramen* (**19**). The medial border of the scapula is sometimes concave and the scapula is then called a **scaphoid scapula**.

Ossification: The scapula develops (E) from several ossification centers. In the 3rd intrauterine month a large bony center develops in the region of the supraspinous and infraspinous fossae and the scapular spine. In the 1st year of life a center develops in the coracoid process, and between the ages of 11 and 18 years smaller centers may appear throughout the scapula. All the centers fuse with each other between the ages of 16 and 22 years. The center that develops in the acromion between 15 and 18 years of age may, in rare instances, remain unfused (os acromiale).

Ligaments of the Scapula

The **coracoacromial ligament** crosses the shoulder joint and extends between the coracoid process and the acromion. The **superior transverse scapular ligament** bridges the scapular notch. (Only in rare cases is there an inferior transverse scapular ligament, which extends from the edge of the scapular spine to the glenoid cavity.)

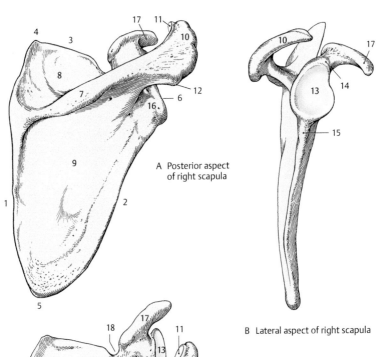

A Posterior aspect of right scapula

B Lateral aspect of right scapula

C Right scapula from above

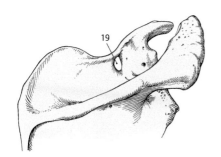

D Foramen scapulae (variant)

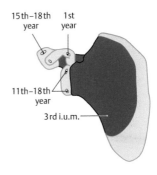

15th–18th year   1st year

11th–18th year

3rd i.u.m.

E Ossification of scapula

Clavicle (A, B, F)

The collar bone or **clavicle** is an S-shaped bone, anteriorly convex in the medial two-thirds of its length, while the lateral third is concave anteriorly. Toward the sternum is the stout **sternal end** (**1**) and toward the scapula the flat **acromial end** (**2**), and between the two lies the **body of the clavicle**. At the sternal end we find a triangular *sternal articular facet* (**3**). The *acromial articular facet* (**4**) is almost oval. Near the sternal end, on the lower surface of the clavicle, is an *impression for the costoclavicular ligament* (**5**). The groove for the subclavian muscle lies on the undersurface of the *clavicular body*. The prominent *conoid tubercle* (**6**) lies near the acromial end close to the *trapezoid line* (**7**).

Ossification: The clavicle develops in connective tissue, and ossification begins in the 6th intrauterine week. The ends are preformed in cartilage but an ossification center does not appear in the sternal end until 16 to 20 years of age. It synostoses with the rest of the clavicle between the ages of 21 and 24 years.

Clinical tip: Cleidocranial dysostosis is an anomaly due to maldevelopment or non-development of the connective tissue part of the clavicle. It is associated with defects of those bones of the skull that are preformed in connective tissue.

Joints of the Shoulder Girdle (C–E)

Connections with the trunk are made through a continuous fibrous (costoclavicular ligament, **8**) and discontinuous synovial (sternoclavicular) joints. In the same way, the parts of the shoulder girdle are connected to each other by continuous fibrous (coracoclavicular ligament) and discontinuous synovial (acromioclavicular) joints.

Sternoclavicular Joint (C)

This is a joint with an *articular disk* (**9**), which divides the joint space in two. The socket is a shallow concave indentation in the sternum, and the head is formed by the sternal end of the clavicle.

The incongruity is offset by the cartilage-like fibrous tissue, which covers both articular facets, and by the disk, which is fixed cranially to the clavicle and caudally to the sternum. The capsule is lax and thick and is strengthened by the *anterior* (**10**) and *posterior sternoclavicular ligaments*. The clavicles are interconnected by the *interclavicular ligament* (**11**). The sternoclavicular joint functions as a ball-and-socket type and has three degrees of freedom.

The **costoclavicular ligament** (**8**) extends between the first rib and the clavicle.

Acromioclavicular Joint (D, E)

This joint consists of two apposing, almost flat articular surfaces covered by cartilage-like fibrous tissue (**12**). The capsule has a strengthening ligament on its superior surface, the *acromioclavicular ligament* (**13**). The **coracoclavicular ligament** extends between the coracoid process and the clavicle. It can be divided into anterolateral and posteromedial parts. The lateral part, the **trapezoid ligament** (**14**), arises from the upper medial margin of the coracoid process and extends to the trapezoid line. The medial part, the **conoid ligament** (**15**), arises from the base of the coracoid process and has a fanlike termination on the conoid tubercle.

Clinical tip: Marked posterior and inferior displacement of the clavicle may compress the subclavian artery, as can be detected by a weakening of the radial pulse.

16 Superior transverse scapular ligament
17 Coracoacromial ligament
18 Subclavius muscle

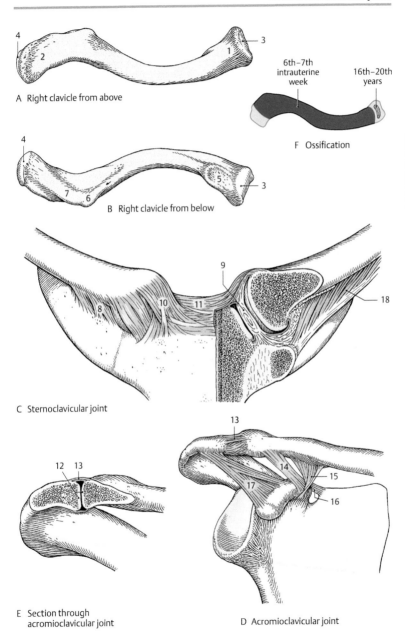

A Right clavicle from above

6th–7th
intrauterine
week   16th–20th
years

F Ossification

B Right clavicle from below

Upper Limb

C Sternoclavicular joint

E Section through
acromioclavicular joint

D Acromioclavicular joint

The Free Upper Limb

The bones of the **free upper limb** are
- The **humerus**
- The **radius** and **ulna**
- The **carpal bones**
- The **metacarpal bones**
- The **phalanges**

Bone of the Arm
Humerus (A–H)

The **humerus** articulates with the scapula, radius, and ulna. It consists of the **body** and **upper (proximal) and lower (distal) ends**. The proximal end is formed by the *head* (**1**), adjoining the *anatomical neck* (**2**). On the anterolateral surface of the proximal end lies laterally the *greater tubercle* (**3**), and medially is the *lesser tubercle* (**4**). Between these tubercles begins the *intertubercular sulcus* (**5**), which is bounded distally by the *crests of the lesser* (**6**) and *greater* (**7**) *tubercles*. The *surgical neck* (**8**) lies proximally on the body of the humerus. In the middle of the body lies laterally the *deltoid tuberosity* (**9**). The body may be divided into an *anteromedial surface* (**10**) with a *medial border* (**11**), and an *anterolateral surface* (**12**) with a *lateral border* (**13**), which becomes sharpened distally and is called the *lateral supracondylar ridge*. The *groove for the radial nerve* (**14**) lies on the posterior surface of the body. The distal end of the humerus bears on its medial side the large *medial epicondyle* (**15**) and on the lateral side the smaller *lateral epicondyle* (**16**).

The *trochlea* (**17**) and the *capitulum* (**18**) of the humerus form the *humeral condyles* for articulation with the bones of the forearm. The *radial fossa* (**19**) lies proximal to the capitulum and proximal to the trochlea is the somewhat larger *coronoid fossa* (**20**).

Medial to the trochlea (**D**) is a shallow sulcus, the *groove for the ulnar nerve* (**21**). On the posterior surface above the trochlea is a deep pit, the *olecranon fossa* (**22**).

The humerus is twisted at its proximal end; that is, the head is rotated posteriorly by approximately 20° relative to the transverse axis of the distal end (**torsion**). The angle between the long axis of the humerus and that of the head averages 130°, and at the distal end, between the transverse axis of the joint and the long axis of the shaft of the humerus, there is an angle of 76 to 89°.

The **proximal epiphyseal line** (**23**) runs transversely through the lesser tubercle and inferior to the greater tubercle. It crosses the zone of attachment of the capsule (see p. 117) in such a way that a small part of the shaft comes to lie within the capsule. At the **distal end** there are two epiphyses and **two epiphyseal lines** (**24**). One epiphysis bears the medial epicondyle and the other the articular surfaces and lateral epicondyle.

Ossification: In general, development of the ossification centers and fusion of the epiphyses occur somewhat earlier in females than in males. The perichondral bone anlage in the shaft appears in the 2nd to 3rd intrauterine month. The endochondral ossification centers in the epiphyses appear between the 2nd week of life and the 12th year. Three centers appear proximally soon after birth, and distally four ossification centers develop later. The distal epiphyseal disks fuse during puberty, the proximal disks at the end of puberty.

■ **Variants:** Just above the medial epicondyle a *supracondylar process* (**25**) is occasionally found, and above the trochlea there may be a *supratrochlear foramen* (**26**).

> **Clinical tip:** 50% of fractures of the humerus occur in the **shaft**. There is a risk of damage to the radial nerve!

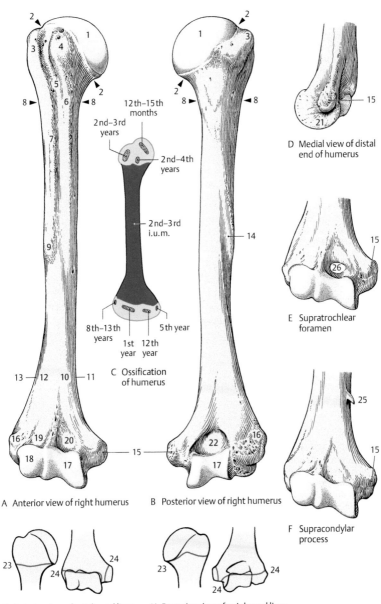

Upper Limb

C Ossification of humerus

12th–15th months
2nd–3rd years
2nd–4th years
2nd–3rd i.u.m.
8th–13th years
1st year
12th year
5th year

D Medial view of distal end of humerus

E Supratrochlear foramen

F Supracondylar process

A Anterior view of right humerus

B Posterior view of right humerus

G Anterior view of epiphyseal lines

H Posterior view of epiphyseal lines

Shoulder Joint (A–G)

The bony socket, the **glenoid cavity**, of the ball-and-socket **shoulder joint** is much smaller than the **head of the humerus**. The hyaline cartilage lining (**1**) of the glenoid cavity is thicker at the margins than in the center. The socket is enlarged by a fibrocartilaginous lip, the **glenoid labrum** (**2**).

The socket is perpendicular to the plane of the scapula, and the position of the scapula determines the attitude of the entire joint. The surface of the glenoid cavity has an area of 6 cm^2 to withstand an atmospheric pressure of 6 kp (approx. 60 N) on the joint. The upper limb weighs approximately 4 kg. As there are no strong ligaments, the shoulder joint is stabilized by the action of the enveloping muscles. It is known as a "muscle-dependent joint."

The head of the humerus (**3**) is ball-shaped. Its hyaline cartilage covering begins at the anatomical neck and extends somewhat farther distally at the intertubercular sulcus. The cartilage gives the head a more oval shape. The **synovial layer of the articular capsule** is attached to the glenoid labrum. It is evaginated pouchlike (**C**) along the intracapsularly coursing tendon of the long head of the biceps (**4**) and surrounds it as the *synovial sheath of the intertubercular groove* (**5**). The **fibrous layer of the joint capsule** in the upper arm forms a connective tissue layer across the intertubercular sulcus and converts it to a fibro-osseous canal. The **articular capsule** is lax and when the arm hangs down it has an outpouching on its medial surface, the *axillary recess* (**6**). The upper portion of the capsule is partly strengthened by the *coracohumeral ligament* (**7**) and three weak *glenohumeral ligaments*. The coracohumeral ligament arises from the base of the coracoid process (**8**) and radiates into the capsule, extending to the greater and lesser tubercles. When the arm is hanging in its normal anatomical position, the upper half of the head of the humerus is in contact with the joint capsule and the lower half with the glenoid cavity.

The shoulder joint is associated with a number of synovial sacs. As a rule, it communicates *with the subcoracoid bursa, the subtendinous bursa of the subscapular muscle* (beneath the tendon of the *subscapular muscle*, **9**), *the intertubercular synovial sac,* and the *coracobrachial bursa.*

Movements of the Shoulder Joint (D–F)

The shoulder joint has **three degrees of freedom of movement**. **Abduction** and **adduction** refer to movements away from the position of rest (**D**) of the head of the humerus in the scapular plane (see p. 110). Pure lateral abduction (**E**) always produces **retroversion** and slight **rotation**, while abduction from the scapular plane is anteriorly directed (frontal abduction).

Flexion (**anteversion**) is forward lifting of the arm. Because of rotary components associated with these other movements, a compound movement, **circumduction**, occurs in which the arm traces the surface of a cone. Abduction (**E**) is *always* associated with movement of the scapula: excessive associated scapular movement occurs with abduction of more than 90° (**F**; elevation), because then the movement of the joint is constrained by the coracoacromial ligament (**10**; see p. 110).

Clinical tip: Dislocation is more common in the shoulder than in any other joint. If associated with a torn capsule, the tear usually has an anteroinferior location.

The palpable and visible **prominence of the shoulder joint** is produced by the greater tubercle, the location of which indicates the position of the humeral head. The protuberance disappears when the shoulder is dislocated, as the humeral head is no longer in its socket. When palpating a dislocated shoulder the finger enters an empty cavity (**G**) below the acromion.

A fracture of the (intracapsular) anatomical neck is uncommon and has a very poor prognosis.

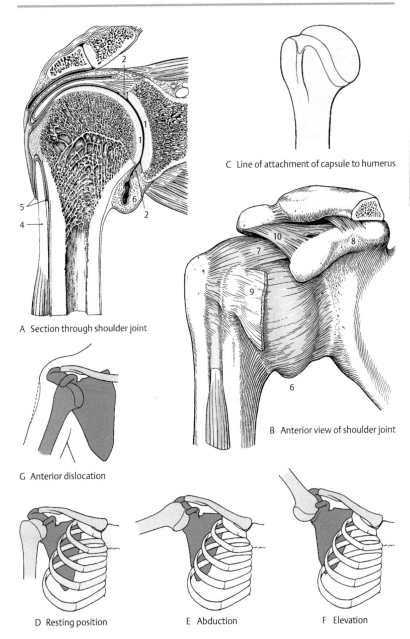

A Section through shoulder joint

C Line of attachment of capsule to humerus

B Anterior view of shoulder joint

G Anterior dislocation

D Resting position

E Abduction

F Elevation

Bones of the Forearm

In the **forearm**, the shorter **radius** lies laterally, the longer **ulna**, medially.

Radius (A–E)

The **radius** comprises a **shaft** (**1**) and a **proximal** and **distal extremity**. The proximal extremity contains the *head* (**2**) with its *articular facet* (**3**), which continues into the *articular circumference* (**4**). Medially, at the junction of the *neck* (**5**) and its shaft, lies the *radial tuberosity* (**6**). The shaft has an approximately triangular shape in cross section with a medially directed *interosseous border* (**7**), an *anterior surface* (**8**), an *anterior border* (**9**), a *lateral surface* (**10**), and a *posterior border* (**11**), which represents the boundary between the lateral and *posterior surfaces* (**12**). The lateral surface of the shaft at approximately its middle third exhibits a distinct, well-developed roughened area, the *pronator tuberosity* (**13**). At the distal end of the radius is the *suprastyloid crest* with the *styloid process* (**14**) and medial to it, the *ulnar notch* (**15**). The *carpal articular surface* (**16**) is directed distally.

Posteriorly are found various distinct **grooves** that transmit the tendons of the long extensors. From lateral (radial) to medial (ulnar), the *first groove* (**17**) resides on the styloid process and transmits the tendons of the abductor pollicis longus and extensor pollicis brevis muscles. The *second groove* (**18**) serves for the passage of the tendons of the extensor carpi radialis longus and brevis, while the *third groove* (**19**) courses obliquely and transmits the tendon of the extensor pollicis longus. In the *fourth groove* (**20**) lie the tendons of the extensor digitorum and extensor indicis muscles. The lateral bony ridge lying by the third groove is usually palpable and is also designated as the **dorsal tubercle** (**21**).

Clinical tip: The styloid process of the radius extends approximately 1 cm farther distally than that of the ulna. This is an important detail to remember in the reduction of fractures.

Ossification: Perichondral ossification of the radial shaft begins in the 7th intrauterine week. The epiphyses are formed endochondrally and postnatally, the distal epiphysis in the 1st and 2nd years of life, the styloid process in the 10th to 12th years, and the proximal epiphysis in the 4th to 7th years. Epiphyseal fusion occurs proximally between ages 14 and 17 years, distally between the 20th and 25th years of life.

Ulna (F–L)

The **ulna** possesses a **shaft** (**22**) and a **proximal** and a **distal extremity**. The proximal end exhibits a curved, hook-shaped process, the *olecranon* (**23**), which has a roughened surface. In front is the *trochlear notch* (**24**), which extends up to the *coronoid process* (**25**).

The *radial notch* (**26**) lies laterally and articulates with the articular circumference of the radial head. The *ulnar tuberosity* (**27**) is located at the transition to the shaft. Lateral to it lies the *supinator crest* (**28**), which appears as an inferior prolongation of the radial notch. The shaft of the ulna is three-sided. The *interosseous border* (**29**) is directed laterally and the *anterior surface* (**30**), which faces anteriorly, is separated from the *medial surface* (**32**) by the *anterior border* (**31**). The medial surface is separated from the *posterior surface* (**33**) by the *posterior border* (**34**). The anterior surface at about the middle of the ulna presents a *nutrient foramen* (**35**) and the *head* (**36**) contains the *articular circumference* (**37**) and the small *styloid process* (**38**) projecting distally.

Ossification: Perichondral ossification of the shaft of the ulna begins in the 7th intrauterine week. The ossification centers in the epiphyses are endochondral in origin and appear distally between the 4th and 7th postnatal years, in the styloid process between the 7th and 8th years, and proximally between the 9th and 11th years of life. Epiphyseal fusion takes place earlier proximally, later distally.

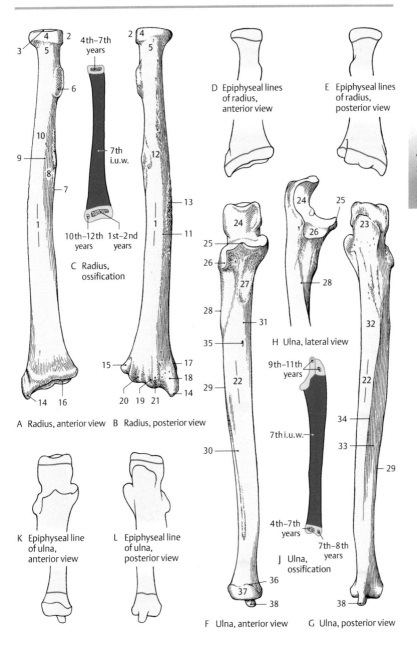

C Radius, ossification

4th–7th years

7th i.u.w.

10th–12th years   1st–2nd years

D Epiphyseal lines of radius, anterior view

E Epiphyseal lines of radius, posterior view

A Radius, anterior view   B Radius, posterior view

K Epiphyseal line of ulna, anterior view

L Epiphyseal line of ulna, posterior view

H Ulna, lateral view

J Ulna, ossification

9th–11th years

7th i.u.w.

4th–7th years

7th–8th years

F Ulna, anterior view   G Ulna, posterior view

Elbow Joint (A–D)

The **elbow joint** is a **compound joint** with three articulating bony surfaces within the joint capsule. It actually consists of three joints:

- The **humeroradial joint**
- The **humeroulnar joint**
- The **proximal radioulnar joint**

It is secured by *bones* and *ligaments*. Bony stability is provided by the trochlea of the humerus and the trochlear notch of the ulna into which it fits. Ligamentous stability is derived from the annular ligament of the radius and the collateral ligaments.

The thin, lax **joint capsule** (**1**) encloses the joint surfaces. In order to prevent pinching of the capsule between these surfaces during movement of the joint, fibers from the brachialis and triceps brachii muscles act as *articular muscles* and radiate into the capsule in order to tense it. Both *humeral epicondyles* (**2**) are outside the capsule (**D**). The synovial membrane surrounds the olecranon fossa and both fossae on the anterior side of the humerus (**D**). Between the **synovial** (**3**) and **fibrous** (**4**) **membranes** of the capsule in the region of the fossa is a large amount of fatty tissue (**5**), which may help to limit extreme movements of the joint. In the ulnar region, the line of attachment of the capsule (**D**) follows the margin of the trochlear notch, so that the tips of the *olecranon* (**6**) and the *coronoid process* (**7**) still project within the capsule. On the radius the capsule extends as a sac below the *annular ligament of the radius* (**8**), the *superior sacciform recess* (**9**). This extension of the capsule makes rotation of the radius possible.

The very strong collateral ligaments are embedded in the sides of the joint capsule. The **ulnar collateral ligament** (**10**) arises from the medial epicondyle of the humerus and usually possesses *two strong fiber bundles*, an *anterior one* (**11**), which is directed to the coronoid process, and a *posterior one* (**12**), which extends to the lateral margin of the olecranon. The ulnar nerve runs under the latter bundle in the groove for the ulnar nerve. Between these two fibrous bundles lies loose connective tissue, which is bounded on the ulnar side by *oblique fibers* (**13**).

The **radial collateral ligament** (**14**) extends from the lateral epicondyle of the humerus to the radial annular ligament and proximal to the latter radiates into the ulna. The radial collateral ligament fuses with the superficial extensors. The **quadrate ligament** connects the neck of the radius to the radial notch of the ulna.

Finally there is the **annular ligament of the radius** (**8**), which is attached at both ends to the ulna and encircles the head of the radius. There is often cartilaginous tissue on its inner surface, which acts as a moveable buttress for the radius during pronation and supination (see p. 122).

Owing to the interaction of these three joints in any flexed or extended position, a simultaneous rotation of the radius around the ulna is possible.

The following movements are possible: flexion, extension, supination, and pronation (see p. 122).

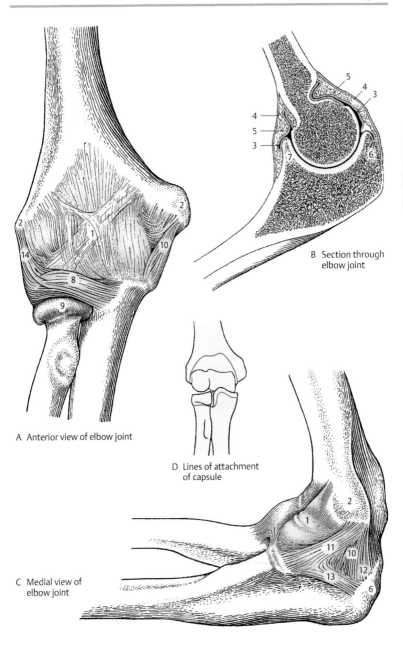

Upper Limb

A Anterior view of elbow joint

B Section through elbow joint

D Lines of attachment of capsule

C Medial view of elbow joint

Upper Limb

Elbow Joint, continued (A–C)

The **humeroradial joint** (**1**) is formed by the **capitulum of the humerus** and the **articular facet** on the **head of the radius**. It corresponds in form to a ball-and-socket joint. The **humeroulnar joint** (**2**), a hinge joint, occurs between the **trochlea of the humerus** and the **trochlear notch of ulna**. On the trochlea there is a *channel* (**3**) that accommodates the leading edge of the trochlear notch. Flexion and extension movements between the upper arm and forearm occur at the humeroradial and humero-ulnar joints. The axis of movement corresponds to the axis of the trochlea of the humerus and its extension through the capitulum of the humerus. The **proximal radioulnar joint** (**4**) is formed between the **articular circumference of the radial head** and the **radial notch of the ulna**, together with the **annular ligament** (**5**). This is a pivot joint and it permits movements of the radius around the ulna together with the distal radioulnar joint. Rotation of the radius around the ulna is called **pronation** (**B**; the bones cross over each other) or **supination** (**C**; the bones are parallel to one another). The axis of this movement runs from the center of the fovea on the radial head to the styloid process of the ulna.

The **"angle of excursion,"** that is, the angle measured anteriorly between the upper arm and forearm at maximal extension, is slightly greater in females (180°) than in males (175°). Hyperextension is possible in children. At maximal flexion the upper arm and forearm form an angle of approximately 35° (soft-tissue restraint). The **"carrying angle,"** that is, the angle, open to the lateral side, between the upper arm and forearm when the limb is fully extended (abduction angle), varies between 158° and 180°, with an average of approximately 168.5°.

Distal Radioulnar Joint (D)

The **distal radioulnar joint** (**6**) is a pivot joint formed by the **head of ulna** and the **ulnar notch of radius**. Between the radius and the styloid process of the ulna lies an *articular disk*, which separates the distal radioulnar joint from the radiocarpal joint. The **capsule** is lax and extends from the *inferior sacciform recess* (**7**) up to the shaft of the ulna. The **proximal** and **distal radioulnar joints** are **necessarily combined joints** to permit pronation and supination.

Continuous Fibrous Joint between Radius and Ulna (D)

The **interosseous membrane of the forearm** (**8**) stretches between the radius and the ulna. Its fibers run laterally and distally to the medial side of the ulna. Fibers of the *oblique cord* (**9**) run in the opposite direction to those of the interosseous membrane. It strengthens the interosseous membrane proximally. The cord begins approximately at the ulnar tuberosity and extends to the interosseous border of the radius distal to the radial tuberosity.

Clinical tip: The interosseous membrane not only prevents parallel displacement of the radius and ulna but also allows tensile and pressure stresses to be transmitted from one bone to the other. It is so strong that during overstrain of the forearm the bones tend to fracture before the fibers are torn.

The most common of all fractures (first described by *Colles* in 1814) is at a **classic site on the radius** and is due to a fall onto the palm of the hand with the arm extended. The weight of the body is transmitted through the humerus and ulna and then passes through the interosseous membrane to the radius. The distal end of the radius resists the counter-pressure, so that maximal stress develops and causes a fracture of the lower radius. The distal fragment is displaced radially and posteriorly as the fibers of the interosseous membrane fix the shaft of the radius to the ulna (bayonet position).

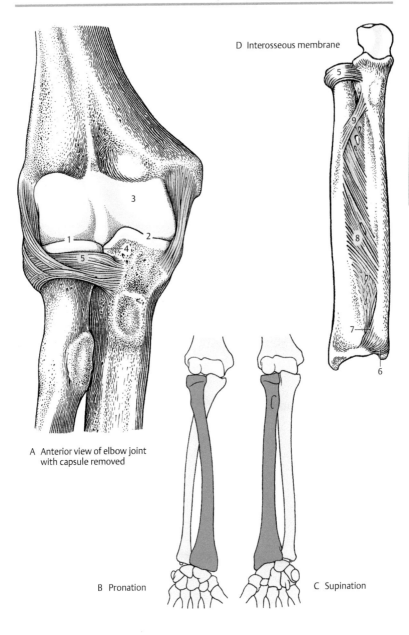

D Interosseous membrane

A Anterior view of elbow joint with capsule removed

B Pronation

C Supination

Upper Limb

Carpus (A–C)

The **carpus** consists of eight **carpal bones** arranged in two rows of four.

In the *proximal row* from lateral to medial are
- The **scaphoid** (**1**)
- The **lunate** (**2**)
- The **triquetrum** (**3**) and, superimposed on it,
- The **pisiform** (**4**)

In the *distal row* from the lateral to the medial side are
- The **trapezium** (**5**)
- The **trapezoid** (**6**)
- The **capitate** (**7**) and
- The **hamate** (**8**)

Each carpal bone has several facets for articulation with the neighboring bones.

Both rows of bones together, that is, the entire carpus, form an arch that is convex proximally and concave distally. The palmar surface of the carpus is also concave and is spanned by the *flexor retinaculum*, which forms the fibro-osseous **carpal tunnel**. The flexor retinaculum stretches from the scaphoid and trapezium to the hamate and triquetrum. Projections on these named bones are palpable through the skin. With the hand hanging freely, the pisiform is easily moved and is readily palpable, as is the tendon of the flexor carpi ulnaris, which inserts into the pisiform. The scaphoid and trapezium form the floor of the radial notch, erroneously called the "anatomical snuffbox" (see p. 392).

Clinical tip: The scaphoid (**1**) is of particular interest clinically since it is the most frequently fractured carpal bone. Ulnar abduction (see p. 132) brings about a divergence of the fragments, whereas with radial abduction (see p. 132) the fragments are compressed. Palmar and dorsal flexion (see p. 132) open the fracture gap toward the dorsal or palmar aspect, respectively.
Inadequate treatment of a scaphoid fracture can lead to a pseudoarthrosis or necrosis of a fragment. Seventy percent of all scaphoid fractures occur in its middle third. **Carpal tunnel syndrome** is described on page 388.

■ **Variants:** Sometimes small accessory bones are found between the carpal bones. More than 20 of such accessory bones have been described so far. However, apart from the **central bone** (**9**), only the **styloid** (**10**), the **secondary trapezoid** (**11**), and the **secondary pisiform** (**12**) are considered to be proven accessory bones.

Clinical tip: The possible presence of accessory carpal bones should always be considered when examining radiographs of the wrist. The most common accessory bone is the os centrale (**9**). Its cartilaginous anlage is believed to be consistently present in humans, but it almost always synostoses with the scaphoid (**1**). Fusion of carpal bones has also been described; the most frequent fusion is between the lunate and triquetrum.

The scaphoid, triquetrum, and pisiform bones may also be divided in two. This may be confused with fractures of these bones.

Upper Limb

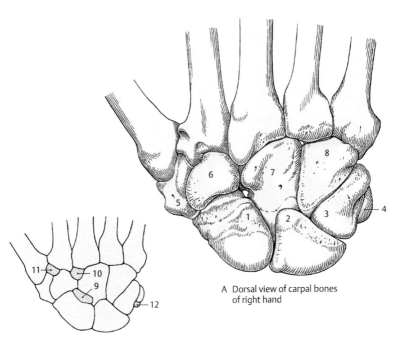

A Dorsal view of carpal bones
of right hand

C Accessory carpal bones

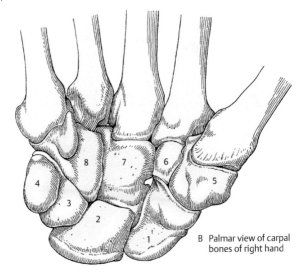

B Palmar view of carpal
bones of right hand

Individual Bones of the Carpus (A, B)

Proximal Row

The **scaphoid** (**1**) is the largest bone in the proximal row. On its palmar surface is a *tubercle* (**2**), which is easily palpable through the skin. The scaphoid articulates proximally with the radius, distally with the trapezium and trapezoid, and medially with the lunate and capitate. Blood vessels enter along the entire roughened surface of the bone. In one-third of cases, blood vessels reach the scaphoid bone only on its distal face; in this case, a fracture of the scaphoid bone (see p. 124) may be followed by necrosis of the proximal fragment.

The crescent-shaped **lunate** (**3**) articulates proximally with the radius and the articular disk, medially with the triquetrum, laterally with the scaphoid, and distally with the capitate, and sometimes also with the hamate.

The **triquetrum** (**4**) is almost pyramidal in shape with its apex pointing medially. The base faces laterally and articulates with the lunate. It articulates proximally with the articular disk and distally with the hamate. The palmar surface has a small articular facet (**5**) for the pisiform.

The **pisiform** (**6**), the smallest carpal bone, is round and possesses on its dorsal surface an articular surface for the triquetrum. It is readily palpable and is inserted as a sesamoid bone in the tendon of the flexor carpi ulnaris.

Distal Row

The **trapezium** (**7**) possesses a *tubercle* (**8**) that is palpable on dorsiflexion on the hand, and medial to it there is a groove (**9**) for the tendon of the flexor carpi radialis. Distally it has a saddle-shaped articular facet (**10**) for the first metacarpal bone. A facet for articulation with the trapezoid lies medially, and between the distal and medial articular facets there is a further small facet for the joint with the second metacarpal bone. Proximally the trapezium articulates with the scaphoid.

The **trapezoid** (**11**) is wider dorsally than on its palmar surface. It articulates proximally with the scaphoid, distally with the second metacarpal, laterally with the trapezium, and medially with the capitate.

The **capitate** (**12**) is the largest carpal bone. It has facets proximally for articulation with the scaphoid and the lunate, laterally for the trapezoid, medially for the hamate, and distally mainly for the third metacarpal bone, as well as partly for the second and fourth metacarpals.

The **hamate** (**13**) is readily palpable. On its palmar aspect is the *hamulus* (**14**), which is curved laterally. The latter is related to the flexor digiti minimi brevis and the **piso-hamate ligament**. It articulates distally with the fourth and fifth metacarpal bones, laterally with the capitate, proximally and medially with the triquetrum, and proximally and laterally with the lunate.

Ossification: The ossification centers arise endochondrally and appear only after birth. In the 1st year of life (usually in the 3rd month) they develop in the capitate and hamate, in the 2nd to 3rd years in the triquetrum. In girls, the bony center appears in the triquetrum at the beginning of the 2nd year, whereas in boys the earliest appearance is seen only after $2\frac{1}{2}$ years. The ossification center for the lunate develops between the 3rd and 6th years, that for the scaphoid between the 4th and 6th years, those for the trapezium and trapezoid between the 3rd and 6th years. The pisiform arises between years 8 and 12.

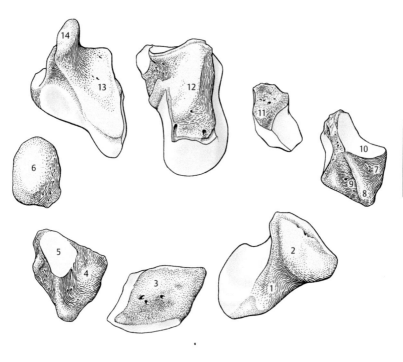

A Carpal bones of right hand, anterior (palmar) view

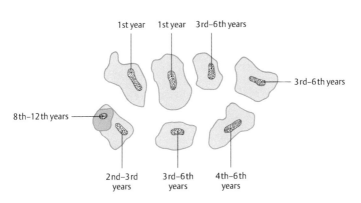

B Development of carpal bones

Bones of the Metacarpus and Digits (A–C)

The five **metacarpals** of the hand each have a *head* (**1**), a *shaft* (**2**), and a *base* (**3**). On all of these there are articular facets at one end (base) for articulation with the carpals and at the other end (head) for the phalanges. The palmar surface is slightly concave and the dorsal surface slightly convex. The dorsal surface exhibits a characteristic triangular configuration toward the head. The proximal articular facet of the **first metacarpal** is saddle-shaped; the **second metacarpal** has a notched base proximally for articulation with the carpus, and on the medial side with the third metacarpal. On the dorso-radial side of the base of the **third metacarpal** is a *styloid process* (**4**) and radially an articular facet for the second metacarpal. Proximally, for junction with the carpus, there is one articular facet, and on the ulnar side there are two articular facets for articulation with the **fourth metacarpal**. The fourth metacarpal has two articular facets radially but only one on its ulnar side for articulation with the **fifth metacarpal**.

The **bones of the digits**: Each digit (i.e., the index, middle, ring, and little finger) consists of more than one bone, namely, a **proximal** (**5**), a **middle** (**6**), and a **distal phalanx** (**7**). The sole exception is the **thumb**, which has only two phalanges.

Each **proximal phalanx** has a flattened palmar surface. Dorsally and transversely it is convex and has roughened sharpened borders for the attachment of the fibrous tendon sheaths of the flexor muscles. It has a *shaft* (**8**), a distal *phalangeal head* (also called a "trochlea") (**9**), and a proximal *base* (**10**). The base has a transverse oval socket, an articular facet for the metacarpals.

The base of the **middle phalanx** has two convex facets separated by a smooth ridge to conform to the shape of the head of the proximal phalanx.

The base of the **distal phalanx** also bears a ridge. At the distal end there is a rough palmar surface for insertion of the tendon of the flexor digitorum profundus as well as a palmar-facing, roughened, spade-shaped *plate* (**11**) at its terminus, *the tuberosity of the distal phalanx.*

Sesamoid bones are consistently found in the joints between the metacarpals and the proximal phalanx of the thumb, one placed medially and the other laterally. Sesamoid bones are also found in variable numbers in the other fingers.

Ossification: In both the metacarpals and the phalanges there is only one epiphyseal ossification center in addition to the perichondral diaphysis (3rd intrauterine month). In the metacarpals the distal epiphyseal centers develop in the 2nd year of life, except for the first metacarpal, in whose proximal end the center appears in the 2nd to 3rd year. In the phalanges epiphyseal ossification centers occur only proximally.

Clinical tip: Pseudoepiphyses may develop in the metacarpal bones. In radiographs they may be distinguished from true epiphyses, as they are attached to the diaphysis by a piece of bone. The first metacarpal bone may have a pseudoepiphysis at its distal end, but all other metacarpals have them at the proximal end: they must be distinguished from fractures. Pseudoepiphyses are found more commonly in certain diseases.

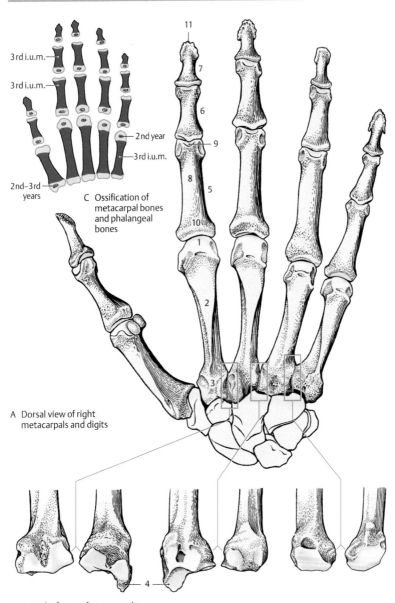

3rd i.u.m.

3rd i.u.m.

2nd year

3rd i.u.m.

2nd–3rd years

C Ossification of metacarpal bones and phalangeal bones

A Dorsal view of right metacarpals and digits

B Articular facets of metacarpals on their opposing surfaces

Upper Limb

Radiocarpal and Midcarpal Joints (A–E)

The **radiocarpal** or **wrist joint** is an ellipsoid joint formed on one side by the **radius** (**1**) and the **articular disk** (**2**) and on the other by the **proximal row of carpal bones**. Not all the carpal bones of the proximal row are in continual contact with the socketshaped articular facet of the radius and the disk. The *triquetrum* (**3**) makes close contact with the disk only during ulnar abduction and loses contact on radial abduction.

The **capsule** of the wrist joint is lax, relatively thin dorsally, and reinforced by numerous ligaments. The joint space is unbranched and sometimes contains *synovial folds*. Often the wrist joint is in continuity with the midcarpal joint.

The **midcarpal joint** is formed by the **proximal** and **distal rows of carpal bones** and has an s-shaped joint space. Each row of carpal bones can be considered as a single articular body, and they interlock with each other. Although there is a certain limited degree of mobility between members of the proximal row of carpal bones, this is not true of the distal row because they are joined one to another (**4**) as well as to the metacarpal bones by strong ligaments. Thus the distal row of carpal bones and the metacarpals form a functional unit.

The **joint capsule** is tense on the palmar surface and lax dorsally. The joint space is branched and has connections with the radiocarpal joint, and around the *trapezium* (**5**) and *trapezoid* (**6**) there are also connections with adjacent carpometacarpal joints.

Sometimes the joint space contains numerous *synovial folds* (**7**). The space between the lunate and triquetrum and the capitate and hamate is padded by synovial folds, which may be visible in radiographs.

Ligaments about the Wrist (A–E)

Four groups of ligaments can be distinguished:

Ligaments that unite the forearm bones with the carpal bones (violet). These include the *ulnar collateral ligament* (**8**), the *radial collateral ligament* (**9**), the *palmar radiocarpal ligament* (**10**), the *dorsal radiocarpal ligament* (**11**), and the *palmar ulnocarpal ligament* (**12**).

Ligaments that unite the carpal bones with one another, or **intercarpal ligaments** (red). These comprise the *radiate carpal ligament* (**13**), the *pisohamate ligament* (**14**), and the *palmar intercarpal* (**15**), *dorsal intercarpal* (**16**), and *interosseous intercarpal* (**4**) *ligaments*.

Ligaments between the carpal and metacarpal bones, or **carpometacarpal ligaments** (blue). To this group belong the *pisometacarpal ligament* (**17**), the *palmar carpometacarpal ligaments* (**18**), and the *dorsal carpometacarpal ligaments* (**19**).

Ligaments between the metacarpal bones, or **metacarpal ligaments** (yellow). These are organized into *dorsal* (**20**), *interosseous* (**21**), and *palmar* (**22**) *metacarpal ligaments*.

Almost all of these ligaments strengthen the joint capsules and partly guide the movements of the joints of hand.

The joints between the carpal bones of a row are known as **intercarpal joints**. Only the joint between the triquetrum and the *pisiform*, the **pisiform joint**, deserves special attention.

Clinical tip: Several more ligaments are described in hand surgery. They are important in cases of surgical intervention.

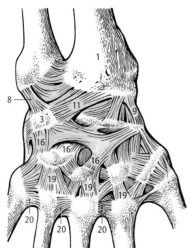

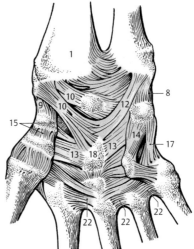

A Ligaments of right wrist,
   dorsal surface

B Ligaments of right wrist,
   palmar surface

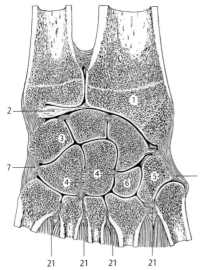

C Section through right wrist,
   dorsal view

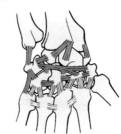

D Diagram of ligaments
   of right wrist, dorsal surface

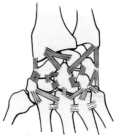

E Diagram of ligaments
   of right wrist, palmar surface

Movements in the Radiocarpal and Midcarpal Joints (A–C)

Starting from the midposition (**A**), we distinguish

- **Marginal movements** of **radial abduction** (**B**) and **ulnar adduction** (**C**)
 from
- **Movements in the plane of the hand**, that is, **flexion** (**palmar flexion**) and **extension** (**dorsiflexion**) as well as
- **Intermediate or combined movements**

Marginal Movements

Pure radial abduction: Radial abduction is carried out by the synergistic cooperation of the following muscles: extensor carpi radialis longus, abductor pollicis longus, extensor pollicis longus, flexor carpi radialis, and flexor pollicis longus. *The scaphoid* (red) *is tilted toward the palmar surface*, where it becomes palpable through the skin. Tilting of this bone allows the *trapezium* (blue) and *trapezoid* (green) to approach the radius. Since the *trapezoid* and the second *metacarpal bone* are rigidly joined together and the flexor carpi radialis and extensor carpi radialis longus are inserted into the second metacarpal bone, radial abduction represents a pulling action on this functional unit. The trapezoid glides along the scaphoid and, as the latter bone is not fixed, it can be moved, and since it cannot free itself from its other articulations, it is forced to tilt.

This tilting movement occurs along a radioulnar transverse axis. In addition to tilting of the scaphoid, there is palmar displacement of the other proximal carpal bones. **Radial abduction occurs around a dorsopalmar axis**, which runs through the *head of the capitate* (violet). In this movement the *pisiform* (dotted line) traverses the greatest path, as can be seen in radiographs.

Pure ulnar abduction: *Ulnar abduction involves a tilting or dorsal shifting of the proximal row of carpal bones.* The muscles collaborating in this action are especially the extensor carpi ulnaris and the flexor carpi ulnaris, in addition to the extensor digitorum and extensor digiti minimi. **Movement toward the ulnar side takes place around a dorsopalmar axis** through the head of the capitate, *the tilting movement around a radioulnar axis.*

Magnitude of Abduction Movements

Abduction movements are equally possible on either side of the **midposition**. The midposition or neutral position corresponds to an ulnar abduction of 12° and must not be confused with the straight position of the hand.

The **straight position** is one in which the long axis of the third finger runs over the capitate bone and is in a straight line with the long axis of the forearm. Starting from the straight position radial abduction is smaller, namely 15°, while ulnar abduction is approximately 40°. These values are only true when the arm is in strict supination; in strict pronation they are slightly greater. The angle is much larger if the forearm is pronated and the humerus rotated around the elbow joint. Possibly the various muscles are able to function more effectively in the latter position.

The radiographs from which Figures **A–C** were drawn were taken with the arm in pronation.

Hamate	Orange
Lunate	Black
Triquetrum	Yellow

A Straight position of right hand
  (from a radiograph)

B Radial abduction of right hand
  (from a radiograph)

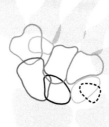

C Ulnar abduction of right hand
  (from a radiograph)

Movements in the Radiocarpal and Midcarpal Joints, continued (A–C)

Movements in the Plane of the Hand

Palmar flexion and dorsiflexion: The *proximal carpal bones are displaced toward the palmar side during dorsiflexion, and toward the dorsal side during palmar flexion.* This becomes particularly evident at the scaphoid (red), which protrudes toward the palmar side during dorsiflexion and is palpable through the skin. *The axes of movements run transversely, through the lunate* (black) *for the proximal row* and *through the capitate* (violet) *for the distal row.* **Flexion and extension consist of movements that take place around both these axes.** The magnitude of the angle between maximum dorsiflexion and palmar flexion is approximately 170°. **Palmar flexion** *occurs mainly at the radiocarpal (wrist) joint,* **dorsiflexion** *predominantly at the midcarpal joint.* Palmar flexion takes place by the action of the long flexors of the fingers, as well as by the wrist flexors and abductor pollicis longus. Dorsiflexion is carried out by the radial extensors of the wrist and by the extensors of the fingers (p. 172).

Intermediate or Combined Movements

These result from the directions in which the involved muscles work, and through them and the movements of the various joints, including the elbow and the shoulder, it is possible to produce movements that approximate those of a ball-and-socket joint. One focus of all joint and movement axes runs through the capitate. The structure of the wrist necessitates certain restrictions of mobility; for example, the wrist cannot be abducted during maximum palmar flexion, because in the latter position the proximal row of carpal bones cannot be either shifted or tilted.

Carpometacarpal and Intermetacarpal joints

Carpometacarpal Joint of the Thumb

This joint is a **saddle joint**, which allows *abduction* and *adduction* of the thumb, as well as *opposition, reposition,* and *circumduction.*

Carpometacarpal Joints

All other joints between the carpal and metacarpal bones are **amphiarthroses**. They are held in place by taut ligaments, the palmar and dorsal carpometacarpal ligaments.

Intermetacarpal Joints

These, too, are **rigid joints** and are stabilized by dorsal, palmar, and interosseous ligaments.

Metacarpophalangeal and Digital Joints (D, E)

The **metacarpophalangeal joints** are **ball-and-socket joints in shape** with *lax capsules. The palmar side of the capsule is strengthened by palmar ligaments and fibrocartilage.* The articulation is between the head of the metacarpal (**1**) and the base of the proximal phalanx (**2**). Movements are constrained by the *collateral ligaments* (**3**), whose origin (**4**) is dorsal to the axis of motion of the joint of the heads of the metacarpals. The greater the movement, the tighter the ligaments become. In flexion, movements of abduction are almost impossible. The joints may be rotated passively by up to 50°. The joints between the bones of the fingers, the **interphalangeal joints of the hand,** are **hinge joints** which can be flexed and extended. They, too, have collateral (**5**) and palmar ligaments.

Trapezoid	Green
Triquetrum	Yellow
Trapezium	Dark blue
Hamate	Orange
Pisiform	Black dotted line

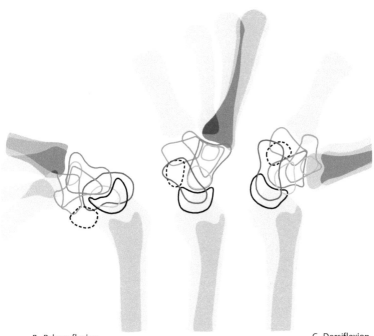

B Palmar flexion of right hand (from a radiograph)

A Midposition of right hand, seen from side (from a radiograph)

C Dorsiflexion of right hand (from a radiograph)

D Lateral view of digital joints

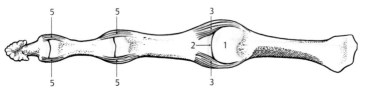

E Palmar view of metacarpophalangeal and digital joints with capsules removed

Muscles, Fasciae, and Special Features

Muscles of the Shoulder Girdle and Arm

Classification of the Muscles (A–C)

Ontogenetically the limb muscles arise from the ventral body wall musculature. Their division into dorsal (posterior) and ventral (anterior) muscle groups results from a consideration of their topography and innervation. The nerves arise from ventral or dorsal parts of the plexus (see Vol. 3). The immigration into the shoulder girdle region of various muscles that ontogenetically stem from other regions, for instance from the branchial musculature, has obscured the simple principle underlying this classification. Further information should be sought in textbooks of embryology. In any description of the musculature, it is important to retain the genetic principle as far as possible and by this to prove the relationship of the individual muscles.

Another method of classification is that of functional relationship. Here muscles are grouped together according to their actions on individual joints.

Shoulder Girdle Muscles

The shoulder girdle muscles may be grouped ontogenetically into those which have migrated from the trunk into the upper limb, those which extend secondarily from the arm into the trunk, and those which have immigrated as craniothoracic muscles from the head to the shoulder girdle.

Shoulder Girdle Muscles with Insertions on the Humerus

Posterior Muscle Group (see p. 138)

- Supraspinatus (**1**)
- Infraspinatus (**2**)
- Teres minor (**3**)
- Deltoid (**4**)
- Subscapularis (**5**)
- Teres major (**6**)
- Latissimus dorsi (**7**)

Anterior Muscle Group (see p. 142)

- Coracobrachialis (**8**)
- Pectoralis minor
  (exception: insertion on the scapula)
- Pectoralis major (**9**)

Trunk Muscles that Insert on the Shoulder Girdle

Posterior Muscle Group (see p. 144)

- Rhomboideus major
- Rhomboideus minor
- Levator scapulae
- Serratus anterior

Anterior Muscle Group (see p. 146)

- Subclavius
- Omohyoid

Cranial Muscles that Insert on the Shoulder Girdle
(see p. 146)

- Trapezius
- Sternocleidomastoid

Muscles of the Arm

The muscles of the upper limb are separated according to their position into those of the arm and those of the forearm (see p. 158). The arm muscles are divided into anterior and posterior groups, which are separated by intermuscular septa.

Anterior Muscle Group (see p. 154)

- Brachialis (**10**)
- Biceps brachii (**11**) with its long (**12**) and short (**13**) heads

Posterior Muscle Group (see p. 156)

- Triceps brachii with its long (**14**), medial (**15**), and lateral (**16**) heads
- Anconeus

17 Axillary artery and vein
18 Brachial artery
19 Brachial veins
20 Basilic vein
21 Cephalic vein
22 Radial nerve
23 Median nerve
24 Ulnar nerve
25 Medial antebrachial cutaneous nerve
26 Musculocutaneous nerve
27 Axillary or circumflex nerve

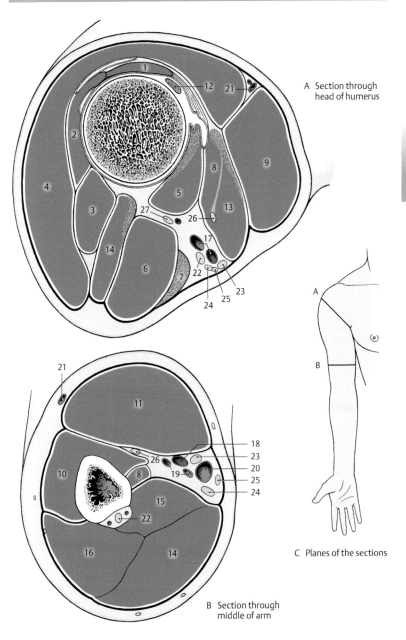

A Section through head of humerus

B Section through middle of arm

C Planes of the sections

Upper Limb

Shoulder Muscles Inserted on the Humerus

Posterior Muscle Group (A–C)

Insert on the greater tubercle of the humerus and on the crest of the greater tubercle and its continuation (the supraspinatus, infraspinatus, teres minor, and deltoid).

The **supraspinatus** (**1**) *arises from the supraspinous fascia and the supraspinous fossa* (**2**). It passes over the joint capsule, with which it is fused, to reach the *upper facet of the greater tubercle* (**3**). It holds the humerus in its socket, tenses the capsule, and abducts the arm. Sometimes there is a synovial bursa near the glenoid cavity.
Nerve supply: suprascapular nerve (C4–C6).

> **Clinical tip:** Tendinopathy of the supraspinatus caused by excessive strain or trauma is common. It is associated with calcification in the tendon near the greater tubercle and causes severe pain on abduction. Tendon ruptures also occur and are most common after 40 years of age.

The **infraspinatus** (**4**) *arises from the infraspinous fossa* (**5**), *the spine of scapula* (**6**), *and the infraspinous fascia, and runs to the greater tubercle* (**7**; middle facet). The infraspinatus reinforces the capsule of the shoulder joint. Its main function is external rotation of the arm. Near the joint socket there is often the subtendinous bursa of the infraspinatus muscle.
Nerve supply: suprascapular nerve (C4–C6).

■■ **Variant:** It is frequently fused with the teres minor.

The **teres minor** (**8**) *arises from the lateral border of the scapula* (**9**) superior to the origin of the teres major, and is *inserted on the lower facet of the greater tubercle* (**10**). It acts as a weak external rotator of the arm.
Nerve supply: axillary (circumflex) nerve (C5–C6).

■■ **Variant:** It may be fused with the infraspinatus.

The **deltoid** (**11**) is divided into three parts: clavicular (**12**), acromial (**13**), and spinal (**14**). The **clavicular part** arises from the *lateral third of the clavicle* (**15**), the **acromial part** from *the acromion* (**16**), and the **spinal part** from the *inferior border of the scapular spine* (**17**). *All three parts are attached to the deltoid tuberosity* (**18**). In the region of the greater tubercle of the humerus, there is a subdeltoid bursa.

The three components of the deltoid muscle act partly as synergists and partly as antagonists. The deltoid is the principal **abductor** of the shoulder joint. Abduction to approximately 90° is performed mainly by the deltoid, at first only by the acromial fibers. Only after the first two-thirds of the movement of abduction have been completed do the clavicular and spinal fibers become responsible for the movement. The clavicular and spinal fibers are able to **adduct** the arm after it has been lowered to a third of its range of movement. The clavicular fibers, aided by some of the acromial fibers, can produce **anteversion**, and the spinal fibers, helped by other acromial fibers, produce **retroversion**. These movements are superimposed on the framework of basic movements of the arm (swinging of the arm while walking). The clavicular and spinal sections of the deltoid exert a rotary action on these movements. The clavicular fibers can produce **internal rotation** of an arm that is adducted and externally rotated, while the spinal fibers can produce **external rotation** of an arm that is rotated internally.
Nerve supply: axillary (circumflex) nerve (C4–C6); clavicular fibers also by pectoral branches (C4–C5).

■■ **Variants:** Fusion with neighboring muscles; absence of the acromial fibers; presence of supernumerary groups of muscle fibers.

19 Teres major
20 Long head of the triceps
21 Lateral head of the triceps
22 Trapezius
23 Levator scapulae

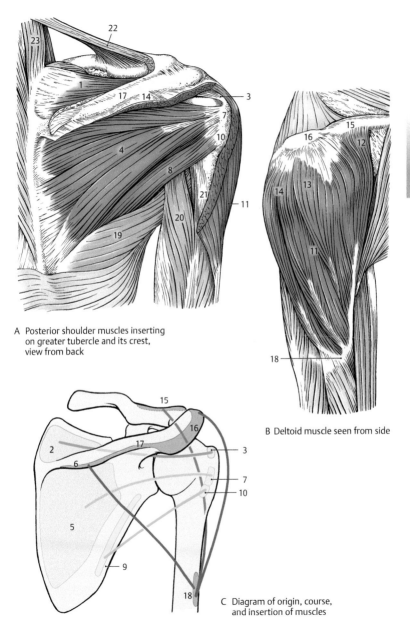

A Posterior shoulder muscles inserting on greater tubercle and its crest, view from back

B Deltoid muscle seen from side

C Diagram of origin, course, and insertion of muscles

Shoulder Muscles
Inserted on the Humerus, continued

Posterior Muscle Group, continued
(A–D)

Inserting on the lesser tubercle and its crest (subscapularis, teres major, and latissimus dorsi).

The **subscapularis** (**1**) *arises in the subscapular fossa* (**2**) *and inserts on the lesser tubercle* (**3**) *and the proximal part of its crest.* Near to its attachment between the subscapularis and the joint capsule is the *subtendinous bursa of the subscapularis* (**4**), and between it and the base of the coracoid process lies the *subcoracoid bursa* (**5**). Both bursae communicate with the joint space. The subscapularis produces internal rotation of the arm.
Nerve supply: subscapular nerve (C5–C8).

■ **Variant:** The presence of accessory bundles.

> **Clinical tip: Paralysis** of the subscapularis produces maximal external rotation of the limb, which indicates that it is a particularly strong medial rotator of the arm.
> The term **"rotator cuff"** is often incorrectly used for the subscapularis, supraspinatus (**6**), infraspinatus (**7**), and teres minor (**8**) muscles. It is more correct to use the term "muscle–tendon cuff" or "tendon hood."

The **teres major** (**9**), which *arises from the lateral border* (**10**) *of the scapula near the inferior angle, is inserted on the crest of the lesser tubercle* (**11**), near the subtendinous bursa of the teres major. Its main function is retroversion of the arm toward the midline, a movement requiring retroversion and a small degree of simultaneous medial rotation. It is particularly prominent if the arm has previously been anteverted and slightly abducted. The muscle also helps in adduction.
Nerve supply: thoracodorsal nerve (C6–C7).

■ **Variants:** Fusion with the latissimus dorsi or complete absence of the muscle.

The **latissimus dorsi** (**12**) is broad and flat, and is the largest muscle in the human body. *It arises from the spinous processes of the seventh to 12th thoracic vertebrae* (**13**) as the **vertebral part,** from the *thoracolumbar fascia* (**14**) *and the posterior third of the iliac crest* (**15**) as the **iliac part,** *from the 10th to 12th ribs* (**16**) as the **costal part,** and, in addition, very often from the *inferior angle of the scapula* as the **scapular part** (**17**). The latissimus dorsi thus usually arises in four parts that have different functions. It develops embryologically with the teres major, with which it is inserted on the *crest of the lesser tubercle* (**18**). The subtendinous bursa of the latissimus dorsi lies just proximal to the junction of both muscles. The latissimus dorsi provides the muscular framework of the posterior axillary fold. It lowers the raised arm and adducts it. When the arm is adducted, it pulls it backward and medially, and rotates it so far medially that the back of the hand can cover the buttock. The latissimus dorsi is often called the "dress coat pocket" muscle. Both latissimi can act together to pull the shoulders backward and downward. They function, too, during forced expiration and in coughing (coughing muscle).
Nerve supply: thoracodorsal nerve (C6–C8).

■ **Variant:** The presence of aberrant muscle fibers that run into the pectoralis major as a muscular arch across the axilla.

19 Long head of triceps muscle
20 Long head of biceps muscle
21 Coracoacromial ligament
22 Glenoid cavity
23 Glenoid labrum
24 Joint capsule
25 Bursa of supraspinatus muscle
26 External oblique abdominal muscle
27 Trapezius muscle (partly resected)

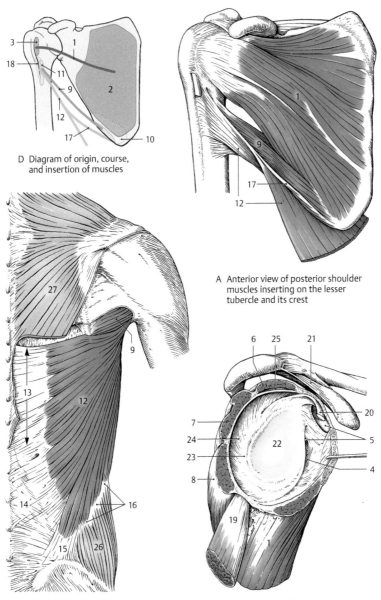

D Diagram of origin, course, and insertion of muscles

A Anterior view of posterior shoulder muscles inserting on the lesser tubercle and its crest

B Posterior view of latissimus dorsi muscle

C Muscle–tendon cuff

Shoulder Muscles Inserted on the Humerus, continued

Anterior Muscle Group (A, B)

The **coracobrachialis** (**1**) *arises from the coracoid process* (**2**) together with the short head of biceps brachii. *It is inserted on the medial surface of the humerus in line with the crest of the lesser tubercle* (**3**). It anteverts the arm and also retains the humeral head in its articular socket.

Nerve supply: musculocutaneous nerve (C6–C7).

The **pectoralis minor** (**4**) is the only shoulder girdle muscle that is not inserted on the bone of the free limb. *It arises from the third to fifth ribs* (**5**) *and is inserted on the coracoid process* (**6**). It lowers and rotates the scapula.

Nerve supply: pectoral nerves (C6–C8).

■■ **Variants:** Arises by variable numbers of slips.

The **pectoralis major** (**7**) consists of three parts: **clavicular**, **sternocostal**, and **abdominal**.

The **clavicular part** *arises from the medial half of the anterior surface of the clavicle* (**8**), while the **sternocostal part** *comes from the sternal membrane and the cartilages of the second to sixth ribs* (**9**). There are additional deep origins (**10**) of the sternocostal part from the third (fourth) to fifth costal cartilages. The weaker **abdominal part** *arises from the anterior layer of the uppermost part* (**11**) *of the rectus sheath. The pectoralis major is inserted onto the crest of the greater tubercle* (**12**) in such a manner that the fibers are twisted, so that the abdominal part is attached most proximally and forms a pocket that is open superiorly.

The pectoralis major is a strong muscle that presents four sides when the arm is hanging freely at the side. When the arm is raised, its borders form a triangle. It forms the muscular framework of the anterior axillary fold.

With the arm abducted, the clavicular and sternal parts can produce anteversion, a movement that is familiar from swimming. All parts of the pectoralis major act together as they forcibly and rapidly lower the raised arm. In addition the whole muscle can adduct the arm and rotate it medially. The sternocostal and abdominal parts together lower the shoulder anteriorly.

Finally the muscle can act as an accessory muscle during inspiration if the arms are fixed. Exhausted athletes after a race may be seen to prop up their arms on their trunk, so that the pectorales majores can be brought into action as accessory muscles of respiration to move the thorax.

Nerve supply: pectoral nerves (C5–T1).

■■ **Variants:** Individual portions may be absent. The sternocostal part may be divided into a sternal and a costal part. Sometimes the clavicular part is in direct contact with the deltoid muscle when there is no clavipectoral trigone (see p. 370). A muscular axillary arch may be formed which is related to the latissimus dorsi. A variant form is encountered in approximately 7% of cases.

13 Short head of the biceps
14 Long head of the biceps
15 Deltoid (partly resected)

Upper Limb

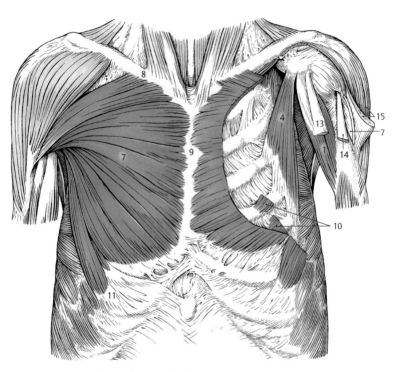

A Anterior shoulder muscles, anterior view

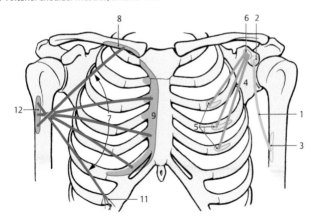

B Diagram of origin, course, and insertion of muscles

Trunk Muscles
Inserting on the Shoulder Girdle

Posterior Muscle Group (A–D)

The **rhomboid minor** (**1**) *originates from the spinous processes of the sixth to seventh cervical vertebrae* (**2**) *and inserts on the medial border of the scapula* (**3**).

The **rhomboid major** (**4**), situated caudal to the rhomboid minor, *arises from the spinous processes of the first to fourth thoracic vertebrae* (**5**) and likewise *inserts on the medial border of the scapula* (**3**), caudal to the insertion of rhomboid minor.

Both muscles have the same actions: they can press the scapula against the chest wall, and they can retract the scapula toward the vertebral column.

The two muscles are sometimes merged to form a single rhomboid muscle.
Nerve supply: dorsal scapular nerve (C4–C5).

The **levator scapulae** (**6**) *arises from the posterior tubercles of the transverse processes of the first to fourth cervical vertebrae* (**7**) and *is inserted on the superior angle of the scapula and the adjacent part of the medial border* (**8**). It elevates the scapula while rotating the inferior angle medially.
Nerve supply: dorsal scapular nerve (C4–C5).

The **serratus anterior** (**9**) *usually arises by nine* (**10**) *slips from the first to ninth ribs* (**10**) and sometimes from ribs 1 to 8. The number of slips is greater than the number of ribs from which they arise, as there are usually two slips from the second rib. *The insertion of the muscle extends from the superior to the inferior angles along the entire medial border of the scapula* (**3**). The muscle is divided into three sections according to the points of insertion: a **superior part** (**11**) inserted near the superior angle of the scapula, an **intermediate part** (**12**) inserted along the medial border of

the scapula, and an **inferior part** (**13**) attached on or near the inferior angle of the scapula.

All three parts pull the scapula anteriorly, a movement essential for anteverting the arm. It is the opposite of that produced by its antagonists, the rhomboid muscles. The superior and inferior parts together press the scapula onto the thorax, and in this movement they act synergistically with the rhomboid muscles. The inferior part rotates the scapula laterally and pulls the inferior angle lateral and forward. This movement makes elevation of the arm possible. All three parts may act to lift the ribs when the shoulder girdle is fixed and can thus serve as accessory muscles of respiration.
Nerve supply: the long thoracic nerve (C5–C7).

▇ **Variants:** Arises by variable numbers of slips.

Clinical tip: Paralysis of the serratus anterior caused by damage to the long thoracic nerve produces a **winged scapula** on the affected side, making it impossible to raise the arm laterally past 90° ("**rucksack paralysis**").
The possibility of damage to the rhomboid muscles must be considered in the differential diagnosis, as this may also produce a winged scapula but does not limit arm elevation (see also pp. 148 and 150).

14 Subscapular
15 Teres major
16 Teres minor
17 Infraspinatus
18 Supraspinatus
19 Clavicle
20 Subclavius
21 External oblique muscle of the abdomen
22 Section through the scapula

Upper Limb

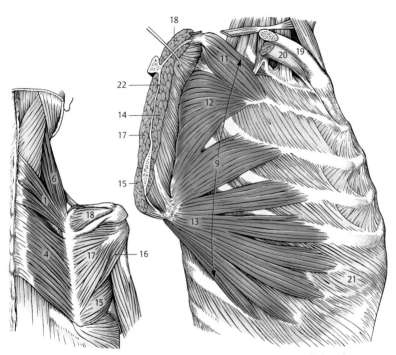

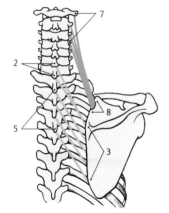

A Rhomboid muscles and levator scapulae muscle, posterior view

C Serratus anterior muscle, lateral view

B Diagram of origin, course, and insertion of muscles

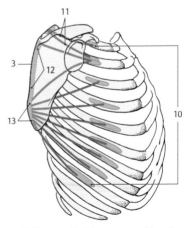

D Diagram of origin, course, and insertion of serratus anterior muscle

Trunk Muscles Inserting on the Shoulder Girdle, continued

Anterior Muscle Group (A–C)

The **subclavius** (**1**) *arises at the osteochondral junction of the first rib and is inserted into the subclavian groove on the inferior surface of the clavicle.* It pulls the clavicle toward the sternum and so stabilizes the sternoclavicular joint.
Nerve supply: subclavian nerve (C5–C6).

▬ **Variant:** This muscle may be absent.

The **omohyoid** is a two-bellied muscle; its **inferior belly** (**2**) *arises from the upper margin (border) of the scapula* near the scapular notch (**3**), and its **superior belly** (**4**) *inserts at the lateral one-third of the lower border of the body of the hyoid bone* (**5**). It is, among others, a fascia tensor and, as a vascular muscle, dilates the internal jugular vein lying beneath it (see also p. 326).
Nerve supply: ansa cervicalis "profunda" (C1–C3).

▬ **Variant:** The muscle may arise from the clavicle instead of the scapula, in which case it is known as the **cleidohyoid muscle**.

Cranial Muscles Inserted on the Shoulder Girdle (A–C)

The **trapezius** (**6**) is divided into **descending**, **transverse**, and **ascending parts**.

The **descending part** *arises from the superior nuchal line, the external occipital protuberance, and the nuchal ligament and is inserted on the lateral third of the clavicle* (**7**).
The **transverse part** *arises from the seventh cervical to third thoracic vertebrae* (from their spinous processes and supraspinous ligaments) and *is inserted on the acromial end of the clavicle, the acromion* (**8**), *and part of the spine of the scapula* (**9**). The **ascending part** *arises from the second or third to 12th thoracic vertebrae* (from the spinous processes and supraspinous ligaments) and *is inserted on the spinal trigone or the adjacent part of the scapula* (**10**; see also figures on p. 329).

The primary action of the trapezius is a static one, namely to stabilize the scapula and thus to fix the shoulder girdle. Active contraction of the trapezius pulls the scapula and the clavicle posteriorly toward the vertebral column. The descending and ascending parts rotate the scapula, and the former, in addition to adduction, also produces a slight elevation of the shoulder and so assists the serratus anterior. If the latter is paralyzed, the action of the descending part of the trapezius may still permit some elevation of the arm above the horizontal.
Nerve supply: accessory nerve and trapezius branch (cervical plexus C2–C4).

▬ **Variants:** The attachment to the clavicle may be widened to extend to the origin of the sternocleidomastoid muscle. In these cases there is a tendinous arch for passage of the supraclavicular nerves (see p. 358).

One head of the **sternocleidomastoid** (**11**) *arises from the sternum* (**12**) and the other *from the clavicle* (**13**). *It is inserted on the mastoid process* (**14**) and *the superior nuchal line* (**15**), where there is a tendinous junction with the origin of the trapezius.

As its action on the shoulder girdle is of minor importance, it is not discussed here but will be described in connection with the muscles of the head (see p. 328).
Nerve supply: accessory nerve and cervical plexus (C1–C2).

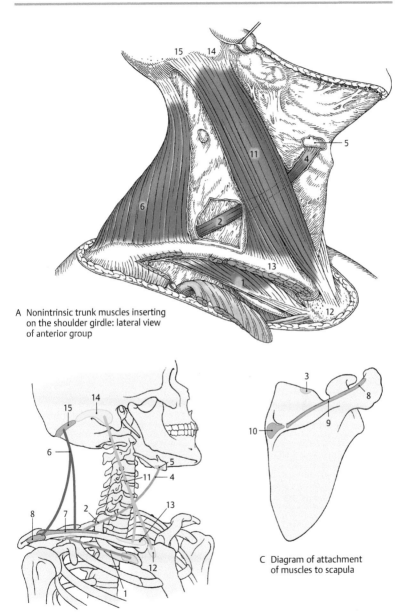

A Nonintrinsic trunk muscles inserting
on the shoulder girdle: lateral view
of anterior group

B Diagram of origin, course, and insertion of muscles

C Diagram of attachment
of muscles to scapula

Upper Limb

Function of the Shoulder Girdle Muscles (A–C)

We distinguish **adduction**, drawing of the arm toward the body, and **abduction**, lateral raising of the arm through 90° around a **sagittal axis** that runs through the humeral head. **Elevation**, which may be a continuation of abduction, is not due to movement within the shoulder joint but is produced by **rotation of the scapula**, the inferior angle of which is moved forward and laterally.

In addition, there is **anteversion** or forward lifting of the arm, and **retroversion** or backward lifting of the arm. Both movements occur around a **frontal axis** that runs through the humeral head.

Finally there is **rotation** of the upper limb. This is accomplished by pivoting the (freely hanging) arm around an **axis that runs from the humeral head through the ulnar styloid process**. It corresponds to the axis of pronation and supination of the forearm, so we may say that rotation leads to reinforcement of the movements of pronation and supination. We distinguish between **external** and **internal rotation**. The compound movement of **circumduction** may also be either a **lateral or medial circumduction**. In it the movement of the humerus is cone-shaped. Obviously the same muscles that are active in rotation of the arm also function in circumduction.

Adductors (A) include
- The pectoralis major (red, pectoral nerves)
- The long head of the triceps brachii (blue, radial nerve, see p. 156)
- The teres major (yellow, thoracodorsal nerve)
- The latissimus dorsi (orange, thoracodorsal nerve)
- The short head of the biceps brachii (green, musculocutaneous nerve)
- The clavicular and spinal parts of the deltoid (brown, broken line, pectoral branches and axillary nerve)

Abduction (B) is produced by
- The deltoid (red, axillary nerve and pectoral branches)
- The supraspinatus (blue, suprascapular nerve)
- The long head of the biceps brachii (yellow, musculocutaneous nerve)

The serratus anterior and trapezius may aid this movement by producing slight rotation of the scapula.

Elevation (C) of the upper limb is produced by
- The serratus anterior (red, long thoracic nerve)

Before the arm can be elevated, it must be abducted by the deltoid, the long head of the biceps brachii and the supraspinatus. In the transition from abduction to elevation, the trapezius (blue, accessory nerve) supports the action of the serratus anterior. The effect of the latter depends on its action on the clavicular joints (acromioclavicular and sternoclavicular joints).

> **Clinical tip:** If the **serratus muscle is paralyzed**, elevation of the arm is limited to the 15° produced by action of the trapezius.
> In **fractures of the humerus**, the level is an important determinant of the displacement of the bony fragments. If the fracture is proximal to the insertion of the deltoid muscle, the greater adductor force causes the proximal bone fragment to be pulled medially. If the bone is broken distally to the deltoid insertion, the overpowering force of the deltoid muscle pulls the proximal part laterally and anteriorly (see p. 380).

The color of the arrows shows the order of importance of the muscles in specific movements:

red
blue
yellow
orange
green
brown

The nerve that innervates each muscle is shown in parentheses.

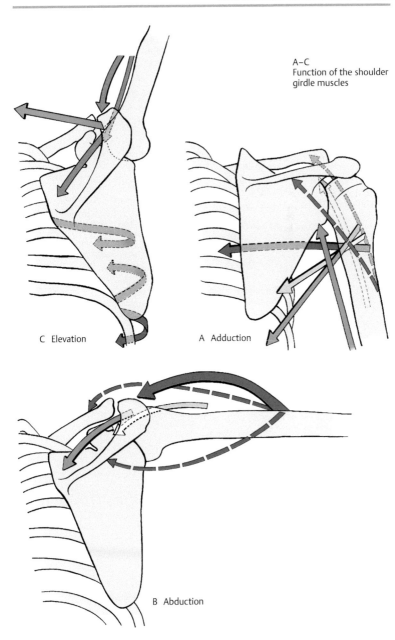

A–C
Function of the shoulder
girdle muscles

C Elevation

A Adduction

B Abduction

Function of the Shoulder Girdle Muscles, continued (A–D)

The muscles that are active during **anteversion** (**A**) include:
- The clavicular and some of the acromial fibers of the deltoid (red, pectoral branches and axillary nerve)
- The biceps brachii (blue, musculocutaneous nerve, see p. 154)
- The pectoralis major (yellow, pectoral nerves)
- The coracobrachialis (orange, musculocutaneous nerve)
- The serratus anterior (green, long thoracic nerve)

Clinical tip: Anteversion is still possible with paralysis of the serratus anterior, but it is accompanied by marked elevation of the scapula from the chest wall (winged scapula).

Retroversion (**B**) is brought about by
- The teres major (red, thoracodorsal nerve)
- The latissimus dorsi (blue, thoracodorsal nerve)
- The long head of the triceps brachii (yellow, radial nerve)
- The deltoid (orange, axillary nerve)

There is always some associated movement at the acromioclavicular joint.

External rotation (**C**) is produced by
- The infraspinatus (red, suprascapular nerve)
- The teres minor (blue, axillary nerve)
- The spinal part of the deltoid (yellow, axillary nerve)

The strongest external rotator, the infraspinatus, performs much more work than all the others combined. With external rotation, the scapula and clavicle are simultaneously pulled backward by the trapezius and rhomboid muscles. Thus this action also involves movements at the sternoclavicular and acromioclavicular joints.

Clinical tip: During sudden external rotation, the antagonistic pulling force of the most powerful medial rotator, the subscapularis, may result in avulsion of the lesser tubercle.

Internal rotation (**D**) is carried out by
- The subscapularis (red, subscapular nerve)
- The pectoralis major (blue, pectoral nerves)
- The long head of the biceps (yellow, musculocutaneous nerve)
- The clavicular part of the deltoid (orange, pectoral branches)
- The teres major (green, thoracodorsal nerve)
- The latissimus dorsi (brown, thoracodorsal nerve)

By far the strongest action is produced by the subscapularis and the weakest by the latissimus dorsi. When the elbow is extended, the short head of the biceps (not shown) also contributes slightly.

The cited movements, however, do not occur exclusively at the shoulder joint. In the living person, an associated movement of the shoulder girdle always takes place, as well as that of the trunk with certain movements.

The color of the arrows shows the order of importance of the muscles in the individual movements:

red
blue
yellow
orange
green
brown

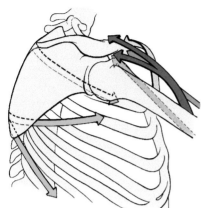

A Anteversion

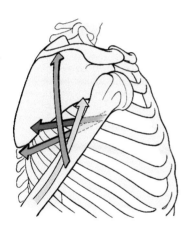

B Retroversion

A–D
Function of the shoulder
girdle muscles
(continued)

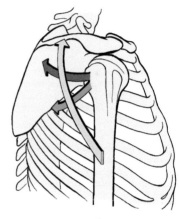

C External rotation

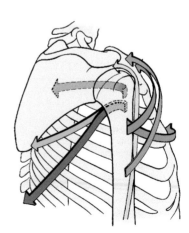

D Internal rotation

Fasciae and Spaces in the Shoulder Girdle Region

Fasciae (A, B)

Each shoulder girdle muscle is surrounded by its own fascia to permit free movement of the muscles relative to each other. Particularly strong fasciae are the **deltoid fascia** (**1**), the **pectoral fascia** (**2**), and the **clavipectoral fascia** (**3**).

The **deltoid fascia** covers the deltoid muscle and sends numerous septa deeply between the individual muscle bundles. Anteriorly it is attached to the pectoral fascia and posteriorly, where it is especially strong, it blends with the fascia that covers the infraspinatus muscle. It continues distally as the *brachial fascia* (see p. 180). Additionally it is fixed to the spine of the scapula, the acromion, and the clavicle.

The **pectoral fascia** covers the superficial surface of the pectoralis major muscle and extends from there over the *deltopectoral sulcus* (**4**) to the deltoid muscle. It is attached to the **axillary fascia** (**5**), which has a mixed loose and dense consistency.

The **clavipectoral fascia** surrounds the subclavius and the pectoralis minor, and partly extends over the coracobrachialis. It separates the pectoralis major from the pectoralis minor. At the lateral border of the latter it radiates into the axillary fascia.

A special feature of the remaining fasciae is that, in the region of the infraspinatus and teres minor, they may become aponeurotic and muscle fibers may actually arise from them.

The **axillary fascia** forms the continuation of the pectoral fascia as far as the fascia covering the latissimus dorsi. It does not consist of regularly arranged, dense connective tissue, but instead there are zones of loose tissue that may easily be removed. After removal of the loose part of the axillary fascia, an oval zone may be seen, the proximal fascial border of which is called the axillary arch of *Langer*.

Special Spaces in the Shoulder Girdle Region (Axillary Spaces and Axilla)

Axillary spaces (see p. 374). There is a **medial** and a **lateral axillary space**. These spaces are called the *triangular* and *quadrangular* spaces, respectively, because of their shapes. The medial or triangular space is bounded by the teres minor, the teres major, and the long head of the triceps brachii, and the lateral or quadrangular space by the long head of the triceps brachii, the teres minor, the teres major, and the humerus.

Axilla. The axilla is *pyramidal* in shape. It is bounded anteriorly by the *anterior axillary fold* (**6**), the muscular basis of which is the pectoralis major, and also deep in the anterior wall are the pectoralis minor and clavipectoral fascia. The posterior wall of the axilla consists of the *posterior axillary fold* (**7**), which is basically formed by the latissimus dorsi. Moreover, the subscapularis participates with the scapula and teres major in forming the posterior wall. The medial wall is formed by the thorax and the serratus anterior and its fascial covering. The lateral wall consists of the upper part of the arm. (The contents of the axilla are described on p. 372.)

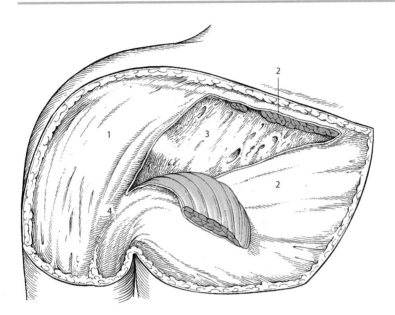

A Fasciae in the region of the clavipectoral triangle

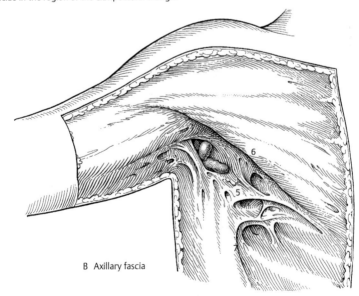

B Axillary fascia

Arm Muscles

According to their position the muscles of the upper limb may be divided into arm and forearm muscles. In the arm, the anterior group of muscles is divided from the posterior group by the intermuscular septa.

Anterior Muscle Group (A–C)

The **brachialis** (**1**) *arises from the distal half of the anterior surface of the humerus* (**2**) *and the intermuscular septa. It is inserted into the ulnar tuberosity* (**3**) *and the joint capsule* (as the articular muscle). It is a single-joint muscle and is the most important flexor of the elbow joint independent of pronation or supination of the forearm. Its full power is exerted in lifting a heavy load. In such a movement there is also slight retroversion at the shoulder joint. Nerve supply: musculocutaneous nerve (C5–C6). A small, lateral part of the muscle is supplied by the radial nerve (C5–C6).

▰ **Variant:** Inserts into the oblique cord or into the radius.

The **biceps brachii** (**4**) arises by its **long head** (**5**) from the *supraglenoid tubercle* (**6**) and by its **short head** (**7**) from the *coracoid process* (**8**). Both heads usually join, at the level of insertion of the deltoid, into the biceps muscle, which again terminates with two tendons. The stronger tendon is *inserted into the radial tuberosity* (**9**), with a bicipitoradial bursa enclosed. The other flattened tendon, the *bicipital aponeurosis* (**10**), whose fibers form the continuation of part of the short head, *radiates into the antebrachial fascia on the ulnar side*. The long head traverses the shoulder joint and, covered by a synovial sheath, it extends along the intertubercular groove (**11**) of the humerus. In its action it uses the head of the humerus as a fulcrum.

The biceps brachii acts on two joints. With its long head it abducts the arm and rotates it medially. The short head is an adductor.

Both heads are active during anteversion of the shoulder joint. The biceps brachii is also a flexor and strong supinator of the elbow joint. Its supinator action is increased during flexion of the elbow joint. It should be pointed out that, on the whole, the supinators are more strongly developed than the pronators. Therefore, the most essential rotary movements of the forearm are supinating movements (e.g., turning a screw). Its aponeurosis spans the fascia of the forearm.

Nerve supply: musculocutaneous nerve (C5–C6).

▰ **Variant:** In 10% of cases a third head may arise from the humerus to join to the belly of the biceps.

Clinical tip: The tendon of the long head of the biceps is especially susceptible to muscle or tendon tears. Rupture of this tendon is marked by a high position of the humeral head.

12 Long head of the triceps brachii
13 Lateral head of the triceps brachii
14 Medial head of the triceps brachii
15 Lateral intermuscular septum
16 Medial intermuscular septum
17 Latissimus dorsi
18 Subscapularis
19 Pectoralis minor
20 Coracobrachialis

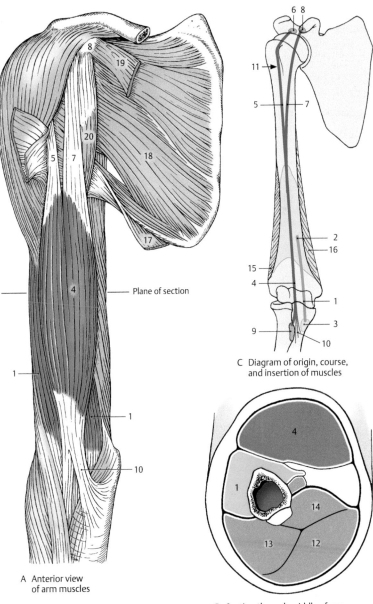

A Anterior view of arm muscles

C Diagram of origin, course, and insertion of muscles

B Section through middle of arm

Arm Muscles, continued

Posterior Muscle Group (A–C)

The **triceps brachii** (**1**) has three heads: long (**2**), **medial** (**3**), and **lateral** (**4**).

The **long head** (**2**) *arises from the infraglenoid tubercle of the scapula* (**5**) and extends distally in front of the teres minor (**6**) and behind the teres major (**7**). The **medial head** (**3**) *arises distally from the groove for the radial nerve* (**8**), *from the posterior surface of the humerus* (**9**), *from the medial intermuscular septum* (**10**), and, in its distal part, also from the *lateral intermuscular septum* (**11**). The medial head is largely covered by the long and lateral heads. It is only visible distally as it lies flattened against the humerus. The **lateral head** (**4**) *arises from the posterior surface of the humerus lateral and proximal to the groove for the radial nerve* (**12**). *Proximally it originates just beneath the greater tubercle* (**13**) *and ends distally in the region of the lateral intermuscular septum* (**11**).

The three heads converge in a common, flattened tendon, which is *inserted on the olecranon of the ulna* (**14**) *and the posterior wall of the capsule.* The long head of the triceps brachii acts on two joints, while with the other heads it acts only on one joint. It is **the** extensor of the elbow joint. At the shoulder the long head is involved in retroversion and adduction of the arm.

Part of the tendon of the triceps brachii radiates into the forearm fascia and may almost completely cover the anconeus. In the region of its attachment to the olecranon there are often bursae: the subcutaneous olecranon bursa and subtendinous bursa of the triceps brachii. Sometimes an intratendinous olecranon bursa can be seen.
Nerve supply: radial nerve (C6–C8).

Variants: A tendinous arch is very often found between the origin of the long head and the tendon of insertion of the latissimus dorsi. Very rarely the long head may arise additionally from the lateral margin of the scapula and from the articular capsule of the shoulder joint.

The **anconeus** (**15**) *arises from the posterior surface of the lateral epicondyle* (**16**) *and the radial collateral ligament and is inserted into the proximal one-fourth of the posterior side of the ulna* (**17**), close to the medial head of the triceps brachii. Its function is to assist the triceps brachii in producing the movement of extension, and it also tenses the capsule of the elbow joint.
Nerve supply: radial nerve (C7–C8).

18 Trapezius
19 Deltoid
20 Infraspinatus
21 Biceps brachii
22 Brachialis
23 Coracobrachialis
24 Humerus

Upper Limb

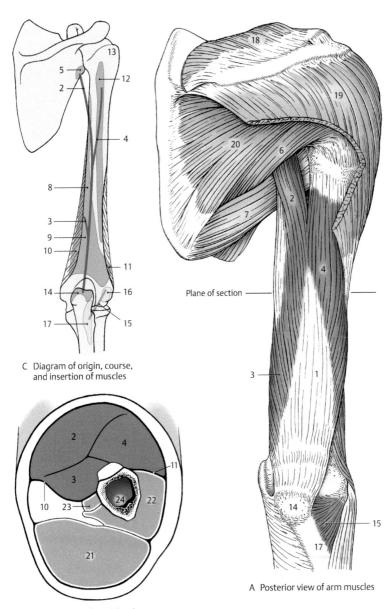

C Diagram of origin, course, and insertion of muscles

Plane of section

B Section through middle of arm

A Posterior view of arm muscles

Muscles of the Forearm

Classification of the Muscles (A–D)

The forearm muscles are divided into three groups according to their relationship to the various joints, their attachments, and their mode of action.

- **The first group** comprises muscles attached to the radius, which are only involved in movements of the bones of the forearm.
- **The second group** of forearm muscles extends to the metacarpus and produces movement at the wrist.
- **The third group** comprises muscles that extend to the phalanges and are responsible for finger movements.

Another system of classification is based on the position of the muscles in relation to each other. The ulna and radius with the interosseous membrane separate an anterior muscle group, the flexors, from a posterior group of extensors. Connective tissue septa between the anterior and posterior muscles separate a radial group. The flexors and extensors can be divided into superficial and deep muscles.

Finally the muscles of the forearm may also be divided into two groups according to their innervation—from either the ventral or dorsal portions of the plexus.

From the practical point of view, the muscles will be classified according to their positions relative to one another. This also provides the most comprehensive functional subdivision.

Anterior Group of Forearm Muscles

Superficial Layer (see p. 160)

- Pronator teres (**1**)
- Flexor digitorum superficialis (**2**)
- Flexor carpi radialis (**3**)
- Palmaris longus (**4**)
- Flexor carpi ulnaris (**5**)

Deep Layer (see p. 162)

- Pronator quadratus (**6**)
- Flexor digitorum profundus (**7**)
- Flexor pollicis longus (**8**)

Radial Group of Forearm Muscles (see p. 164)

- Extensor carpi radialis brevis (**9**)
- Extensor carpi radialis longus (**10**)
- Brachioradialis (**11**)

Posterior Group of Forearm Muscles

Superficial Layer (see p. 166)

- Extensor digitorum (**12**)
- Extensor digiti minimi (**13**)
- Extensor carpi ulnaris (**14**).

Deep Layer (see p. 168)

- Supinator (**15**)
- Abductor pollicis longus (**16**)
- Extensor pollicis brevis (**17**)
- Extensor pollicis longus (**18**)
- Extensor indicis (**19**)

20 Median nerve
21 Ulnar nerve
22 Superficial branch of radial nerve
23 Deep branch of radial nerve
24 Muscular branch of median nerve
25 Brachialis artery
26 Radial artery
27 Ulnar artery
28 Basilic vein
29 Cephalic vein
30 Radius
31 Ulna
32 Interosseous membrane
33 Common interosseous artery and vein
34 Anterior interosseous artery
35 Posterior interosseous artery

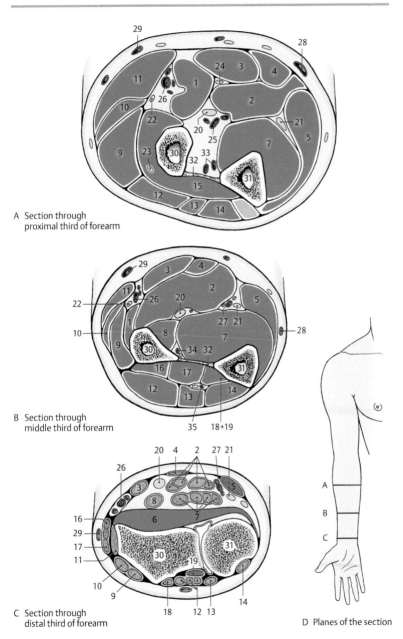

A Section through
   proximal third of forearm

B Section through
   middle third of forearm

C Section through
   distal third of forearm

D Planes of the section

Upper Limb

Anterior Forearm Muscles

Superficial Layer (A–D)

The **pronator teres** (**1**) *originates with its* **humeral head** *from the medial epicondyle of the humerus* (**2**) *and from the medial intermuscular septum* and with its **ulnar head** *from the coronoid process of the ulna* (**3**). It inserts on the pronator tuberosity (**4**) of the radius. Together with the pronator quadratus, it pronates the forearm and contributes to flexion at the elbow joint.
Nerve supply: median nerve (C6–C7).

■ **Variants:** The ulnar head may be absent. If a supracondylar process is present (see p. 114), the humeral head will also arise from it.

The **flexor digitorum superficialis** (**5**) *arises by its* **humeral head** *from the medial epicondyle of the humerus* (**6**), by its **ulnar head** *from the coronoid process of the ulna* (**7**), and by its **radial** head *from the radius* (**8**). Between the heads stretches a tendinous arch that is crossed below by the median nerve and the ulnar artery and vein. Its tendons run in a common sheath (see p. 182) through the carpal tunnel. The muscle ends in four tendons, each inserted onto the lateral *bony crests* (**9**) *in the center of the middle phalanges of the second to fifth fingers*. At this point the tendons divide into two slips (**10**, **perforated muscle**). The tendons of the *flexor digitorum profundus* (**11**) glide between and through them. It is a very weak elbow flexor, but a strong flexor of the wrist and finger joints. Its action on the digits is weakened when the wrist is maximally flexed.
Nerve supply: median nerve (C7–T1).

The **flexor carpi radialis** (**12**) *arises from the medial epicondyle of the humerus* (**6**) *and from the superficial fascia of the forearm. It inserts into the palmar surface of the base of the second metacarpal* (**13**) *and also* in some cases on the third metacarpal. It runs in the carpal tunnel in a groove in the trapezium, which is closed to form a fibroosseous canal. It is a weak flexor and pronator of the elbow joint, participates in

palmar flexion of the wrist, and acts with the extensor carpi radialis longus (see p. 164) to produce radial abduction.
Nerve supply: median nerve (C6–C7).

The **palmaris longus** (**14**) *arises from the medial epicondyle of the humerus and radiates into the palmar surface of the hand* with the **palmar aponeurosis** (**15**; see also p. 178). It flexes the hand toward the palm and tenses the palmar aponeurosis.
Nerve supply: median nerve (C7–T1).

■ **Variant:** It may be absent, but even then the palmar aponeurosis is always present.

The **flexor carpi ulnaris** (**16**) lies on the medial side. Its **humeral head** *arises from the medial epicondyle of the humerus* (**6**) and its **ulnar head** *from the olecranon and the upper two-thirds of the posterior margin of the ulna* (**17**). *It is inserted onto the pisiform bone* (**18**) and extends by the *pisohamate ligament* as far as the *hamate* (**19**) and by the *pisometacarpal ligament* to the *fifth metacarpal* (**20**). Proximal to its attachment to the pisiform bone, the muscle usually gives off descending tendon fibers that pass obliquely distally and radiate into the antebrachial fascia. It runs outside the carpal tunnel. It participates in palmar flexion, where it is more effective than the flexor carpi radialis and also helps in ulnar adduction of the hand.
Nerve supply: ulnar nerve (C7–C8).

21 Brachioradialis
22 Flexor pollicis longus
23 Pronator quadratus
24 Biceps brachii
25 Flexor retinaculum
26 Lumbricales
27 Abductor pollicis brevis
28 Flexor pollicis brevis
29 Palmaris brevis
30 Ulna
31 Radius
32 Vinculum longum
33 Vinculum breve

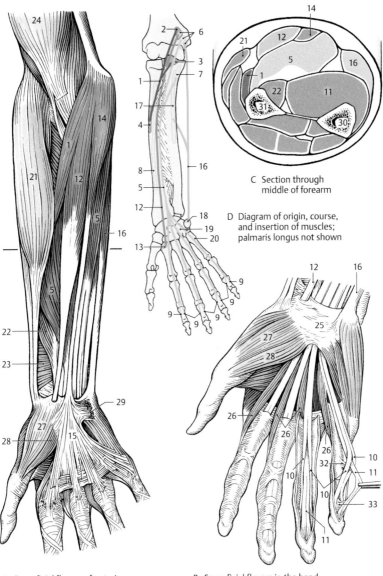

C Section through middle of forearm

D Diagram of origin, course, and insertion of muscles; palmaris longus not shown

A Superficial flexors of anterior group of forearm muscles (plane of section indicated)

B Superficial flexors in the hand, palmar aponeurosis removed

Upper Limb

Anterior Forearm Muscles, continued

Deep Layer (A–C)

The **pronator quadratus** (**1**) *arises from the distal one-fourth of the palmar surface of the ulna* (**2**) *and is inserted on the distal one-fourth of the palmar surface of the radius* (**3**). It pronates the forearm, assisted by the pronator teres.

Nerve supply: anterior interosseous branch of the median nerve (C8–T1).

▮ **Variants:** The muscle may extend farther proximally. It may also reach different carpal bones and, rarely, the muscles of the thenar eminence. The pronator quadratus is sometimes absent.

The **flexor digitorum profundus** (**4**) *arises from the proximal two-thirds of the palmar surface of the ulna* (**5**) *and the interosseous membrane*. In its course through the carpal tunnel, its tendons and those of the superficial flexors of the fingers (see p. 160) are surrounded by a common tendon sheath (see p. 182). *It is attached by four tendons to the base of the distal phalanges of the second to fifth fingers* (**6**). Because of its relationship to the flexor digitorum superficialis whose terminal tendon it pierces, it is also called the **perforating muscle**. In addition, the *lumbrical muscles* (**7**) arise from the radial side of its tendons. It is a flexor of the wrist and midcarpal, metacarpophalangeal, and phalangeal joints.

Nerve supply: anterior interosseous branch of the median nerve and the ulnar nerve (C7–T1).

▮ **Variant:** The tendon that reaches the index finger often has its own muscle belly (see **A**).

The **flexor pollicis longus** (**8**) *arises from the anterior surface of the radius, distal to the radial tuberosity, and from the interosseous membrane* (**9**). Surrounded by its own tendon sheath (see p. 182), it extends through the carpal tunnel, then lies between the heads of the flexor pollicis brevis, and *continues onto the base of the*

distal phalanx of the thumb (**10**). It is a flexor of the distal phalanx of the thumb and is also able to abduct it a little in the radial direction.

Nerve supply: anterior interosseous branch of the median nerve (C7–C8).

▮ **Variant:** In 40% of cases there is also a humeral head arising from the medial epicondyle of the humerus. In these cases there is a tendinous connection with the humeral head of the flexor digitorum superficialis.

11 Brachioradialis
12 Flexor retinaculum
13 Abductor pollicis brevis
14 Flexor pollicis brevis
15 Flexor carpi radialis
16 Palmaris longus
17 Flexor digitorum superficialis
18 Flexor carpi ulnaris
19 Pronator teres
20 Radius
21 Ulna

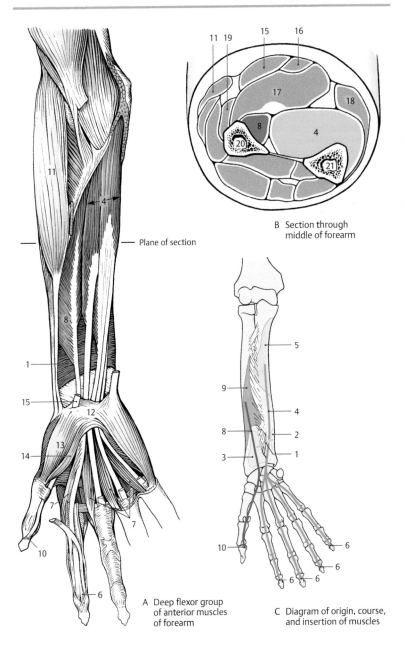

B Section through middle of forearm

Plane of section

A Deep flexor group of anterior muscles of forearm

C Diagram of origin, course, and insertion of muscles

Upper Limb

Radial Forearm Muscles (A–D)

The radial group includes three muscles that act as flexors at the elbow joint.

The **extensor carpi radialis brevis** (**1**) *arises from the common head of the lateral epicondyle of the humerus* (**2**), *from the radial collateral ligament, and from the annular radial ligament, and is inserted on the base of the third metacarpal* (**3**). It runs through the second tendon compartment (see p. 182) on the dorsum of the wrist. The extensor carpi radialis brevis is a weak flexor of the elbow joint. It brings the arm to the midposition from ulnar abduction and flexes posteriorly.

Nerve supply: deep branch of the radial nerve (C7).

The **extensor carpi radialis longus muscle** (**4**) *arises from the lateral supracondylar crest of the humerus* (**5**) and the *lateral intermuscular septum* as far as the lateral epicondyle and runs with the extensor carpi radialis brevis through the second tendon compartment. *It is inserted onto the base of the second metacarpal* (**6**). It is a weak flexor at the elbow joint, a weak pronator in the flexed arm, and a supinator in the extended arm. At the carpal joints it acts with the extensor carpi ulnaris in dorsiflexion and with the flexor carpi radialis in radial abduction.

Nerve supply: deep branch of the radial nerve (C6–C7).

The two muscles just described are called **"fist clenchers,"** as during clenching the hand must be slightly flexed posteriorly to permit maximal action by the flexors.

Clinical tip: Pain may occur in the lateral epicondyle of the humerus when the fist is clenched. This is called **epicondylitis of the humerus** and is thought to result from periosteal irritation at the origin of the two radial extensors due to overuse (tennis elbow).

The **brachioradialis** (**7**) *arises from the lateral supracondylar crest of the humerus* (**8**) *and the lateral intermuscular septum. It is inserted into the radial surface of the styloid process of the radius* (**9**). Unlike the muscles of the forearm described above, this muscle acts only on a single joint. It brings the forearm into the midposition between pronation and supination. In this position it acts as a flexor. It has a minimal flexor action during slow movements and in the supinated forearm.

Nerve supply: radial nerve (C5–C6).

Clinical tip: The radial artery pulse is palpable just proximal to its insertion, between its tendon and the tendon of the flexor carpi radialis (see p. 160; see also p. 386).

10 Extensor digitorum
11 Extensor digiti minimi
12 Extensor carpi ulnaris
13 Extensor pollicis longus
14 Extensor pollicis brevis
15 Abductor pollicis longus
16 Ulna
17 Radius

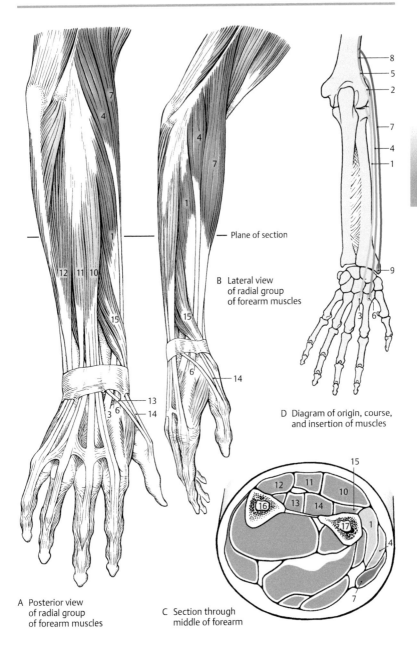

Plane of section

B Lateral view
of radial group
of forearm muscles

D Diagram of origin, course,
and insertion of muscles

A Posterior view
of radial group
of forearm muscles

C Section through
middle of forearm

Upper Limb

Posterior Forearm Muscles

Superficial (Ulnar) Layer (A–C)

The **extensor digitorum** (**1**) *has a flattened origin from the lateral epicondyle of the humerus* (**2**), *the radial collateral ligament, the annular radial ligament and the antebrachial fascia.* It runs through the fourth tendon compartment (see p. 182). *With its tendons it forms the dorsal aponeurosis* (**3**) *of the second to fifth fingers.* In addition, slips of the tendons run to the bases of the proximal phalanges (**4**) and to the capsules of the metacarpophalangeal joints. Between the individual tendons **intertendinous connections** (**5**) are always present, starting from the fourth to the third and fifth fingers. The extensor digitorum extends and spreads the fingers. It is the strongest dorsiflexor of the wrist and midcarpal joints and it acts, too, as an ulnar abductor.
Nerve supply: deep branch of the radial nerve (C6–C8).

■ **Variants:** The tendon of the second finger may have a separate muscle belly. The tendon to the fifth finger can be absent. In other cases the tendons to individual fingers may also be duplicated.

The **extensor digiti minimi** (**6**) *arises by a common head with the extensor digitorum* (**2**) and extends through the fifth tendon compartment in the dorsum of the wrist, usually as two tendons, *to the dorsal aponeurosis of the fifth finger.* Sometimes it is absent and then the extensor digitorum takes over its function with an additional tendon. It extends the fifth digit and helps in dorsiflexion and ulnar abduction of the hand.
Nerve supply: deep branch of the radial nerve (C6–C8).

The **extensor carpi ulnaris** (**7**) *arises from the common head* (**2**), together with the extensor digitorum, *and from the ulna* (**8**) and runs on the posteromedial side of the ulna

through the sixth tendon compartment *to the base of the fifth metacarpal* (**9**).
It is actually misnamed because it acts as a strong ulnar abductor, an action that is most easily understood from the course of its tendon in relation to its axis of movement (see p. 134); the tendon runs to the radiocarpal joint on the dorsal side and to the midcarpal joint on the palmar side. This leads to dorsiflexion of the radiocarpal joint and palmar flexion in the midcarpal joint; that is, the two functions balance one another. Hence the principal action of the muscle is as an abductor. Its antagonist is the abductor pollicis longus.
Nerve supply: deep branch of the radial nerve (C7–C8).

■ **Variant:** An additional tendon that extends to the proximal phalanx is frequently found on the radial side.

10 Extensor carpi radialis longus
11 Extensor carpi radialis brevis
12 Abductor pollicis longus
13 Extensor pollicis brevis
14 Extensor pollicis longus
15 Extensor indicis
16 Radius
17 Ulna
18 Anconeus

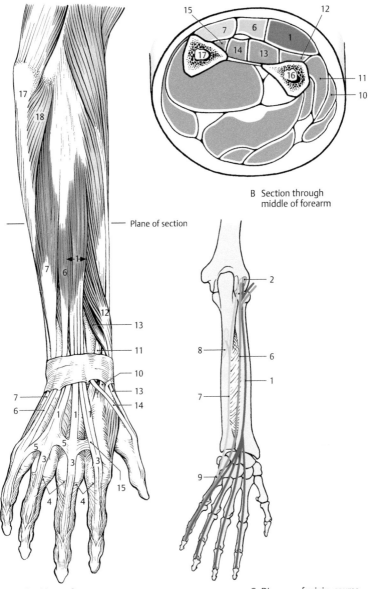

B Section through middle of forearm

Plane of section

A Superficial layer of posterior forearm muscles

C Diagram of origin, course, and insertion of muscles

Upper Limb

Posterior Forearm Muscles, continued

Deep Layer (A–C)

The surfaces from which the **supinator** (**3**) *originates include the supinator crest of the ulna* (**1**)*, the lateral epicondyle of the humerus* (**2**)*, the radial collateral ligament, and the annular radial ligament.* Those fibers originating from the most posterior portion of the radial collateral ligament run superficially and form a distally convex tendinous arch. *The muscle inserts on the radius* (**4**) *between the radial tuberosity and the attachment of pronator teres.* It encircles the radius and supinates the forearm, in contrast to the biceps brachii, in every position of flexion and extension.
Nerve supply: deep branch of the radial nerve (C5–C6).

The **abductor pollicis longus** (**5**) *arises from the posterior surface of the ulna* (**6**) distal to the supinator crest of the ulna, *from the interosseous membrane* (**7**)*,* and *from the posterior surface of the radius* (**8**)*.* It runs through the first tendon compartment (see p. 182) and is *inserted on the base of the first metacarpal* (**9**)*.* Part of the tendon reaches the trapezium and another part often fuses with the tendon of the extensor pollicis brevis and abductor pollicis brevis.

Due to its position it flexes the hand toward the palm and abducts it radially. The main function of this muscle is abduction of the thumb.
Nerve supply: deep branch of the radial nerve (C7–C8).

The **extensor pollicis brevis** (**10**) *arises from the ulna* (**11**) distal to the abductor pollicis longus, *from the interosseous membrane* (**12**)*,* and *from the posterior surface of the radius* (**13**) and *inserts on the base of the proximal phalanx of the thumb* (**14**)*.* It extends and abducts the thumb because of its close relationship to the abductor pollicis

longus, with which it runs in the first tendon compartment.
Nerve supply: deep branch of the radial nerve (C7–T1).

▬ **Variants:** Its terminal tendon is frequently duplicated. In rare cases it may be absent.

The **extensor pollicis longus** (**15**) *arises from the posterior surface of the ulna* (**16**) *and the interosseous membrane* (**17**)*.* It runs through the third tendon compartment on the dorsal side of the wrist. *It is inserted on the base of the distal phalanx* (**18**) *of the thumb.* It uses the dorsal tubercle on the radius, which is situated lateral to the third tendon compartment, as a fulcrum for extending the thumb. At the wrist it dorsiflexes and abducts the hand radially.
Nerve supply: deep branch of the radial nerve (C7–C8).

The distal third of the *posterior surface of the ulna* (**19**) *and the interosseous membrane* (**20**) *are the sites of origin* of the **extensor indicis** (**21**)*.* It runs with the extensor digitorum muscle, through the fourth tendon compartment, and *projects* its tendon *into the dorsal aponeurosis of the index finger.* It extends the index finger and participates in dorsiflexion at the wrist and midcarpal joints.
Nerve supply: deep branch of the radial nerve (C6–C8).

▬ **Variants:** Two or three tendons are frequently present. The muscle is sometimes absent.

22 Extensor digitorum
23 Extensor digiti minimi
24 Extensor carpi ulnaris
25 Ulna
26 Radius

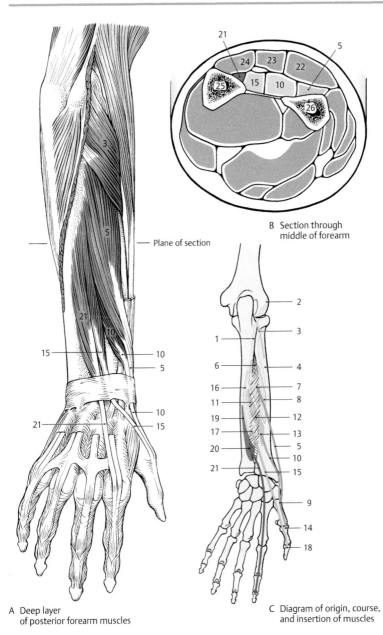

B Section through
   middle of forearm

— Plane of section

Upper Limb

A Deep layer
   of posterior forearm muscles

C Diagram of origin, course,
   and insertion of muscles

Function of Muscles of the Elbow Joint and Forearm (A–D)

The movements at the elbow joint are **flexion** and **extension. The axis of movement runs through the epicondyles of the humerus.** All muscles that pass in front of the axis act as flexors and all those passing behind it act as extensors at the elbow joint. Since many of the muscles act on several joints, their names are not always appropriate for their function in relation to the elbow joint. In addition, their action at the elbow joint depends on the position of the neighboring joints.

The **flexors** (**A**) include
- Biceps brachii (red, musculocutaneous nerve)
- Brachialis (blue, musculocutaneous nerve)
- Brachioradialis (yellow, radial nerve)
- Extensor carpi radialis longus (orange, radial nerve)
- Pronator teres (green, median nerve)

Less important are (not shown) the flexor carpi radialis, extensor carpi radialis brevis, and palmaris longus. Flexion in the position of pronation, performed by contraction of almost all the flexors, is strongest. The exceptions are the brachialis muscle, which is equally strong in all positions, and the biceps brachii muscle, whose flexor power is reduced in pronation.

The only important **extensor** (**B**) is the triceps brachii (red, radial nerve). The most effective parts of it are the medial and lateral heads, while the long head of the triceps is only of secondary importance. The anconeus may be disregarded as an extensor.

The movements of the forearm are **rotational movements** at the proximal and distal radioulnar joints, with associated movements at the humeroradial joint.

These rotational movements are **pronation** and **supination** (see p. 122) and **they occur around an axis that runs from the fovea on the head of the radius to the styloid process of the ulna.**

Pronation and supination are executed with almost equal force but with greater strength when the elbow joint is flexed. The preponderance of pronation is a false impression due to a medial rotation in the shoulder joint (*Lanz* and *Wachsmuth*).

The muscles that act as **supinators** (**C**) are
- Supinator (red, radial nerve)
- Biceps brachii (blue, musculocutaneous nerve)
- Abductor pollicis longus (yellow, radial nerve)
- Extensor pollicis longus (orange, radial nerve)
- Brachioradialis (not shown)

Extensor carpi radialis longus also functions as a supinator when the arm is extended.

Pronation (**D**) is produced by the
- Pronator quadratus (red, median nerve)
- Pronator teres (blue, median nerve)
- Flexor carpi radialis (yellow, median nerve)
- Extensor carpi radialis longus (orange, radial nerve) in the flexed arm
- Brachioradialis (not shown)
- Palmaris longus (not shown)

The color of the arrows shows the order of importance of the muscles in each movement:

red
blue
yellow
orange
green

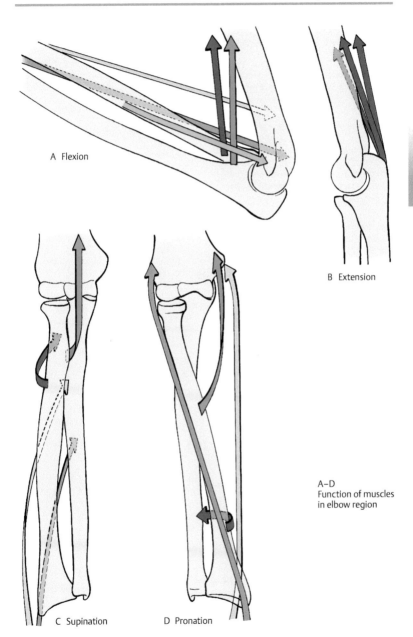

A Flexion

B Extension

C Supination

D Pronation

A–D
Function of muscles
in elbow region

Function of Muscles of the Wrist and the Midcarpal Joint (A–D)

We distinguish **dorsiflexion** (**A**), which lifts the dorsum of the hand, from **palmar flexion** (**B**), which lowers the dorsum of the hand.

These movements take place at the radiocarpal and midcarpal joints **through an imaginary transverse axis that runs through the capitate bone**. We also distinguish **radial abduction** (**C**) and **ulnar abduction** (**D**) **about a dorsopalmar axis through the capitate bone**.

It should be noted here that, in the resting position of the hand, the long axis through the third metacarpal bone, the axis through the capitate, and the main axis of the forearm run parallel to one another. The main axis of the forearm runs from the middle of the radial head to the styloid process of the ulna. This axis corresponds to the axis of movements during pronation and supination.

Palmar flexion is the most powerful of the movements described above. The flexors are considerably stronger than the extensors and, among them, the flexors of the fingers are the most powerful.

> **Clinical tip:** The predominance of the flexors causes the hand to assume a position of palmar flexion after a long period of immobilization (fracture healing). Thus, the hand should always be immobilized in slight dorsiflexion during healing.

The following muscles are active in **dorsiflexion**:
- Extensor digitorum (red, radial nerve)
- Extensor carpi radialis longus (blue, radial nerve)
- Extensor carpi radialis brevis (yellow, radial nerve)
- Extensor indicis (orange, radial nerve)
- Extensor pollicis longus (green, radial nerve)
- Extensor digiti minimi (not shown)

Palmar flexion can be produced by
- Flexor digitorum superficialis (red, median nerve)
- Flexor digitorum profundus (blue, median nerve and ulnar nerve)
- Flexor carpi ulnaris (yellow, ulnar nerve)
- Flexor pollicis longus (orange, median nerve)
- Flexor carpi radialis (green, median nerve)
- Abductor pollicis longus (brown, radial nerve)

The two digital flexor muscles are the strongest flexors at the wrist joint.

Radial abduction is produced by
- Extensor carpi radialis longus (red, radial nerve)
- Abductor pollicis longus (blue, radial nerve)
- Extensor pollicis longus (yellow, radial nerve)
- Flexor carpi radialis (orange, median nerve)
- Flexor pollicis longus (green, median nerve)

Ulnar abduction is produced by
- Extensor carpi ulnaris (red, radial nerve)
- Flexor carpi ulnaris (blue, ulnar nerve)
- Extensor digitorum (yellow, radial nerve)
- Extensor digiti minimi (not shown)

The color of the arrows shows the order of importance of the muscles in each movement:

red
blue
yellow
orange
green
brown

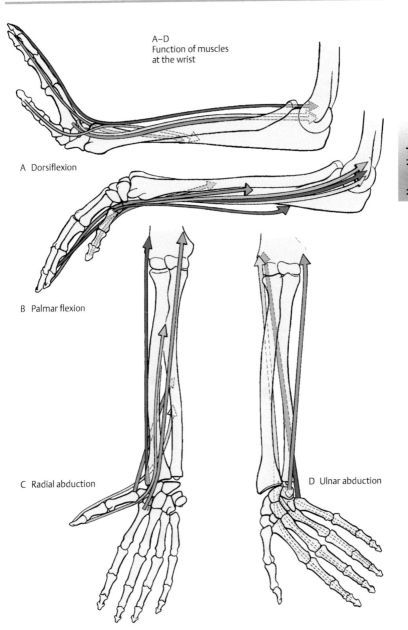

A–D
Function of muscles
at the wrist

A Dorsiflexion

B Palmar flexion

C Radial abduction

D Ulnar abduction

Upper Limb

Intrinsic Muscles of the Hand

The intrinsic hand muscles are divisible into three palmar groups:

- The **central muscles** of the **hand**
- The **thenar muscles** of the thumb
- The **hypothenar muscles** of the little finger

The extensor aponeurosis lies on the dorsum of the digits.

Central muscles of the hand (A–D)

The seven short, pennate **interossei** may be divided into **three palmar muscles arising by one head** and **four dorsal muscles that arise by two heads**.

The **palmar interossei** (1) *arise from the second, fourth, and fifth metacarpal bones* (**2**). *They insert by short tendons on the corresponding proximal phalanges* (**3**) *and they also radiate into the corresponding tendons of the dorsal aponeurosis* (**4**).

Their tendons run dorsal to the *deep transverse metacarpal ligaments* (**5**) and palmar to the axis of the metacarpophalangeal joints. Thus they flex the metacarpophalangeal joints, and by their expansions into the dorsal aponeurosis they are able to extend the interphalangeal joints. Through their relationship to the metacarpal and phalangeal bones, they also adduct in relation to an axis that passes longitudinally through the middle finger; they move the second, fourth, and fifth fingers toward the middle finger.

The **dorsal interossei** (**6**) *arise by two heads from the adjacent sides of the five metacarpal bones* (**2, 7**). Like the palmar interosseous muscles, *they extend to the proximal phalanges and radiate into the dorsal aponeurosis* (**4**). The first dorsal interosseous extends to the proximal phalanx of the second finger on the radial side, the second and third interosseous muscles reach the proximal phalanx of the middle finger on both the radial and ulnar

sides, and the fourth dorsal interosseous muscle extends to the proximal phalanx of the fourth finger on the ulnar side.

Like the palmar interosseous muscles, they flex the metacarpophalangeal joints and extend the interphalangeal joints. They function as abductors in relation to the axis of the middle finger (stretching of the finger).

Nerve supply: deep branch of the ulnar nerve (C8–T1).

The four **lumbricales** (**8**) *arise from the radial sides of the tendons of the flexor digitorum profundus* (**9**). As these tendons are mobile, the sites of origin of the lumbricales are not fixed. Covered by the palmar aponeurosis and palmar to the deep transverse metacarpal ligaments (**5**), *they run to the extensor aponeurosis* (**4**) *and to the capsules of the metacarpophalangeal joints.* They flex the metacarpophalangeal joints and extend the interphalangeal joints.

Nerve supply: the two radial lumbricales are supplied by the median nerve and the two ulnar ones by the deep branch of the ulnar nerve (C8–T1).

10 Flexor retinaculum
11 Abductor pollicis brevis
12 Flexor pollicis brevis
13 Transverse head of the abductor pollicis
14 Abductor digiti minimi
15 Flexor carpi ulnaris
16 Flexor carpi radialis

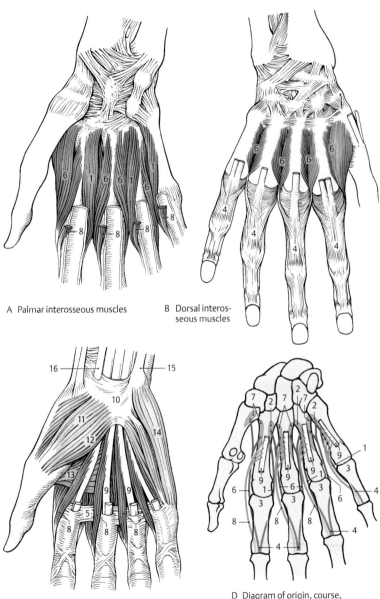

A Palmar interosseous muscles

B Dorsal interosseous muscles

C Lumbricales muscles

D Diagram of origin, course, and insertion of muscles

Intrinsic Muscles of the Hand, continued

Thenar Muscles (A–D)

These include
- Abductor pollicis brevis
- Flexor pollicis brevis
- Adductor pollicis
- Opponens pollicis

The **abductor pollicis brevis** (1) *arises from the scaphoid tubercle* (2) *and the flexor retinaculum* (3). *It is inserted into the radial sesamoid bone* (4) *and the proximal phalanx* (5) *of the thumb.* It abducts the thumb.
Nerve supply: median nerve (C8–T1).

The **flexor pollicis brevis** has a **superficial head** (6) and a **deep head** (7). *The former arises from the flexor retinaculum* (3) *and the latter from the trapezium* (8), *trapezoid* (9), *and capitate* (10). It is *inserted into the radial sesamoid bone* (4) *of the metacarpophalangeal joint of the thumb.* It flexes, adducts, and abducts the thumb and is able to bring the thumb into opposition.
Nerve supply: the superficial head is supplied by the median nerve and the deep head by the ulnar nerve (C8–T1).

The **adductor pollicis** *also has two heads of origin*, the **transverse head** (11) *originating from the entire length of the third metacarpal* (12), and the **oblique head** (13) *originating from the adjacent carpal bones.* It is *inserted into the ulnar sesamoid bone* (14) *of the metacarpophalangeal joint of the thumb.* It produces adduction and assists in opposition and flexion of the thumb.
Nerve supply: deep branch of the ulnar nerve (C8–T1).

The **opponens pollicis** (15) *arises from the tubercle of the trapezium* (16) *and the flexor retinaculum* (3) *and inserts on the radial border of the first metacarpal* (17). It produces opposition of the thumb and assists in adduction.
Nerve supply: median nerve (C6–C7).

In summary, the muscles of the thenar eminence may also be classified according to their function:

Adduction of the thumb is produced by the adductor pollicis with the help of the flexor pollicis brevis and opponens pollicis.

Abduction is produced by the abductor pollicis brevis and partly by the flexor pollicis brevis.

The position of **opposition** is produced principally by the opponens pollicis, assisted by the flexor pollicis brevis and adductor pollicis.

Reposition (return to the neutral position) is effected by the long muscles on the dorsal side: extensor pollicis brevis, extensor pollicis longus, and abductor pollicis longus.

> **Clinical tip:** The "reticular bands" (*Landsmeer*) run from the attachments of the abductor pollicis brevis and adductor pollicis; they reach to the extensor tendons and insert together with them on the distal phalanges. They are important during hand surgery.

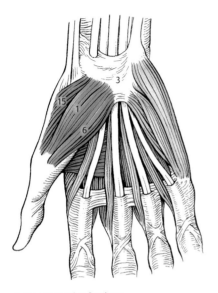

A Thenar muscles, first layer

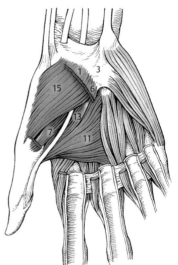

B Thenar muscles, second layer

C Thenar muscles, third layer

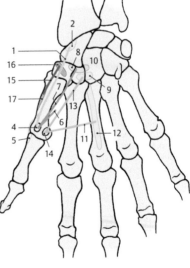

D Diagram of origin, course, and insertion of muscles

Intrinsic Muscles of the Hand, continued

Palmar Aponeurosis (A)

The **palmar aponeurosis** (also see p. 388) consists of *longitudinal* (**1**) and *transverse* (**2**) fascicles. The longitudinal fibers run to the tendon sheaths of the flexor tendons (**3**), the deep transverse metacarpal ligaments (**4**), and the ligaments of the metacarpophalangeal joints. They also radiate into the corium of the palm of the hand (**5**). The palmar aponeurosis is connected to the deep palmar fascia (see p. 180) by nine septa (**6**). Eight of the septa border both sides of the tendons of the superficial and deep flexors of the digits, while the ninth septum lies on the radial side of the first lumbrical muscle (see p. 174). The septa arise from both the longitudinal and transverse fasciculi.

The connection of the deep palmar fascia with the carpal bones corresponds to the anchoring of the palmar aponeurosis to the skeleton of the hand. The longitudinal fasciculi reach the second through the fifth fingers and radiate mostly in the hand and in the fibrous layer of the synovial sheaths (see p. 182). A few of the fibers join the superficial transverse metacarpal ligament. The transverse fasciculi lie proximally deeper than the longitudinal fasciculi. Distally the transverse fasciculi (**2**) are visible, lying in the same layer as the longitudinal fibers.

The palmar aponeurosis forms a functional unit with the ligaments, septa, and fasciae. It is firmly fixed to the skin of the palm of the hand over the carpal bones.

In the hypothenar eminence lies the **palmaris brevis** (**7**), which may be in the process of involution *and whose fibers connect the palmar aponeurosis and the flexor retinaculum* (**8**) *to the skin on the ulnar border of the hand.*
Nerve supply: superficial branch of the ulnar nerve (C8–T1).

Hypothenar Muscles (B–D)

The muscles of the hypothenar eminence consist of
– Abductor digiti minimi (**9**)
– Flexor digiti minimi brevis (**10**)
– Opponens digiti minimi (**11**)

The **abductor digiti minimi** (**9**) *arises from the pisiform* (**12**), *the pisohamate ligament* (**13**), *and the flexor retinaculum* (**8**) and is *inserted into the ulnar margin of the base of the proximal phalanx of the fifth digit* (**14**). In part it also radiates into the extensor aponeurosis of the little finger. It functions as a pure abductor.
Nerve supply: deep branch of the ulnar nerve (C8–T1).

The **flexor digiti minimi brevis** (**10**) *arises from the flexor retinaculum* (**8**) *and also from the hamulus of the hamate* (**15**). At its insertion it blends with the tendon of the abductor digiti minimi and *ends on the palmar surface of the base of the proximal phalanx* (**16**). It flexes the metacarpophalangeal joint.
Nerve supply: deep branch of the ulnar nerve (C8–T1).

■ **Variant:** Very often the muscle is absent.

The **opponens digiti minimi** (**11**), like the flexor digiti minimi brevis, *arises from the hamulus of the hamate* (**15**) *and from the flexor retinaculum* (**8**). *It is inserted into the ulnar margin of the fifth metacarpal* (**17**). It brings the little finger into the position for opposition.
Nerve supply: deep branch of the ulnar nerve (C8–T1).

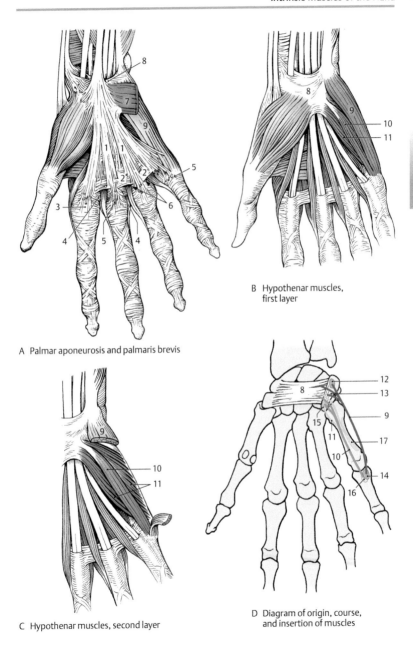

A Palmar aponeurosis and palmaris brevis

B Hypothenar muscles, first layer

C Hypothenar muscles, second layer

D Diagram of origin, course, and insertion of muscles

Fasciae and Special Features of the Free Upper Limb

Fasciae (A–D)

In the arm the **brachial fascia** (**1**) surrounds the flexors and extensors. Between the flexor and extensor groups of muscles on the medial and lateral sides of the humerus are the *medial* (**2**) and *lateral* (**3**) *brachial intermuscular septa*. These septa connect the brachial fascia with the humerus. The medial intermuscular septum begins proximally at the level of the insertion of the coracobrachialis, while the lateral septum begins just distal to the deltoid tuberosity. Both septa are attached to the borders of the humerus and extend to the corresponding epicondyles. The fascia of the arm is continuous with the *axillary fascia* (**4**) and with the *antebrachial fascia* (**5**). On the anterior surface of the arm just above the elbow there is an aperture, the *hiatus basilicus* (**6**; see p. 376).

The **antebrachial fascia** (**5**) is tightly attached to the posterior surface of the ulna. The *bicipital aponeurosis* (**7**) radiates into the forearm fascia, and the latter sends strong septa (**8**) deep between the individual muscle groups (see p. 158). At the distal end of the forearm the fascia is strengthened by transverse bands to form the extensor retinaculum on the dorsal surface, which provides conduits for the tendons of various muscles. Deep to the extensor retinaculum there are six compartments for passage of the extensor tendons. On the palmar surface, descending tendon fibers of the flexor carpi ulnaris muscle spread radially and distally near to the wrist into the antebrachial fascia. A separate space (**ulnar tunnel** also known as **Guyon's box**; see p. 388) is formed by these fiber bundles and the fascia that covers the deep muscles.

The **dorsal fascia of the hand** (**9**) superficially forms a close, dense expansion of the extensor retinaculum (see p. 182), composed of strong transverse fibers. Distally, it becomes the dorsal aponeurosis of the fingers. In addition it is more or less tightly connected to the intertendinous connections (see p. 166). The dorsal fascia of the hand is attached to the metacarpal bones on the ulnar and radial margins of the back of the hand. Between the tendons of the long extensors of the fingers and the dorsal interosseous muscles (see p. 174) there is a deep, delicate sheet (**10**) of this fascia.

The **palmar aponeurosis** (**11**; see p. 178) on the palmar side forms a continuation of the flexor retinaculum (see p. 182), the superficial and lateral boundaries of the central midhand compartment. Via nine septa, it is connected to the **deep palmar fascia** (**12**), which covers the palmar interosseous muscles. The adductor pollicis muscle (**14**) is covered by its own delicate **adductor fascia** (**13**).

The superficial transverse metacarpal ligament is found at the roots of the fingers. It is a thin, transverse ligament into which some of the longitudinal fasciculi of the palmar aponeurosis radiate. There is close contact between this ligament and the subcutaneous tissue.

15 Palmar interosseous muscles
16 Dorsal interosseous muscles

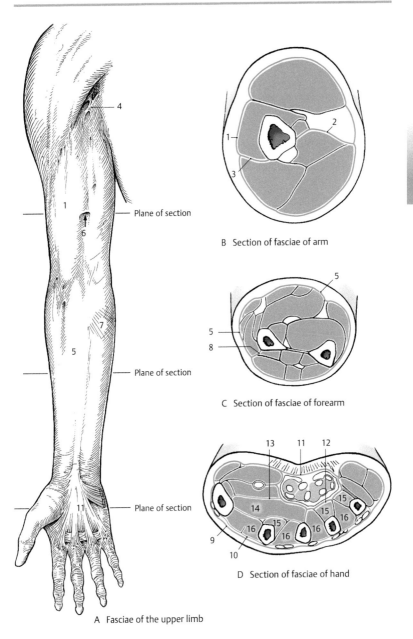

B Section of fasciae of arm

C Section of fasciae of forearm

D Section of fasciae of hand

A Fasciae of the upper limb

Carpal Tendon Sheaths (A–E)

There are dorsal carpal tendon sheaths, palmar carpal tendon sheaths, and palmar digital tendon sheaths.

Dorsal Carpal Tendon Sheaths (A)

The **dorsal synovial sheaths** lie in six tendon compartments formed by the *extensor retinaculum* (**1**) and *septa* (**2**) which arise from the undersurface of the retinaculum and are attached to bony ridges on the radius and ulna. These six fibro-osseous compartments contain the synovial sheaths of variable length for nine tendons. They are counted from the radial to the ulnar side. In the *first compartment* lie the *sheaths containing the tendons of the abductor pollicis longus and extensor pollicis brevis* (**3**). In the *second compartment* lie the sheaths for the tendons of the extensor carpi radialis longus and brevis, the *vagina tendinum musculorum extensorum carpi radialium* (**4**). In the *third compartment*, the slightly obliquely situated *canal* contains the *sheath with the tendon of the extensor pollicis longus* (**5**). The *fourth compartment*, the last compartment attached to the radius, contains the tendon sheath for the *extensor digitorum* and *extensor indicis* (**6**). The *fifth compartment* carries the *tendon sheath for the extensor digiti minimi* (**7**), and the *sixth compartment* contains the *tendon sheath for the extensor carpi ulnaris* (**8**).

Palmar Carpal Tendon Sheaths (B)

The *flexor retinaculum* (**9**) completes the carpal tunnel (see p. 124), through which the median nerve runs and the tendons of various flexor muscles run in three **palmar synovial tendon sheaths**. Most radially the tendon of the flexor carpi radialis runs in the *synovial tendon sheath for the flexor carpi radialis* (**10**) in its own groove in the trapezium bone, thereby dividing the radial attachment of the flexor retinaculum into two parts. Adjacent to it lies the *syn-*ovial sheath of the flexor pollicis longus (**11**), through which runs the digital tendon sheath of the thumb. The flexor digitorum superficialis and flexor digitorum profundus muscles run together in a *common synovial sheath of the flexor muscles* (**12**).

Digital Tendon Sheaths (B)

The five **synovial sheaths for the digits of the hand** are surrounded by **fibrous sheaths**, which consist of *annular* (**13**) and *cruciate* (**14**) fibers. Between the parietal and visceral layers of the synovial sheath (see p. 32) is a mesotendon with blood vessels and nerves. A *mesotendon* in the region of the digital tendon sheaths is called a *vinculum longum* (see p. 160) and *vinculum breve* (see p. 160).

▄▄ **Variants (C–E):** In approximately 72% of the population the *digital tendon sheath of the little finger* (**15**) is directly connected to the carpal tendon sheath (**12**), while the other tendon sheaths usually extend from the metacarpophalangeal joint to the base of the distal phalanx. In approximately 18% of cases there is no connection between the tendon sheath of the little finger (**15**) and the carpal tendon sheaths. In addition to a direct connection of the tendon sheath of the fifth finger to the carpal tendon sheath, the *tendon sheath of the index finger* (**16**) (in 2.5%) or the *tendon sheath of the ring finger* (**17**) (in approx. 3%) may communicate directly with the carpal tendon sheaths.

Clinical tip: Inflammation of the abductor pollicis longus and extensor pollicis brevis tendon sheats is common and causes pain in the region of the radial styloid process.

18 Intertendinous connection

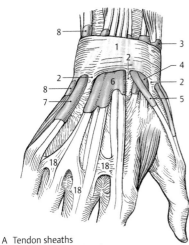

A Tendon sheaths of the back of the hand

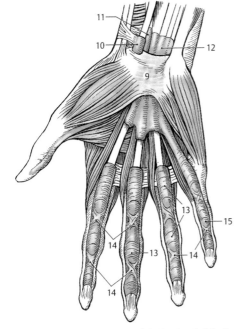

B Tendon sheaths of the palm of the hand and of the fingers

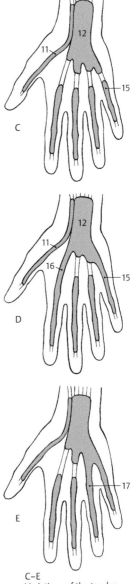

C–E
Variations of the tendon
sheaths of the palm

Upper Limb

Anatomical Terms and their Latin Equivalents

Upper Limb	Membrum superius
Carpal tunnel	Canalis carpi
Elbow joint	Articulatio cubiti
Flexor retinaculum	Retinaculum musculorum flexorum
Forearm	Antebrachium
Humeroradial joint	Articulatio humeroradialis
Humeroulnar joint	Articulatio humeroulnaris
Index finger	Index or digitus secundus
Little finger	Digitus minimus or quintus
Midcarpal joint	Articulatio mediocarpalis
Middle finger	Digitus medius or tertius
Neck of radius	Collum radii
Oblique cord	Chorda obliqua
Palmar (dorsal) carpal tendon sheaths	Vaginae tendinum carpales palmares (dorsales)
Proximal radioulnar joint	Articulatio radioulnaris proximalis
Ring finger	Digitus anularis or quartus
Shaft of radius	Corpus radii
Shoulder girdle	Cingulum membri superioris
Shoulder joint	Articulatio humeri
Sternal end	Extremitas sternalis claviculae
Thumb	Pollex
Ulnar notch	Incisura ulnaris
Wrist joint	Articulario radiocarpalis

Upper Limb

Systematic Anatomy of the Locomotor System

Lower Limb

Bones, Ligaments, and Joints

Pelvis

The bony **pelvis** consists of
- The two hip bones
- The sacrum
- The coccyx (see p. 48)

Hip Bone (A–C)

The **hip bone** consists of three parts, the **pubis**, the **ilium**, and the **ischium**, which synostose in the *acetabular fossa* (**2**), which is bordered by the *acetabular labrum* (**1**) and is surrounded by the *lunate surface* (**3**). The *acetabular notch* (**4**) opens the acetabulum inferiorly and thus limits the *obturator foramen* (**5**).

The **pubis** consists of a *body* (**6**), a *superior ramus* (**7**), and an *inferior ramus* (**8**). The two rami border the obturator foramen anteriorly and inferiorly. Near to the superior end of the medially orientated *symphyseal surface* (**9**) lies the *pubic tubercle* (**10**), from which the *pubic crest* (**11**) extends medially and the *pecten pubis* (**12**) runs laterally toward the *arcuate line of the ilium* (**13**). At the junction of the superior pubic ramus and ilium is the elevation of the *iliopubic eminence* (**14**). The *obturator groove* (**15**) lies inferior to the pubic tubercle and is bordered internally by the *anterior obturator tubercle* (**16**) and the *posterior obturator tubercle* (**17**), which is not always present.

The **ilium** is divided into the *body* (**18**) and the *wing of ilium*. The body forms part of the acetabulum and is delimited externally by the *supra-acetabular groove* (**19**) and internally by the arcuate line (**13**). External to the iliac wing lies the *gluteal surface* (**20**) and internal to it is the *iliac fossa* (**21**). Behind the iliac fossa is the sacropelvic surface with the *iliac tuberosity* (**22**) and the *auricular surface* (**23**). The *iliac crest* (**24**) starts anteriorly at the *anterior superior iliac spine* (**25**) and divides into the *outer* (**26**) and *inner* (**27**) *lips*, and an *intermediate zone* (**28**), which extends upward and backward. There, the outer lip bulges laterally as the *iliac tubercle* (**29**). The iliac crest ends in the *posterior superior iliac spine* (**30**). Beneath the latter lies the *posterior inferior iliac spine* (**31**), while anteriorly beneath the anterior superior iliac spine is the *anterior inferior iliac spine* (**32**). The *inferior gluteal* (**33**), *anterior gluteal* (**34**), and *posterior gluteal* (**35**) lines lie on the gluteal surface. In addition, there are various vascular canals among which at least one corresponds functionally to an emissary vessel.

The **ischium** is divided into the *body* (**36**) and the *ramus of the ischium* (**37**), which together with the inferior pubic ramus forms the inferior border of the obturator foramen. The ischium bears the *ischial spine* (**38**), which separates the *greater sciatic notch* (**39**) from the *lesser sciatic notch* (**40**). The greater sciatic notch is formed partly by the ischium and partly by the ilium, and it extends to the inferior surface of the auricular facies. The *ischial tuberosity* (**41**) develops on the ramus of the ischium.

Ossification: Three anlages appear: in the 3rd intrauterine month (ilium), the 4th to 5th intrauterine month (ischium), and the 5th to 6th intrauterine month (pubis). They fuse at the center of the acetabulum in a γ-shaped junction. Within the acetabulum one or more individual ossification centers develop between the ages of 10 and 12 years. Synostosis of the three bones occurs between the ages of 5 and 7 years, but within the acetabulum itself not until between the ages of 15 and 16 years. Epiphyseal centers of ossification occur in the spines at the age of 16, in the ischial tuberosity and in the iliac crest between the ages of 13 and 15 years.

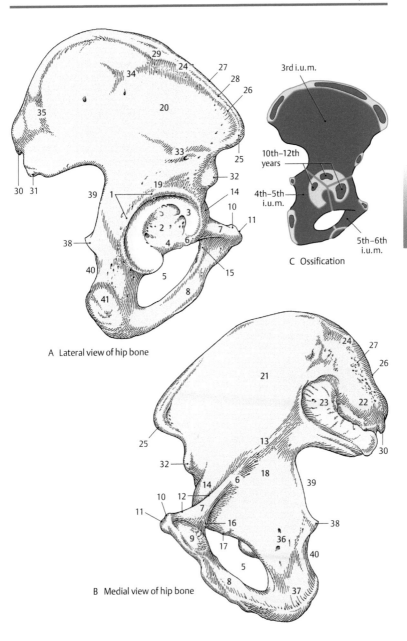

A Lateral view of hip bone

C Ossification

3rd i.u.m.

10th–12th years

4th–5th i.u.m.

5th–6th i.u.m.

B Medial view of hip bone

Connections between the Bones of the Pelvis (A, B)

Symphysis

The two hip bones are joined at the *pubic symphysis* (**1**) by a fibrous cartilage with a hyaline cartilage covering, the *interpubic disk*. Within the disk a small nonsynovial cavity may be present. Cranially and caudally the junction is reinforced by the **superior** (**2**) and the **inferior** (**3**) *pubic ligaments*, respectively.

Sacroiliac Joint (4)

This articulation (**4**) is formed by the auricular surface of the hip bone and the auricular surface of the sacrum. Both are covered by fibrocartilage. A tight capsule encloses the joint, which is almost immobile in the male and slightly mobile in the female (amphiarthrosis). The capsule is reinforced by the *anterior* (**5**), *interosseous* (**6**), and *posterior* (**7**) *sacroiliac ligaments*. It is reinforced indirectly by the *iliolumbar ligament* (**8**), which connects the ilium (**9**) to the lumbar vertebrae (**10**), as well as by the *sacrotuberous* (**11**) and *sacrospinous* (**12**) *ligaments*.

Ligaments in the Pelvic Region

The **obturator membrane** (**13**) closes the obturator foramen, except for the small opening of the **obturator canal** (**14**), which transmits the homonymous blood vessels and nerves.

The **sacrospinous** (**12**) and **sacrotuberous** (**11**) **ligaments** extend like a fan from the lateral margin of the sacral bone (**15**) and coccyx (**16**) to the ischial spine (**17**) and to the ischial tuberosity (**18**). The sacrotuberous ligament is stronger and longer than the sacrospinous ligament.

Owing to these two ligaments, the greater sciatic notch is converted into the *greater sciatic foramen* (**19**) and the lesser sciatic notch into the *lesser sciatic foramen* (**20**). In addition to the sacrospinous ligament, the sacrotuberous ligament also takes part in the delimitation of the greater sciatic foramen.

> **Clinical tip:** Although rare (more common in females than males), an **obturator hernia** of the thigh can extend through the obturator canal covered by the pectineus muscle. **Sciatic hernias**, also rare, pass through the sciatic foramina and protrude below the inferior border of the gluteus maximus.

The **iliolumbar ligament** (**8**) passes from the costal processes of the fourth and fifth lumbar vertebrae (**21**) to the iliac crest (**22**) and to the adjacent region of the iliac tuberosity (**23**). The **transverse acetabular ligament** bridges the acetabular notch and completes the articular surface for the femoral head.

The **inguinal ligament** (lig. of Vesalius) (**24**) is formed by the inferior border of the aponeurosis of the external oblique muscle. It extends between the anterior superior iliac spine (**25**) and the pubic tubercle (**26**). At the latter point of attachment it spreads out along a broad surface in the form of the **lacunar ligament** (**27**). Between the inguinal ligament and the anterior margin of the hip bone are the *muscular* (**28**) and *vascular* (**29**) *spaces*, which are separated from each other by the **iliopectineal arch** (**30**).

Morphology of the Bony Pelvis
(see p. 190)

We distinguish a true and a false, or a lesser and greater pelvis. The region inferior to the terminal line is called the lesser pelvis. The *pelvic inlet* (superior pelvic aperture) leads into the lesser pelvis, which is bordered by the promontory, the arcuate line, the iliopubic eminence, the pecten of the pubis, and the superior edge of the symphysis (*terminal line*). The *pelvic outlet*, the inferior pelvic aperture, is the region between the subpubic angle or pubic arch, the ischial tuberosities, and the coccyx.

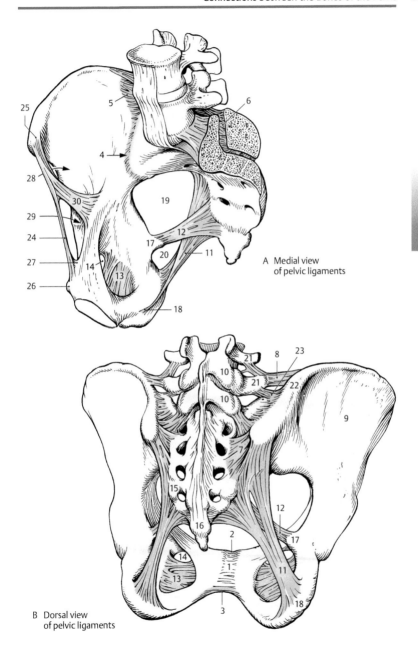

A Medial view
of pelvic ligaments

B Dorsal view
of pelvic ligaments

Morphology of the Bony Pelvis, continued

Orientation of the Pelvis and Sex Differences (A–F)

The plane of the pelvic inlet forms an approximately 60° angle with the horizontal plane. This angle is called the **pelvic inclination**. In an upright posture the anterior superior iliac spines and the pubic tubercles are in the same frontal (coronal) plane.

Classification of Pelvic Types

In females we distinguish various pelvic shapes, of which the most common (50%) is the gynecoid type. Other forms are the android, anthropoid, and platypelloid types. Classification into four main types is achieved by measuring certain pelvic diameters. The pelvic **diameters** or **conjugates** are measured at the pelvic inlet and outlet and as oblique diameters.

Diameters and External Pelvic Dimensions (A–C)

The **transverse diameter** (**1**) (13.5–14 cm) joins the extreme lateral points of the pelvic inlet. The **oblique diameter I** (**2**) (12–12.5 cm) is the line drawn between the right sacroiliac joint and the left iliopubic eminence. The **oblique diameter II** (**3**) (11.5–12 cm) represents a line between the left sacroiliac joint and the right iliopubic eminence.

The **anatomical conjugate** (**4**) (approx. 12 cm) is the line between the symphysis and the promontory. The **true conjugate** (**5**) joins the posterior surface of the symphysis (retropubic eminence) to the promontory. It is the shortest diameter of the pelvic inlet (11.5 cm); because it is of particular importance in parturition, it is also known as the *obstetric conjugate*. As the true conjugate cannot be measured directly, it is derived from the **diagonal conjugate** as an oblique diameter (13 cm). The diagonal conjugate (**6**) extends from the inferior pubic ligament to the promontory and is measured transvaginally.

The **straight conjugate** (**7**) at the pelvic outlet represents the connection between the inferior border of the symphysis and the tip of the coccyx (9.5–10 cm). As its length is variable due to the flexibility of the coccyx, the **median conjugate** (**8**) of the pelvic outlet, which connects the inferior border of the symphysis to the inferior border of the sacrum (11.5 cm), is a more important longitudinal diameter. An additional measure is the **transverse diameter of the pelvic outlet** (10–11 cm) between the two ischial tuberosities.

Two distances on the pelvis can be measured using a pelvic caliper. The **interspinous distance** (**9**) between the anterior superior iliac spines is approximately 26 cm in the female, and the **intercristal distance** (**10**) between the most lateral points of the two iliac crests is 29 cm in the female. The **external conjugate**, the distance between the spinous process of the fifth lumbar vertebra and the upper edge of the symphysis (about 20 cm), can also be measured with a caliper. In some instances the **intertrochanteric distance** (31 cm) between the two femurs is also measured.

The **female pelvis** (**D**, red) has wider projecting iliac wings, transversely directed obturator foramina, and a definite **pubic arch**. The lesser pelvis is larger than in the male.

The **male pelvis** (**D**, light gray), has more upright iliac wings, longitudinally orientated obturator foramina, and a **subpubic angle**.

E *Pubic arch* demonstrated by placing the hand on it; the arch lies between the thumb and the index finger.

F *Subpubic angle*, demonstrated by placing the hand on it; the angle lies between the index and middle fingers.

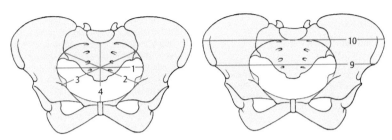

B Pelvic diameters

C External pelvic dimensions

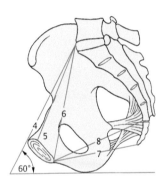

A Pelvic inclination

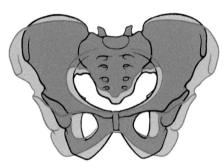

D Comparison of a male pelvis and female pelvis

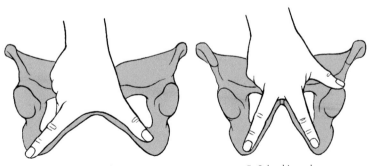

E Pubic arch

F Subpubic angle

The Free Part of the Lower Limb

Femur (A–C)

The thigh bone, or **femur**, is the largest tubular bone in the body and comprises a **shaft** (**1**) with a **neck** (**2**) and **two ends**, **proximal** and **distal**. The angle between the femoral shaft and neck is called the **neck-shaft angle** or CCD angle (see also p. 196).

The shaft has three surfaces: *anterior* (**3**), *lateral* (**4**), and *medial* (**5**). The lateral and medial surfaces are separated on the posterior side by a two-lipped roughened line, the *linea aspera* (**6**), which represents a thickening of the cortical bone. A nutrient foramen is found near this line. The *medial* (**7**) and *lateral* (**8**) *lips* of the linea aspera diverge proximally and distally, the lateral lip becoming continuous proximally with the *gluteal tuberosity* (**9**). This tuberosity can often develop very strongly and is then called a *third trochanter* (**10**). The medial lip extends up to the undersurface of the femoral neck.

Somewhat lateral to this lip is a ridge, the *pectineal line* (**11**), descending from the lesser trochanter. Both proximally and distally the femoral shaft loses its triangular form and becomes more quadrangular.

The *head of the femur* (**12**), with its umbilicate pit *or fovea* (**13**), presents an irregular border with the neck. The junction of the neck and shaft is marked on the anterior surface by the *intertrochanteric line* (**14**) and on the posterior surface by the *intertrochanteric crest* (**15**). At the boundary between the middle and proximal thirds of the intertrochanteric crest is a slight elevation, the *quadrate tubercle* (**16**). Directly below the *greater trochanter* (**17**) is a pit-like depression, the *trochanteric fossa* (**18**). The *lesser trochanter* (**19**) protrudes backward and medially.

The *medial* (**20**) and *lateral* (**21**) *condyles* form the distal end of the femur. Both are united on the anterior surface by the *patellar surface* (**22**), whereas they are separated on the posterior surface by the *intercondylar fossa* (**23**). This fossa is delimited from the posterior surface of the shaft by the *intercondylar line* (**24**), which forms the base of a triangle (*popliteal surface*, **25**). The sides of this triangle represent the continuation of the lips of the linea aspera and are also known as the *medial* and *lateral supracondylar lines*.

The *medial epicondyle* (**26**) protrudes medially above the medial condyle and bears an elevation, the *adductor tubercle* (**27**). The *lateral epicondyle* (**28**), situated on the lateral side, is separated from the lateral condyle by the *popliteal groove* (**29**).

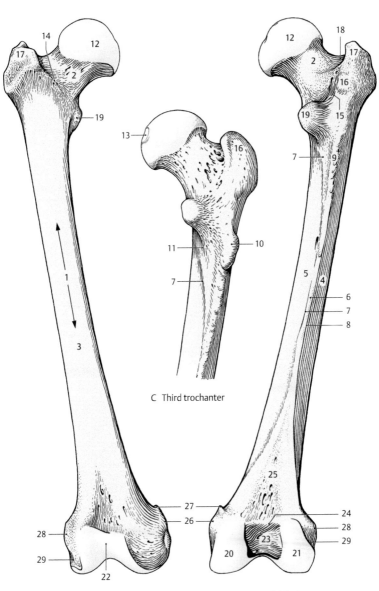

C Third trochanter

A Anterior view of right femur

B Posterior view of right femur

Femur (A–C), continued

The *medial* (**1**) and *lateral* (**2**) *condyles* differ in both size and shape. They diverge distally and posteriorly. The lateral condyle is wider in front than at the back, while the medial condyle is of uniform width. The oblique orientation of the femoral shaft means that in the upright position both condyles are in the horizontal plane despite their different sizes.

In the transverse plane both condyles are only slightly and almost equally curved (**3**) about the *sagittal axis* and in the sagittal plane there is a curvature (**4**) that increases posteriorly. This means that the radius of curvature decreases posteriorly. The midpoints of the curve thus lie on a spiral line (an involute), that is, on a curve the midpoints of which follow another curve. This produces not one but innumerable *transverse axes*, which produce a simultaneous gliding and rolling of the condyles during flexion of the knee joint (see p. 212). This mechanism also ensures that the collateral ligaments become sufficiently lax to permit rotation of the knee. The medial condyle has an additional curvature about a *vertical axis*, the "rotatory curvature" (**5**).

Ossification: The perichondral bony cuff of the shaft appears in the 7th intrauterine week. In the 10th month of fetal life an endochondral center becomes visible in the distal epiphysis (**sign of maturity**). Further ossification centers develop in the head of the femur in the 1st year of life, in the greater trochanter in the 3rd year, and in the lesser trochanter at approximately 11 to 12 years of age. The proximal epiphysis fuses at an earlier age (17–19 years) than the distal epiphysis (19–20 years).

Patella (D–H)

The patella is the largest sesamoid bone in the human body. It is triangular in shape with its base facing proximally and its tip, the *patellar apex* (**6**), facing distally. It has two surfaces, one facing the joint with the femur and the other directed anteriorly. These two surfaces join at a lateral (thinner) and a medial (thicker) margin.

The anterior surface may be divided into three parts and incorporates the tendon of the quadriceps femoris muscle.

In the upper third there is a coarse, flattened, rough surface that often has exostoses and serves largely for attachment of the quadriceps tendon. The middle third is characterized by numerous vascular canaliculi, while the lower third includes the apex, which serves as the origin of the patellar ligament.

The inner surface may be divided into an articular surface covering about three-quarters and a distal surface with vascular canaliculi. This is filled by fatty tissue, the infrapatellar fat pad.

The articular surface is divided into a lateral (**7**) and a medial (**8**) facet by a variably developed vertical ridge. Four types may be distinguished: Type 1, the most common, has a larger lateral and a smaller medial articular surface; Type 2 has two almost equally large articular facets; Type 3 has a particularly small, hypoplastic medial articular face, and in Type 4 there is little or no visible ridge separating the facets.

The total articular surface area of the patella in the adult is approximately $12\,cm^2$ and, especially in the center, is covered by cartilage of up to 6 mm in thickness. Maximal cartilage thickness is found at approximately 30 years of age and then continually decreases with aging.

Ossification (F): An ossification center develops in the 3rd to 4th year of life.

■ **Variants:** The lateral proximal edge of the patella often shows an emargination (indentation), producing a **patella emarginata** (**G**). A **bipartite patella** is the result of the ossification of an additional cartilaginous layer in the same area in which there has been an emargination. The old idea that several ossification centers occur in the patella which then fail to fuse is not accepted today (Olbrich). In addition to a bipartite patella (**H**) there are **multipartite patellas**. Partite patellas occur almost exclusively in males. They can be distinguished from fractures by their position and their shape.

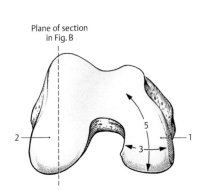

Plane of section
in Fig. B

2 —

5

3 — 1

A Distal view of the femoral condyles

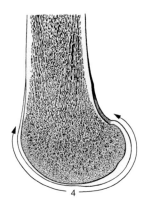

4

B Section through the lateral condyle

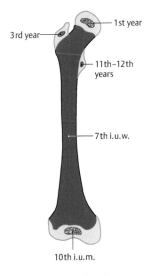

1st year

3rd year

11th–12th
years

7th i.u.w.

10th i.u.m.

C Ossification of the femur

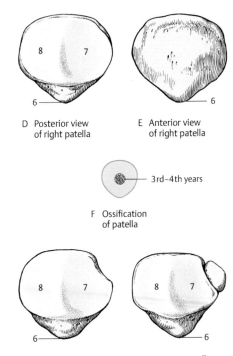

8 7

6

D Posterior view
of right patella

6

E Anterior view
of right patella

3rd–4th years

F Ossification
of patella

8 7

6

G Patella emarginata

8 7

6

H Bipartite patella

Positions of the Femur (A–G)

The angle formed between the neck and the shaft of the femur is often erroneously called the **collodiaphyseal angle**, or more correctly the **neck–shaft angle**. It is approximately 150° in newborns, decreasing to 145° by 3 years of age (**A**). In adults (**B**) the angle varies between 126° and 128°, and in old age (**C**) it reaches 120°.

> **Clinical tip:** In certain **bone diseases** (e.g., rickets), the neck–shaft angle may be reduced to 90°. The neck–shaft angle critically affects the strength and stability of the femur: the smaller the angle, the greater the risk of a **femoral neck fracture**. The incidence of femoral neck fractures in the elderly is related to the loss of elasticity of the bony tissues as well as to a decrease in the neck–shaft angle.

The neck–shaft angle influences the relationship of the femoral shaft to the mechanical axis of the leg. The **mechanical axis** of the (healthy) leg forms a straight line passing from the center of the femoral head through the center of the knee joint to the center of the calcaneus. The plane that passes through the lower surface of the femoral condyles is at right angles to this vertical line. This produces an angle between the axis of the femoral shaft and the mechanical axis of the leg. This angle is related in part to the neck–shaft angle and is important in determining the correct position of the lower limb (see also p. 214).

> **Clinical tip:** Pathologic changes in the neck–shaft angle lead to abnormalities in the position of the legs. An abnormally small neck–shaft angle produces **coxa vara** (**D**), and an abnormally large angle leads to **coxa valga** (**E**). The latter is usually combined with genu varum or bowleggedness (see p. 214), as any change in the shape of the femur naturally must affect the knee joint. Coxa vara leads to genu valgum (see p. 214).

The femur also has **torsion** (**F**), meaning that it is twisted about its long axis. If a line drawn through the femoral neck is superimposed on a line drawn transversely through the condyles, an angle will be produced. The mean angle in Europeans is 12°, with a range from 4 to 20°. The torsion angle, which is associated with the inclination of the pelvis, makes it possible for flexion movements of the hip joint to be transposed into rotatory movements of the femoral head.

Abnormal values for the torsion angle result in atypical leg positions. If the torsion angle is increased, the limb is turned inward, and if it is decreased or absent, the limb is turned outward. Both positions result in a decreased range of motion to one side.

> **Clinical tip:** In the moderately flexed hip, the tip of the greater trochanter does not rise above a line that joins the anterior superior iliac spine to the ischial tuberosity. This theoretical line is known as the **Roser–Nélaton line** (**G**). With a fracture or dislocation of the femoral neck, these three points no longer lie on a straight line. Thus, the Roser–Nélaton line may be of help in the diagnosis of fractures, although its practical value is disputed.

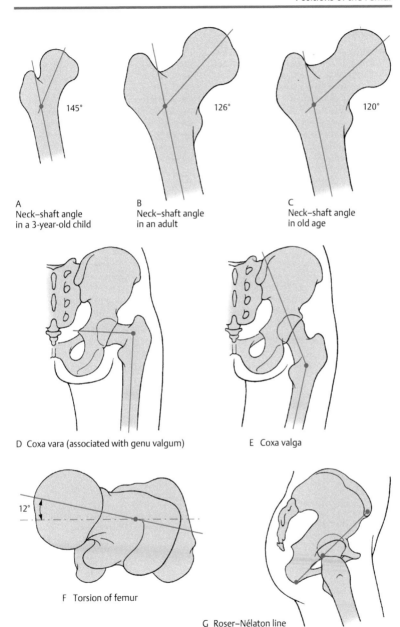

Lower Limb

A
Neck–shaft angle
in a 3-year-old child

145°

B
Neck–shaft angle
in an adult

126°

C
Neck–shaft angle
in old age

120°

D Coxa vara (associated with genu valgum)

E Coxa valga

12°

F Torsion of femur

G Roser–Nélaton line

Hip Joint (A–D)

The articular surfaces of the **hip joint** are formed by the **lunate surface of the acetabulum** (**1**) and the **femoral head** (**2**). The lunate surface of the joint cavity presents a section of a hollow sphere and is extended beyond the equator by the **acetabular labrum** (**3**). The acetabular labrum consists of fibrocartilaginous material. The lunate surface and labrum cover two-thirds of the femoral head. The bony socket is incomplete and is closed inferiorly by the **transverse acetabular ligament** (**4**). The acetabular labrum is found on the free margin of this ligament. The *ligament of the head of the femur* (**6**), which is covered by synovial membrane, extends from the acetabular fossa, where there is a fatty cushion (**5**), to the head of the femur. This ligament contains the artery to the head of the femur, which arises from the acetabular branch of the obturator artery. The femoral head is also supplied by branches of the medial and lateral circumflex femoral arteries.

The central portion of the superior rim of the acetabulum appears thickened in radiographs and is called the **acetabular roof**.

The **joint capsule** is attached to the hip bone outside the acetabular labrum, so that the latter projects freely into the capsular space. The capsular attachment (**8**) to the circumference of the head of the femur lies at about the same distance from the cartilaginous rim of the femoral head. Therefore, the extracapsular part of the neck is shorter in front than at the back. Anteriorly the line of attachment is in the region of the *intertrochanteric line* (**7**), while posteriorly the line of attachment (**8**) is a finger-width away from the *intertrochanteric crest* (**9**).

Hip joint ligaments. Among these ligaments is the strongest in the human body, the *iliofemoral ligament* (**10**), which has a tensile strength of 350 kg. Often this ligament

is incorrectly termed the **Bertini ligament**, but actually it was first described by **Bellini**.

There are five ligaments, of which four are extracapsular and one is intracapsular.

The **extracapsular ligaments** are the *zona orbicularis* (**11**), the *iliofemoral ligament* (**10**), the *ischiofemoral ligament* (**12**), and the *pubofemoral ligament* (**13**). The last three ligaments strengthen the capsule and, at the same time, prevent an excessive range of movement. The zona orbicularis lies like a collar around the narrowest part of the femoral neck. On the inner surface of the capsule it is to be seen as a distinct circular elevation, and externally it is covered by the other ligaments, which partly blend with the capsule. The head of the femur projects into the zona orbicularis like a button in a buttonhole. Together with the acetabular labrum and atmospheric pressure, the zona orbicularis provides an additional mechanism for maintaining contact between the femoral head and acetabulum.

The *ligament of the head of the femur* is entirely **intracapsular**.

Areas of the capsule that are not strengthened by ligaments represent areas of weakness. The *iliopectineal bursa* lies between the capsule and the iliopsoas muscle. In 10 to 15% of people it communicates with the hip joint.

Clinical tip: In inflammatory processes with joint effusion, the weaker areas are pushed outward and become very pressure-sensitive. Dislocations tear the capsule, and the ligament of the head of the femur with the artery of the head of the femur may be severed. This may produce nutritional deficiencies in the femoral head. **Femoral neck fractures** are classified as **medial** or **lateral**. Medial fracture lines are within the joint capsule, while lateral fracture lines are extracapsular.

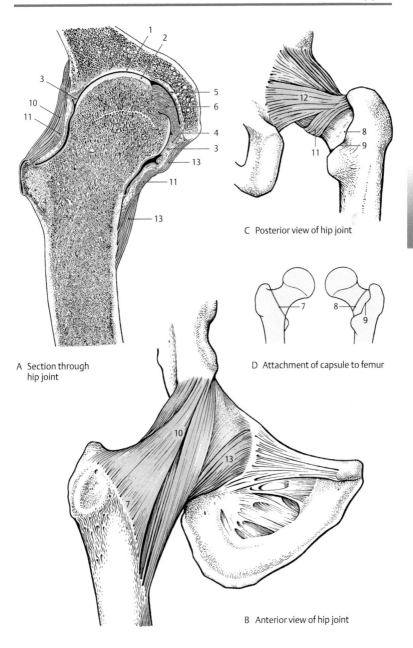

A Section through hip joint

C Posterior view of hip joint

D Attachment of capsule to femur

B Anterior view of hip joint

Lower Limb

Hip Joint (A–D), continued

Ligaments of the Hip Joint (A, B)

The **iliofemoral ligament** (**1**) arises from the *anterior inferior iliac spine* (**2**) and the *rim of the acetabulum* and extends to the *intertrochanteric line* (**3**). It has a strong **transverse part** (**4**), which lies farther cranially and runs parallel to the axis of the neck, and a weaker **descending part** (**5**) lying farther caudally and running parallel to the axis of the shaft.

The two parts, of which the lateral portion is twisted like a screw, act differently and form roughly the outline of an inverted Y. In the upright position, with the pelvis tilted posteriorly, the twist and tension of this ligament permits the stance to be maintained without muscular activity and prevents the trunk from tilting backward. In addition, the iliofemoral ligament keeps the femoral head in contact with the acetabulum. When the thighs are flexed, there is a reduction in tension in both iliofemoral ligaments, which allows the pelvis to tilt a little farther back, so that the sitting posture becomes possible. The thicker, transverse (lateral) part of the ligament prevents external rotation and adduction of the femur. The descending (medial) part restricts internal rotation. When the thigh is flexed, the entire ligament becomes lax, so that a much greater degree of rotation is possible.

The **ischiofemoral ligament** (**6**) arises from the *ischium* below the acetabulum and runs almost horizontally over the neck of the femur to the attachment of the lateral part of the iliofemoral ligament. In addition it also blends with the **zona orbicularis** (**7**). It constrains internal rotation of the thigh.

The **pubofemoral ligament** (**8**), the weakest of the three ligaments, arises from the *obturator crest* and the adjacent part of the *obturator membrane* (**9**). It radiates into the capsule, specifically into the **zona orbicularis** (**7**), and continues by way of this into the femur. It restricts movements of abduction.

The intracapsular **ligament of the head of the femur** extends from the *acetabular notch to the fovea of the femoral head*. It does not help to maintain contact between these structures. When the hip is dislocated, it may prevent further displacement to a certain degree, since only then does it become stretched.

Movements of the Hip Joint

In life, muscle tone restricts joint movement, most noticeably when the extended limb is raised forward.

Movements of the hip joint include **flexion (anteversion)** and **extension (retroversion)**, **abduction** and **adduction**, **circumduction** and **rotation**. Flexion and extension occur about a **transverse axis through the femoral head**. With the knee flexed, the thigh can be raised against the abdomen. This range of flexion is much greater than that of extension, which can only be executed slightly beyond the vertical.

Abduction and **adduction** occur about an **anterior–posterior axis through the femoral head**.

Rotation of the femur occurs around a (**vertical**) **axis through the femoral head and the medial femoral condyle**. With the leg extended, the range of rotation is approximately 60°.

Circumduction is a compound movement in which the leg describes the surface of an irregular cone, the apex of which lies in the femoral head.

10 Acetabular labrum
11 Ischial tuberosity
12 Greater trochanter

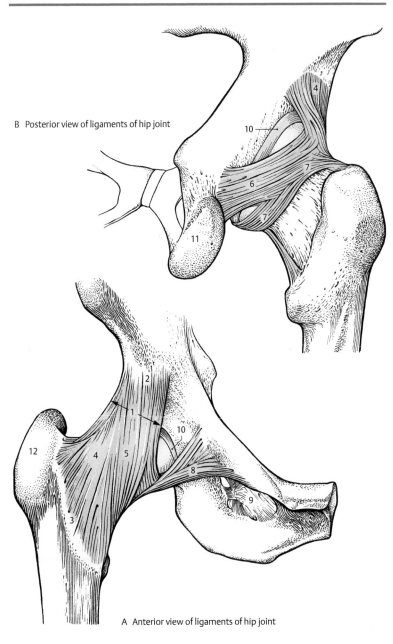

B Posterior view of ligaments of hip joint

A Anterior view of ligaments of hip joint

Bones of the Lower Leg

The bones of the lower leg are the tibia and fibula. The tibia is the stronger bone, which alone provides the connection between the femur and the bones of the ankle and foot.

Tibia (A–D)

The **tibia** has a somewhat triangular **shaft** (**1**) and **proximal** and **distal ends**. At the **proximal end lie** the *medial* (**2**) and *lateral* (**3**) *condyles*. The proximal surface, the *superior articular surface*, is interrupted by the *intercondylar eminence* (**4**). This elevation is subdivided into a *medial* (**5**) and a *lateral* (**6**) *intercondylar tubercle*. In front of and behind the eminence lie the *anterior* (**7**) and *posterior* (**8**) *intercondylar areas*. On the outward-facing overhang of the lateral condyle there is a small *articular facet*, directed laterally and distally, for articulation with the fibula (**9**).

The three-sided **shaft of the tibia** has a sharp *anterior border* (**10**), which proximally becomes the *tibial tuberosity* (**11**) and is flattened distally. It separates the *medial surface* (**12**) from the *lateral surface* (**13**). The lateral surface joins the *posterior surface* (**15**) at the *interosseous border* (**14**). The posterior surface is separated from the medial surface by the *medial border* (**16**). Proximally on the posterior surface of the shaft of the tibia is a slightly roughened area, the *soleal line* (**17**), extending obliquely from the distomedial side to the proximolateral side. Lateral to this there is a *nutrient foramen* (**18**) of varying size.

The **distal end** is prolonged medially to form the *medial malleolus* (**19**) with its *malleolar articular facet*. The *malleolar groove* (**20**) runs along its posterior surface. The *inferior articular surface of the tibia*, which lies on the lower surface of the distal end of the tibia, articulates with the talus. On the lateral side, in the *fibular notch* (**21**), there is a syndesmotic connection, that is, a fibrous joint, with the fibula.

In the adult the proximal end of the tibia is bent slightly backward. We speak of **retroversion** or an actual backward tilting of the tibia. The angle between the superior articular facet of the tibial condyle and the horizontal averages 4 to 6°. In the last gestational months this initially very small angle increases to about 30°. In the first months after birth, and more especially when learning to stand upright, the angle becomes smaller.

The superior articular surface lies behind the long axis of the tibia. This means that the proximal end of the tibia is shifted posteriorly. This shift is referred to as **retroposition**.

The tibia also shows **torsion**, that is, a twist between its proximal and distal ends. This is often present in adults and is attributed to increased growth of the medial tibial condyle.

Ossification: In the shaft of the tibia perichondral ossification begins in the 7th intrauterine week, an endochondral ossification center develops at the proximal end in the 10th intrauterine month or in the 1st year, and an endochondral ossification center in the distal epiphysis appears at the beginning of the 2nd year. The distal epiphysis fuses first, between the ages of 17 and 19 years, and the proximal epiphysis fuses later, between the ages of 19 and 20 years.

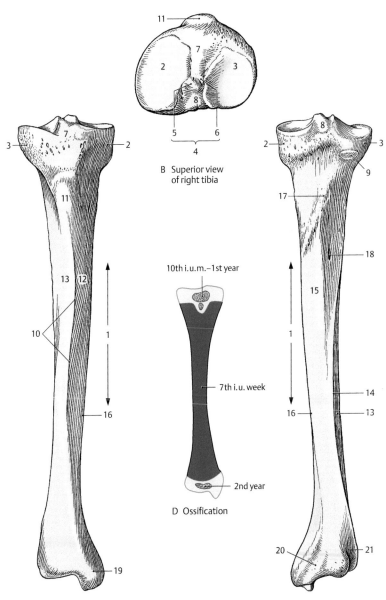

B Superior view of right tibia

10th i.u.m.–1st year

7th i.u. week

2nd year

D Ossification

A Anterior view of right tibia

C Posterior view of right tibia

Lower Limb

Bones of the Leg, continued

Fibula (A–D)

The **fibula** corresponds approximately in length to the tibia, but is a slimmer and therefore more flexible bone. It, too, consists of **two extremities and a shaft**.

The **proximal end** is the *head of the fibula* (**1**) with its *articular facet* (**2**) and a small protuberance, the *apex of the fibular head* (**3**).

The **shaft of the fibula** (**4**) is approximately triangular in its midportion and has three borders and three surfaces. In the distal third there is a fourth border. The sharpest edge is the forward-facing *anterior border* (**5**), which separates the *lateral* (**6**) from the *medial* (**7**) *surface*. The *medial crest* (**8**) separates the medial surface from the *posterior surface* (**9**). It is separated from the *lateral surface* (**6**) by the *posterior border* (**10**). On the medial surface there is a low but very sharp bony ridge, the *interosseous border* (**11**), to which the *interosseous membrane* (**12**) is attached. Approximately in the center of the posterior surface or on the posterior border, there is a nutrient foramen.

On the lateral surface of the **distal end**, which expands distally, there is the large, flat *lateral malleolus* (**13**) with a *facet for articulation with the talus on its inner surface* (**14**). Behind it there is a deep groove, the *lateral malleolar fossa* (**15**), to which the posterior talofibular ligament is attached. A variable, well-developed groove, the *malleolar groove* (**16**), is present on the lateral surface behind the lateral malleolus. The tendons of the peronei muscles (see p. 260) course in this groove.

Ossification: The perichondral bony cuff develops in the region of the shaft in the 2nd intrauterine month. An endochondral ossification center develops in the malleolus in the 2nd year and in the head of the fibula in the 4th year.

The distal epiphysis fuses earlier, between the ages of 16 and 19 years, and the proximal somewhat later, between 17 and 20 years. The junction line of the proximal epiphysis runs below the head of the fibula, and that of the distal epiphysis above the malleolus.

Clincal tip: Care must be taken not to confuse these epiphyseal plates, particularly that of the distal epiphysis, with fracture lines.

Lower Limb

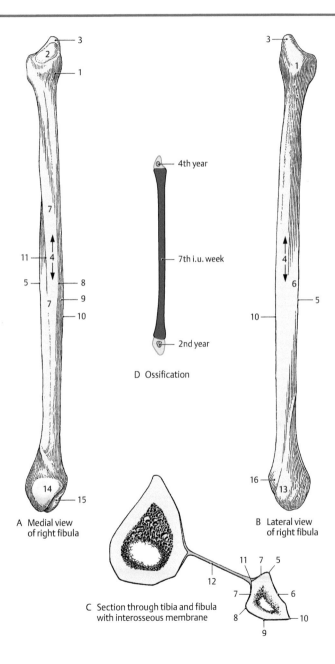

4th year

7th i.u. week

2nd year

D Ossification

A Medial view
of right fibula

B Lateral view
of right fibula

C Section through tibia and fibula
with interosseous membrane

Knee Joint (A–C)

The **knee joint** is the largest joint in the human body. It is a hinge joint, a special type of mobile trochoginglymus. Flexion of the knee joint combines rolling and gliding movements. In the flexed position some rotation is possible.

The articular bodies of the knee joint consist of the **femoral condyles** and the **tibial condyles**. The incongruence of these joint surfaces is compensated for by a relatively thick cartilaginous covering and by the **menisci**. In addition to the tibia and femur, the **patella** also forms part of the knee joint. The clinician also uses the term *femoropatellar joint*, meaning that region of the knee joint in which the patella is in contact with the femur.

The femoral condyles diverge to some extent distally and posteriorly. The *lateral condyle* is wider in front than at the back, while the *medial condyle* is of more constant width. In the transverse plane the condyles are only slightly bent on a sagittal axis. In the sagittal plane, the curvature increases toward the back; that is, the radius of curvature becomes smaller (see p. 194). In addition, the medial condyle curves about a vertical axis (curvature of rotation). The *superior tibial articular surface* is formed by the condyles, which are separated by the intercondylar eminence and both intercondylar areas.

The wide, lax **capsule** (**1**) is thin in front and at the side and is strengthened by ligaments. The patella is inserted into the anterior wall of the capsule.

At various points the knee joint possesses **ligaments**, **menisci**, and **communicating bursae**.

Ligaments. The **patellar ligament** (**2**) is a continuation of the *quadriceps tendon* (**3**), which extends from the *patella* to the *tibial tuberosity* (**4**). The **lateral patellar retinaculum** (**5**) is formed by fibers of the vastus lateralis muscle and some fibers from the rectus femoris. Some fibers of the iliotibial

tract also radiate into it. Laterally it joins the tibial tuberosity of the tibia. The **medial patellar retinaculum** (**6**) is formed to a large extent by fibers from the vastus medialis, which runs distally, medial to the patellar ligament, and is attached to the tibia in front of the medial collateral ligament. Transverse fibers (**8**), which arise from the *medial epicondyle* (**7**) radiate into the medial patellar retinaculum. Two lateral ligaments act as guidance ligaments for flexion and extension of the joint. The **medial collateral ligament** (**9**) is a flattened, triangular ligament, which is integrated into the fibrous membrane of the capsule and is fused to the medial meniscus (see p. 208). It contains three groups of fibers. The *anterior long fibers* (**10**) extend from the medial epicondyle (**7**) to the *medial border of the tibia* (**11**). The *short, upper, posterior fibers* (**12**) radiate into the medial meniscus, and the *inferior, posterior fibers* (**13**) extend from the medial meniscus to the tibia. It is covered partly by the superficial pes anserinus and is crossed inferiorly by the portion of the *semimembranosus tendon* (**14**) that is attached to the tibia. The round **lateral collateral ligament** (**15**) is not fused with the capsule or with the lateral meniscus. It arises from the *lateral epicondyle* (**16**) and is attached to the *head of the fibula* (**17**).

On the posterior surface, the **oblique popliteal ligament** (**18**) represents a lateral expansion of the semimembranosus tendon (**14**). It extends laterally and proximally. The **arcuate popliteal ligament** (**19**) arises from the *apex of the fibular head* (**20**) and passes into the capsule, crossed by the *popliteus tendon* (**21**).

22 Suprapatellar bursa
23 Subtendinous bursa of medial gastrocnemius
24 Medial head of gastrocnemius
25 Lateral head of gastrocnemius

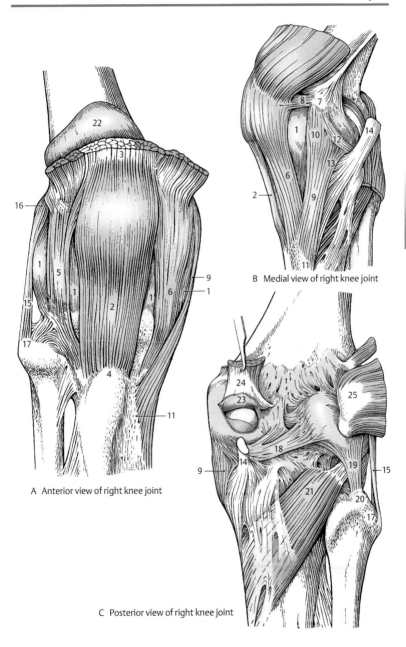

B Medial view of right knee joint

A Anterior view of right knee joint

C Posterior view of right knee joint

Lower Limb

Knee Joint, continued (A–C)

A further group of **ligaments** of the knee joint is that of the *cruciate ligaments*. They serve in particular to maintain contact during rotary movements. They are intra-capsular but extra-articular (see p. 210).

The **anterior cruciate ligament** (**1**) runs from the anterior intercondylar area of the tibia to the inner surface of the lateral condyle of the femur. Fibers arising from the lateral side extend farther dorsally than those from the medial side.

The **posterior cruciate ligament** (**2**) is stronger than the anterior cruciate ligament. It passes from the lateral surface of the medial condyle of the femur to the posterior intercondylar area.

The **menisci** consist of connective tissue with extensive collagen fiber material and interspersed cartilagelike cells. The collagen fibers run in two principal directions. The strong fibers follow the shape of the menisci between their attachments, while the weaker fibers pass radially to an imaginary midpoint and interlace between the longitudinally running fibers. This arrangement means that curved longitudinal tears (see below) can occur more easily than transverse tears. The cartilagelike cells are more abundant near the surface of the menisci.

In transverse section the menisci are seen to be flattened medially. On the external surface they fuse with the synovial membrane of the joint capsule. They may move over the underlying tibia. They receive their blood supply from the middle genicular and inferior lateral and medial genicular arteries of the knee, which together form the perimeniscal marginal arterial arcades.

The **medial meniscus** (**3**) is semicircular in shape and is fused with the *medial collateral ligament* (**4**). Their points of attachment are relatively widely separated. The medial meniscus is wider posteriorly than anteriorly, so the *anterior crus* (**5**) is much thinner than the *posterior crus* (**6**). Its attachment makes it far less mobile than the lateral meniscus. External rotation of the leg causes the greatest displacement and pulling stress on it. Internal rotation relaxes it.

The **lateral meniscus** (**7**) is almost circular; its points of attachment lie close together, and it is of uniform width. It is more mobile than the medial meniscus, as it does not fuse with the *lateral collateral ligament* (**8**), and therefore it is less stressed by the different movements. From its posterior horn arise one or two ligaments. The **anterior meniscofemoral ligament** (**9**) anteriorly and the **posterior meniscofemoral ligament** (**10**) posteriorly pass behind the posterior cruciate ligament to the medial femoral condyle. The posterior meniscofemoral ligament is present more often than the anterior (about 30%). Less often (see **C**) both ligaments are present. The **transverse ligament of the knee** (**11**) joins the two menisci in front. In 10% of cases it is divided into several strips.

Clinical tip: Clinicians distinguish an **anterior** and a **posterior horn** in each meniscus. Menisci may be torn by continuous excessive force or by uncoordinated movements (e.g., flexion in external rotation with the foot planted). **Damage to the medial meniscus is approximately 20 times more frequent than to the lateral meniscus** because of its more limited mobility and its thin anterior crus. *Longitudinal tears* (*bucket handle tear*) or *avulsions of the anterior* or *posterior horn* may occur. After surgical removal of a meniscus, with preservation of the marginal zone of the capsule, meniscoid tissue may be formed, which takes over the function of the meniscus. The meniscofemoral ligaments may cause difficulties during operations on the posterior horn.

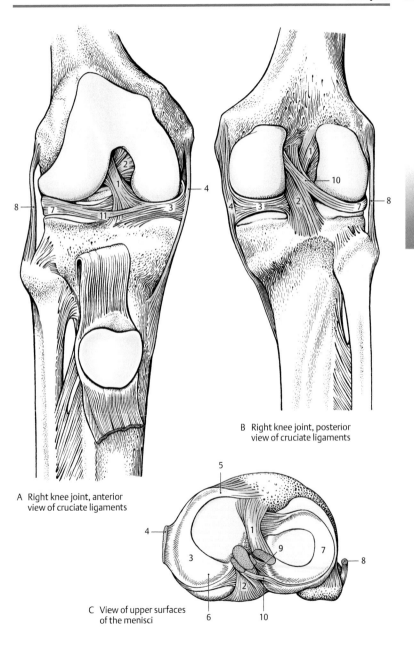

A Right knee joint, anterior
view of cruciate ligaments

B Right knee joint, posterior
view of cruciate ligaments

C View of upper surfaces
of the menisci

Knee Joint, continued (A–D)

The *synovial* (**1**) and *fibrous* (**2**) *membranes* of the **joint capsule** are separated by fatty deposits on their anterior and posterior surfaces. The reflection of the **synovial membrane** anteriorly lies on the *femur* (**3**), usually at some distance from the margin of the cartilage where the synovial membrane arises (**4**). This is due to the presence of the *suprapatellar bursa* (**5**), which communicates with the joint space. It should be noted that at this site of reflection (**6**), the synovial membrane appears slightly lifted from the bone by periosteal connective tissue (**7**). On the *tibia* (**8**) the attachment and the reflection of the synovial membrane anteriorly lie close to the cartilaginous margin. Posteriorly the attachment of the synovial membrane to the femur is at the *cartilage margin* (**9**) of the *femoral condyles*, which produce two dorsally directed extensions (**10**) in the joint space. In the center, the synovial membrane passes in front of the *anterior cruciate* (**11**) and *posterior cruciate* (**12**) ligaments, so that although the ligaments are intracapsular they are extra-articular between the synovial (**1**) membranes. Their posterior attachment to the tibia is exactly on the cartilage margin (**13**). The *menisci* (**14**) are incorporated into the synovial membrane.

The **joint space** itself has a complicated structure. Anteriorly, in the exposed joint, there is a wide fatty pad, the *infrapatellar fat pad* (**15**), inserted between the synovial and fibrous membranes. It extends from the inferior border of the *patella* (**16**), which is enclosed in the anterior wall of the capsule, to the *infrapatellar synovial fold* (**17**) dividing the remnant of the original subdivision of the joint into two chambers.

The infrapatellar synovial fold extends through the joint space with a free upper margin and continues on the cruciate ligaments, which it surrounds from the anterior side (see above). The *alar folds* (**18**) lie lateral to the infrapatellar fat pad and to the infrapatellar synovial fold.

There are numerous **bursae** about the knee joint, some of which communicate with the joint cavity. The largest of the **communicating bursae** is the *suprapatellar bursa* (**5**), which lies anteriorly and extends the joint space proximally. Posteriorly lie the *subpopliteal recess* and the *semimembranosus bursa*, which are much smaller. At the origin of the two heads of the gastrocnemius are the *lateral* and *medial subtendinous bursae of gastrocnemius*.

The **noncommunicating synovial bursae** include the *subcutaneous prepatellar bursa*, which is located directly in front of the patella, as well as the *deep infrapatellar bursa* (**19**), which is situated between the *patellar ligament* (**20**) and the fibrous membrane of the joint capsule. In particular cases the latter bursa can also be in communication with the articular cavity. Additional smaller bursae that are not consistently present include the *subfascial prepatellar bursa*, the *subtendinous prepatellar bursa*, and the *subcutaneous prepatellar bursa*.

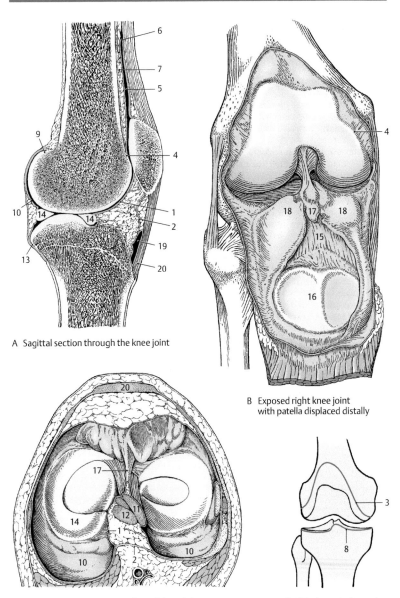

A Sagittal section through the knee joint

B Exposed right knee joint
with patella displaced distally

C Transverse section through knee joint,
proximal view of distal part

D Attachment of capsule

Movements of the Knee Joint (A–E)

The knee may be **flexed** and **extended** about an almost transverse axis, and in the flexed position **rotation** is possible about the axis of the lower leg.

In the **extended knee** (A) both *collateral ligaments* (**1, 2**) and the anterior part of the *anterior cruciate ligament* (**3**) are taut. During extension the femoral condyles glide into the almost extreme position in which the *medial collateral ligament* (**1**) is completely unfolded. During the last 10° of movement before complete extension there is an **obligatory terminal rotation** of about 5° (**the joint is "screwed home"**). *This is caused by stretching of the anterior cruciate ligament and is permitted by the shape of the medial femoral condyle (see p. 194), assisted by the iliotibial tract (see p. 254).* Both lateral ligaments become taut and at the same time there is a slight unwinding of the cruciate ligaments (**3, 4**). Final rotation of the nonweight-bearing active leg is produced by external rotation of the tibia, and of the weight-bearing (standing) leg by internal rotation of the thigh. In the position of extreme extension the collateral (**1, 2**) and cruciate ligaments are tensed (**A**).

The normal range of extension is to 180°, although in children and adolescents the leg can be hyperextended by approximately 5°. In the newborn, maximal extension is impossible due to the physiologic occurrence of tibial retroversion (see p. 202).

In the **flexed knee** (B) the *lateral collateral ligament* (**2**) is completely relaxed, and the *medial collateral ligament* (**1**) is largely lax, while the *anterior* (**3**) and *posterior* (**4**) cruciate ligaments are taut. In flexion, rotation is possible under the control of the cruciate ligaments. The range of **internal rotation** (**C**) **of the leg** is less than that of **exter-**

nal rotation. *During internal rotation of the tibia on the femur, the cruciate ligaments are twisted around each other and so prevent any appreciable internal rotation. In the same way the posterior fibers of the medial collateral ligament* (**1**) *are tensed at extreme internal rotation. During external rotation, the cruciate ligaments become unwound.* The limit of external rotation is primarily determined by the *medial collateral ligament* (**2**); its maximal range is 45 to 60°. The amount of rotation can be verified by movement of the head of the fibula (**5**) when the leg is lifted from the ground.

Because of the oblique position of the cruciate ligaments, in every position one cruciate ligament or part of one is always tense. In any case, these ligaments come to control the joint as soon as the collateral ligaments become inadequate; that is, the cruciates maintain stability when the collaterals relax.

During rotation the femur and *menisci* (**6**) move over the tibia, and during flexion and extension the femur rolls and glides on the menisci, so that we may consider the knee to be a **"mobile joint."**

Clinical tip: The relatively large and incongruent joint surfaces are subject to considerable stress and they often show damage to the cartilaginous covering in old age, as well as bony changes. In a case of ruptured anterior cruciate ligament (**D**), the so-called **anterior drawer sign** (**E**) is observed; that is, in the flexed position (with the collateral ligaments relaxed) the lower leg can be pulled forward 2 to 3 cm (arrow).

Rupture of the posterior cruciate ligament and the lateral collateral ligament results in the **posterior drawer sign**; that is, the lower leg can be pushed backward. Abnormal lateral movements occur if there is a torn lateral ligament (**wobbly joint**).

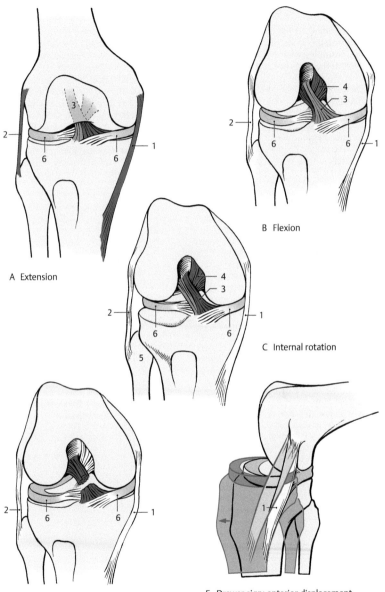

Lower Limb

A Extension

B Flexion

C Internal rotation

D Torn anterior cruciate ligament

E Drawer sign: anterior displacement
of tibia when cruciate ligament is severed

Alignment of the Lower Limb (A–C)

Irrespective of the neck–shaft angle of the femur (see p. 196), the alignment or shape of the lower limb depends on the correct development of the knee joint. A malalignment of the lower limb will cause abnormal loading and early signs of wear and tear in the knee joint.

If the knee joint has developed normally, the limb is straight (**genu rectum**, **A**). In that case the *mechanical axis of the leg* (**1**) runs *through the center of the femoral head* (**2**), the *center of the knee joint*, and, when extended, also through the *center of the calcaneus* (**3**).

When the *mechanical axis* is displaced laterally (**1**), that is, it runs *through the lateral femoral condyle* (**4**) or *the head of the fibula* (**5**), the condition is known as **genu valgum** or "*knock-knee*" (**B**). In this case the medial collateral ligament (**6**) will be overstretched and there is excessive stress on the lateral meniscus (**7**), the cartilage-covered articular surface of the lateral femoral condyle (**4**), and the lateral condyle of the tibia (**8**). The joint space is larger on the medial than on the lateral side. In genu valgum we have increased terminal rotation. In a case of knock-knees the medial surfaces of the legs near the knee joints touch, while the medial malleoli elsewhere have no contact.

When the *mechanical axis* (**1**) runs *through the medial femoral condyle* (**9**) or medial to it, the condition is known as **genu varum** (**C**) or "*bowleggedness*." The lateral collateral ligament (**10**) is overstretched and there is increased wear and tear on the medial meniscus (**11**) and on the cartilage covering of the articular surfaces. In the region of the knee joint the legs cannot be made to touch. In genu varum the legs cannot be completely extended, so terminal rotation cannot occur.

Connections between the Tibia and the Fibula (D)

The **tibiofibular joint** (**12**) is an almost immobile synovial joint (**amphiarthrosis**) between the *head of the fibula* (**13**) and *the fibular articular facet of the lateral tibial condyle* (**14**). It possesses a **tense capsule** that is reinforced by the *anterior and posterior ligaments of the head of the fibula*. It is also known as a **compensation joint** because, during maximal forward dorsiflexion in the ankle (talocrural) joint, there is a widening of the ankle mortise that results in a compensatory movement in the tibiofibular joint.

In addition to the synovial joint between the leg bones, the **interosseous membrane of the leg** (**15**), as a **fibrous joint**, stabilizes the two bones. The fibers in the interosseous membrane run inferiorly from the tibia to the fibula and are very tense.

At the distal end of the two bones is the **tibiofibular syndesmosis** (**16**). This consists of an *anterior tibiofibular ligament*, a relatively flat ligament that runs obliquely over the anterior surfaces of the distal ends of both bones, and the *posterior tibiofibular ligament* on their posterior surfaces. The fiber direction of the posterior ligament is more horizontal. Both ligaments are slightly compliant, allowing a small degree of relative movement between the tibia and fibula during dorsiflexion.

17 Semitendinosus, gracilis, and sartorius, strongly loaded
18 Biceps femoris and iliotibial tract, strongly loaded

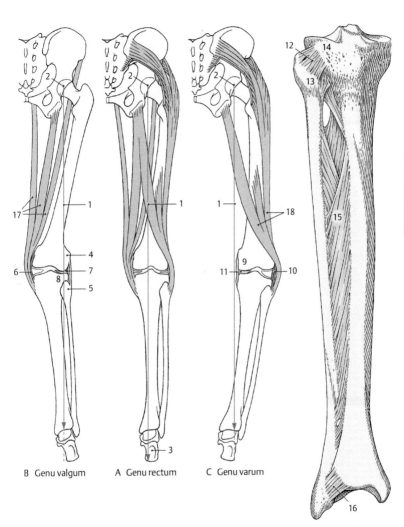

B Genu valgum   A Genu rectum   C Genu varum

A–C Positions of the lower limb and knee joint
(according to *Lanz–Wachsmuth*)

D Connections between
tibia and fibula

Bones of the Foot (A–G)

The skeleton of the foot may be divided into

- The **tarsus** (ankle)
- The **metatarsus** (midfoot)
- The **digits** (toes)

The **tarsus** consists of seven bones—the **talus, calcaneus, navicular, cuboid**, and the three **cuneiforms**. The **metatarsus** consists of **five metatarsals**, and the **digits** are formed by the **phalanges**.

Tarsal Bones

The **talus** (**A–C**) transmits the weight of the entire body to the foot. We distinguish in it a **head** (**1**), a **body** (**2**), and a **neck** (**3**). The head of the talus carries the *navicular articular surface* for articulation with the navicular bone, and the neck of the talus has small vascular channels and roughened areas. On the body of the talus we distinguish the *trochlea* (**4**) and behind this a *posterior talar process* with *lateral* (**5**) and *medial tubercles* (**6**). Immediately adjacent to the medial tubercle is the *groove for the tendon of the flexor hallucis longus* (**7**). The trochlea of the talus and its superior surface are wider in front than at the back. This is more pronounced in the right talus than in the left. On the lateral side, the superior surface blends with the *lateral malleolar facet* (**8**), which extends onto the *lateral talar process* (**9**). Medially lies the smaller *medial malleolar facet* (**10**). The three joint surfaces serve for articulation with the ankle mortise. As an inferior continuation of the navicular articular surface, we find the *anterior facet for the calcaneus* (**11**). Continuous with the anterior calcaneal facet (infrequently there is an intermediate cartilage-free zone) lies the *middle calcaneal facet* (**12**). Posterior to the latter, the *talar sulcus* (**13**) and the large posterior calcaneal articular facet (**14**) are found.

The talus also articulates with ligaments that have cartilage deposits (see p.224). Variably developed articular surfaces are therefore present on its inferior surface. These are referred to

as the (larger) *articular facet for the plantar calcaneonavicular ligament* and the (smaller) *articular surface for the calcaneonavicular part of the bifurcate ligament.*

Ossification: An ossification center appears in the talus in the 7th to 8th intrauterine month.

■ **Variant:** In exceptional cases, the lateral tubercle of the posterior talar process forms an independent bone, the **os trigonum** or **accessory talus.**

The **calcaneus** (**D–G**) is the largest tarsal bone. Posteriorly it bears the large calcaneal tuberosity, or **tuber calcanei** (**15**), which has two forward-facing processes at the point of transition onto its lower surface, the *lateral* and *medial processes of the tuber calcanei*. The Achilles tendon is inserted into the roughened area on the tuber calcanei. Anteriorly there is the *surface for articulation with the cuboid bone* (**16**). On the upper surface of the calcaneus, there are normally three articular surfaces, the *anterior* (**17**), *middle* (**18**), and *posterior* (**19**) *talar articular surfaces*. Between the latter two lies the *calcaneal sulcus* (**20**), which, together with the talar sulcus (see above), forms the **sinus tarsi**. The two anterior articular surfaces may be joined together. On the medial surface, the **talar shelf, sustentaculum tali** (**21**), projects outward. It bears the middle talar articular facet. Inferiorly lies the *groove for the tendon of the flexor hallucis longus* (**22**). In most cases there is a slightly elevated bony tubercle on the lateral surface of the talus, the *peroneal trochlea* (**23**), under which runs the *groove for the peroneus longus tendon* (**24**).

Ossification: A bony center develops in the calcaneus in the 4th to 7th intrauterine month.

Clinical tip: In some cases there is an anteriorly directed bony process, the **calcaneal spur**, arising from the medial tuberal process, from which various muscles of the sole of the foot arise. A calcaneal spur may be very painful.

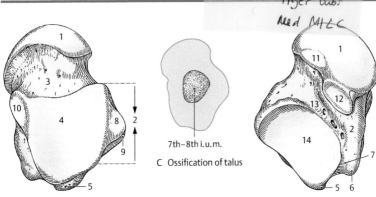

A Superior view of right talus

C Ossification of talus

7th–8th i.u.m.

B Inferior view of right talus

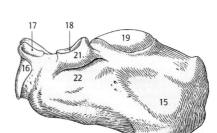

D Superior view of right calcaneus

E Medial view of right calcaneus

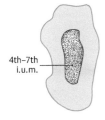

4th–7th i.u.m.

G Ossification of calcaneus

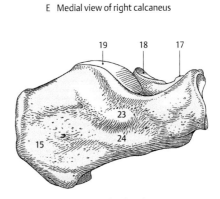

F Lateral view of right calcaneus

Lower Limb

Bones of the Foot, continued

Tarsal Bones, continued (A–P)

The **navicular** (**A–C**) articulates with the talus and with the three cuneiform bones. A concave articular surface faces the head of the talus. The *tuberosity of the navicular* (**1**) is directed *plantarly and medially*. Distally there are three joint surfaces separated only by small crests for the three cuneiform bones.

Ossification: An ossification center develops in the 3rd to 4th year of life.

The **cuboid** (**D–F**) is shorter laterally than medially. Distally there are joint surfaces for the fourth and fifth metatarsal bones separated by a ridge. Medially lies the joint surface for articulation with the lateral cuneiform bone, and sometimes, behind it, we find a small area for articulation with the navicular. The *calcaneal process* (**2**), with its surface for articulation with the calcaneus, is directed posteriorly. On the inferior surface runs the *groove for the peroneus longus tendon* (**3**), posterior to which is a transverse ridge, the *tuberosity of the cuboid* (**4**).

Ossification: The ossification center in the cuboid develops in the 10th intrauterine month (**sign of maturity**).

The three **cuneiform bones** (**G–P**) differ from each other in size and in their position in the skeleton of the foot. The **medial cuneiform** (**G, H**) is the largest and the **intermediate cuneiform** (**J, K**) is the smallest. The broad surface of the medial cuneiform faces the sole of the foot, while the intermediate and **lateral** (**L, M**) cuneiforms have their sharp edges directed plantarly.

All three cuneiform bones have articular surfaces proximally for articulation with the navicular (**5**). Distally and directed toward the digits are articulations for the metatarsals. The medial cuneiform articulates with the first metatarsal and, to a small extent, with the second metatarsal (**6**), while the lateral cuneiform has sur-

faces for articulation with the third metatarsal, a small facet for the second metatarsal (**7**), and sometimes an equally small facet for the fourth metatarsal. The intermediate cuneiform articulates distally only with the second metatarsal. The three cuneiforms also articulate with one other. In addition, the lateral cuneiform has a joint surface (**8**) for articulation with the cuboid.

Ossification: Ossification centers appear in the medial cuneiform (**N**) in the 2nd to 3rd year, in the intermediate cuneiform (**O**) in the 3rd year, and in the lateral cuneiform (**P**) in the 1st to 2nd year.

A Posterior view
of right navicular

B Anterior view
of right navicular

C Ossification
of navicular

3rd–4th years

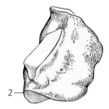

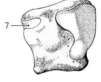

D Dorsal view
of right cuboid

E Plantar view
of right cuboid

F Ossification
of cuboid bone

10th i.u.m.

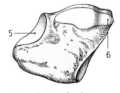

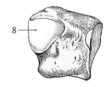

G Medial view of right
medial cuneiform

J Medial view of right
intermediate cuneiform

L Medical view of right
lateral cuneiform

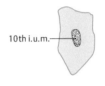

H Lateral view of right
medial cuneiform

K Lateral view of right
intermediate cuneiform

M Lateral view of right
lateral cuneiform

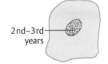

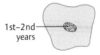

2nd–3rd
years

3rd year

1st–2nd
years

N Ossification of medial
cuneiform

O Ossification of inter-
mediate cuneiform

P Ossification of lateral
cuneiform

Bones of the Foot, continued

Metatarsals (A, B)

The five **metatarsals** are long bones that are convex dorsally. They all possess a *base* (**1**), a *shaft* (**2**), and a *head* (**3**). The **first metatarsal** is the shortest and thickest. There is a *tuberosity* at the base of the first metatarsal on its plantar surface. In the region of this tuberosity and lateral to it, the bone articulates laterally with the base of the second metatarsal and posteriorly via a curved surface with the medial cuneiform (**4**). On its anterior end the head carries, on its plantar surface, a small ridge, and on either side of it there are two small grooves. In these are regularly found two small **sesamoid bones** (**5**). The **second, third,** and **fourth metatarsals** are slimmer and their bases are wider dorsally than on their plantar sides. On the facing sides there are joint surfaces for articulation with each other, and they have proximal posterior facets for the cuneiform and cuboid bones. The heads of these three metatarsals are compressed from side to side so that they resemble rollers. The **fifth metatarsal** differs in that it has a *tuberosity* (**6**) on its lateral side.

Bones of the Toes

The second to fifth digits each have a **proximal, middle,** and **distal phalanx,** while the first digit has only two phalanges. Each phalanx has a *base* (**7**), a *shaft* (**8**), and a *head* (**9**). The distal phalanx (**10**) has a *distal tuberosity*. There are small grooves on the proximal and middle phalanges.

■■■ **Variant:** Occasionally the middle and distal phalanges of the fifth digit may be joined. This may already be the case in the cartilaginous stage before birth.

Sesamoid Bones

Numerous sesamoid bones are found near the metatarsophalangeal joints, although they are consistently present only in the region of the head of the first metatarsal.

Ossification: The cartilaginous metatarsal anlages develop a perichondral bony cuff in the shaft in the 2nd to 3rd intrauterine month, and occasionally there is also an epiphyseal ossification center. Like the metacarpals, the epiphyseal bony center of the first metatarsal is in its base; in the other metatarsals it is always in the head. The epiphyseal endochondral ossification centers develop in the 2nd to 4th years. In some instances there may be an additional second epiphyseal anlage in the first and fifth metatarsal bones.

Epiphyseal centers appear in the base of the phalanges in the 1st to 5th year, while perichondral ossification in the shaft develops in the 2nd to 8th intrauterine month. They fuse during puberty. The individual bony anlages are relatively variable and their times of appearance can be different, so the figures quoted here should only be taken as a general guide.

11 Intermediate cuneiform
12 Lateral cuneiform
13 Cuboid
14 Navicular

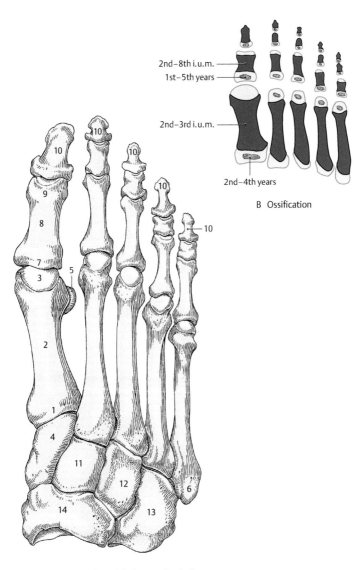

2nd–8th i.u.m.

1st–5th years

2nd–3rd i.u.m.

2nd–4th years

B Ossification

A Dorsal view of metatarsals and phalanges of right foot

Joints of the Foot (A–C)

The **joints of the foot** include the ankle joint, or **talocrural joint**, and the **subtalar** and **talocalcaneonavicular joints**.

In addition we have **cuneonavicular**, **calcaneocuboid**, **cuneocuboid**, and **intercuneiform articulations**.

The **tarsometatarsal joints** are the articulations between the tarsal and metatarsal bones.

Articular connections between the bases of the metatarsals are the **intermetatarsal joints** and those between the metatarsals and the phalanges of the foot are the **metatarsophalangeal joints**.

Also present are joints between the phalanges, or **interphalangeal articulations** of the foot.

Ankle Joint

The **articular surfaces** of the talocrural joint are formed by the *ankle mortise* (**1**) and the superior surface of the *talar trochlea* along with its medial and lateral malleolar facets. The tibia and fibula form a mortise, or clasp, for the trochlea of the talus (see p. 216). The articular surface of the fibula extends farther distally than the tibia.

The **joint capsule** (**2**) is attached to the margins of the cartilage-covered articular surfaces. The joint cavity contains anterior and posterior synovial folds.

Ligaments of the ankle joint. The largest ligament on the medial side is the *deltoid (or medial) ligament* (**3**), which consists of tibionavicular (**4**), tibiocalcaneal (**5**), and anterior and posterior (**6**) tibiotalar parts. The tibionavicular part (**4**) extends from the tibia (**7**) to the navicular (**8**) and covers the anterior tibiotalar part. The tibiocalcaneal part (**5**) runs to the sustentaculum tali (**9**) and partly covers the tibionavicular part (**4**). Other ligaments include the *anterior talofibular ligaments,* (**10**), the *posterior talofibular ligament* and the *calcaneofibular*

ligament (**11**). The anterior talofibular ligament connects the lateral malleolus to the neck of the talus. The posterior talofibular ligament runs almost horizontally from the lateral malleolar fossa to the posterior talar process. The joint capsule bulges distal and proximal to this ligament. The malleolar mortise is fixed by the *anterior* (**12**) and *posterior tibiofibular ligaments*. These bands and the calcaneofibular ligament are collectively known as the lateral collateral ligaments.

Movements. Both **plantar flexion** and **dorsiflexion** are possible. In plantar flexion, as the trochlea of the talus is narrower posteriorly, which leaves more free play in the mortise, slight side-to-side movement is possible. The ankle joint is a **hinge joint** with a **transverse axis**, *beginning just beneath the tip of the medial malleolus and running through the thickest part of the lateral malleolus*. The range of movement between maximal dorsiflexion and plantar flexion is up to 70°.

Clinical tip: Two joint lines are available for amputation of the forefoot or of the forefoot and midfoot. **Chopart's joint line** (**C**, red) is incorrectly called the "transverse tarsal joint." It first runs between the talus (**13**) and calcaneus (**14**) and then between the navicular (**8**) and cuboid (**15**). The *bifurcate ligament* (Chopart's ligament) (**16**, see p. 226) is a key landmark, as it must be divided in opening Chopart's joint line. **Lisfranc's joint line** (**C**, blue) lies between the tarsals and the metatarsals. It should be noted that the second metatarsal (**17**) projects proximally, so the line is not straight.

18 Plantar calcaneocuboid ligament
19 Long plantar ligament
20 Medial cuneiform
21 Intermediate cuneiform
22 Lateral cuneiform
23 Medial tubercle of posterior process of the talus
24 Plantar calcaneonavicular ligament

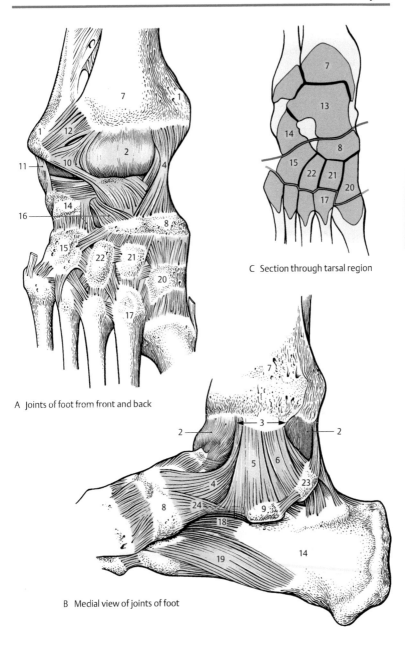

A Joints of foot from front and back

C Section through tarsal region

B Medial view of joints of foot

Joints of the Foot, continued

Subtalar and Talocalcaneonavicular Joints (A, B)

Although separate, these joints act in unison. The **subtalar joint** (**1**) forms the posterior part and the **talocalcaneonavicular joint** (**2**) forms the anterior part of the joint. The **articular surfaces of the subtalar joint** are formed by the *talus* (**3**) and the *calcaneus* (**4**). The **capsule** is loose and thin and is strengthened by the *medial* and *lateral* (**5**) *talocalcaneal ligaments.*

The **talocalcaneonavicular joint** is made up of three bones. In addition to the **articular surfaces** of the *talus*, *calcaneus*, and *navicular* (**6**), there is an additional articular surface covered by cartilage on the *plantar calcaneonavicular ligament* (**7**). This ligament connects the calcaneus in the region of the medial articular surface with the navicular bone, and together with the latter forms the articular cavity for the head of the talus (**spring ligament**).

The **capsule** of the talocalcaneonavicular joint (anterior part) is attached directly at the edge of the cartilage or it extends as far as the plantar calcaneonavicular ligament. The tense *bifurcate ligament* (see **8**, p. 226), which binds the calcaneus (**4**), navicular (**6**), and cuboid (**9**) together, strengthens the capsule. The *interosseous talocalcaneal ligament* (**10**), lying in the sinus tarsi, divides the subtalar from the talocalcaneonavicular joint.

In summary the ankle joint permits **hinge movements** while the subtalar and the talocalcaneonavicular joints permit **rotation**. The ankle joint is a hinge joint, a **ginglymus**, and the others are pivot joints, **trochi**, and together they function as a **trochoginglymus**. Movements of rotation are known as **pronation** (**eversion**) and **supination** (**inversion**), corresponding to the pronating and supinating movements of the hand.

Supination is the elevation of the medial (inner) edge of the foot, and pronation is the elevation of the lateral edge of the foot with simultaneous external rotation. The full range of movement of pronation and supination between their extreme limits amounts to 60°.

Joints between the Other Tarsal and Metatarsal Bones (A, B)

The **calcaneocuboid joint** (**11**) is an amphiarthrosis. The joint cavity is a part of Chopart's joint line (see p. 222). The **cuneonavicular** and the **tarsometatarsal joints** as well as the **cuneocuboid joint** are also amphiarthroses. The ligaments that reinforce the joint capsules will be discussed on page 226. These amphiarthroses include the **intertarsal joints** and the **intermetatarsal joints**, which lie between the adjacent sides of the bases of the second through fifth metatarsals.

Joints of the Toes

The **metatarsophalangeal** and **interphalangeal joints** of the foot may be divided into the proximal and the middle and distal joints. The proximal metatarsophalangeal joints are ball-and-socket joints, although their mobility is restricted by collateral ligaments. The middle and distal joints are pure hinge joints.

12 Dorsal calcaneocuboid ligament
13 Dorsal cuboideonavicular ligament
14 Talonavicular ligament
15 Dorsal tarsometatarsal ligaments
16 Dorsal metatarsal ligaments
17 Long plantar ligament
18 Plantar metatarsal ligaments
19 Tendon of peroneus longus
20 Tendon of tibialis anterior
21 Tendon of tibialis posterior
22 Tendon of peroneus brevis
23 Plantar calcaneocuboid ligament
24 Plantar cuboideonavicular ligament

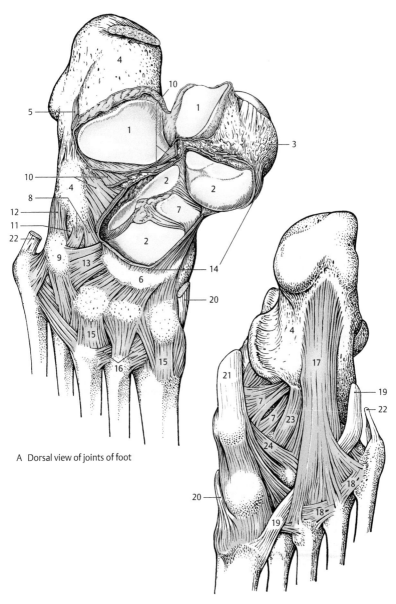

A   Dorsal view of joints of foot

B   Plantar view of joints of foot

Ligaments of the Joints of the Foot (A, B)

The ligaments of the tarsus are divided into several groups.

Ligaments that join the lower leg bones to each other and to the tarsals (red) include the *deltoid ligament* (**1**), the lateral ligament consisting of the *anterior talofibular ligament* (**2**) and the *posterior talofibular ligament* (**3**), the *calcaneofibular ligament* (**4**), the *anterior tibiofibular ligament* (**5**), and the *posterior tibiofibular ligament* (**6**).

Ligaments that join the talus to the other tarsals (green) include the *talonavicular ligament* (**7**), the *interosseous talocalcaneal ligament* (**8**), the *lateral* (**9**) and *medial* (**10**) *talocalcaneal ligaments*, and the *posterior talocalcaneal ligament* (**11**).

The remaining dorsal tarsal ligaments (yellow) include the *bifurcate ligament* (**12**) with its calcaneonavicular and calcaneocuboid fibers, the *dorsal intercuneiform ligaments* (**13**), the *dorsal cuneocuboid ligament* (**14**), the *dorsal cuboideonavicular ligament* (**15**), the *dorsal cuneonavicular ligaments* (**16**), and the *dorsal calcaneocuboid ligaments* (**17**).

The plantar tarsal ligaments (blue) **connect the individual tarsals on their plantar surfaces.** They include the *long plantar ligament* (**18**) extending from the calcaneal tuberosity to the cuboid and metatarsal bones. The *plantar calcaneonavicular or spring ligament* (**19**, see p. 228) is important for the stability of the foot. The medial part of the long plantar ligament, the *plantar calcaneocuboid ligament* (**20**), is particularly important. In addition, there are the *plantar cuneonavicular ligaments*, the *plantar cuboideonavicular ligament*, the *plantar intercuneiform ligaments*, the *plantar cuneocuboid ligament*, and the interosseous ligaments, namely the *interosseous cuneocuboid ligament* and the *interosseous intercuneiform ligaments*.

Ligaments between the tarsus and metatarsus (violet). These may be divided into the *dorsal* and *plantar tarsometatarsal ligaments* and the *interosseous cuneometatarsal ligaments*.

Ligaments between the metatarsals (pink). They include the *dorsal* and *plantar interosseous metatarsal ligaments*, all of which are near the bases of the metatarsals.

Morphology and Function of the Skeleton of the Foot (C, D)

Examination of the skeleton of the foot reveals that in the posterior segment the bones lie over one another, whereas in the middle and anterior regions they lie side by side. By this means the foot becomes arched with the formation of **sagittal** (longitudinal) and **transverse arches**.

It is incorrect to describe the foot as having a longitudinal or transverse "vault." All types of vault (barrel, cloister, cross) have a **keystone-type** construction that makes them *inherently* stable. An arch does not possess this property.

Starting from the talus, a medial row of bones (light gray) continues straight on, while a lateral row (dark gray) fans out from the calcaneus toward the front. The **medial row** consists of the *talus* (**21**), the *navicular* (**22**), the *cuneiforms* (**23**), and the *three medial metatarsals* with their *associated phalanges*. The **lateral row** contains the *calcaneus* (**24**), the *cuboid* (**25**), and the *two lateral metatarsals* with their *corresponding phalanges*.

This results in the foot being wide in front and narrower at the back; it is also higher behind than in front. Finally the foot also has an arch that faces medially and is curved both longitudinally and transversely. The longitudinal curvature is more pronounced on the medial side of the foot than on the lateral side. The transverse arch is well developed only in the midfoot and forefoot.

Clinical tip: Clinically the talus and calcaneus are considered part of the hindfoot, while the other tarsals are regarded as the midfoot and the metatarsals and phalanges as the forefoot.

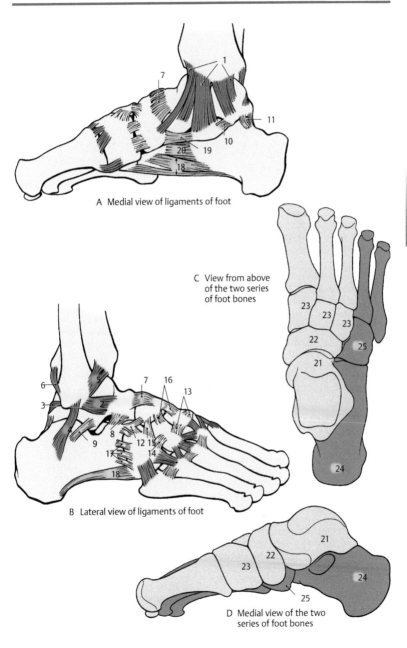

A Medial view of ligaments of foot

C View from above of the two series of foot bones

B Lateral view of ligaments of foot

D Medial view of the two series of foot bones

The Plantar Arch and Its Function (A–C)

The plantar arch is normally in a position of supporting the weight of the body. The **bony points of support of the arch** on a level ground surface are the *calcaneal tuberosity* (**1**), the *head of the first metatarsal* (**2**), and the *head of the fifth metatarsal* (**3**). Thus, the supporting surface is in the form of a triangle (**A**, dotted red). If a **footprint** (**B**) is examined, a somewhat larger weight-bearing surface is found, which is produced by the soft tissues. The **line of transmission** of the weight of the body runs from the *tibia* (**4**) to the *calcaneus* (**5**) and to the *midfoot* and *forefoot* (**6**). The transmission of pressure to the arch in both directions tends to flatten its curvature, and this is opposed by the ligaments and the plantar muscles.

Ligaments. *Ligaments cannot fatigue and have a greater resistance to stress than muscles.* Their resistance does not vary, but if they are overstretched they are unable to return to their previous shape.

The ligaments may be divided into the **plantar aponeurosis** (**7**), the **long plantar ligament** (**8, 9**), the **plantar calcaneonavicular ligament** (**10**), and the **short plantar ligaments**.

The **superficial plantar aponeurosis** (**7**) joins the calcaneal tuberosity to the plantar surface of the digits. It acts especially in the standing (static) position. In the metatarsal part of the foot, tension in the transverse fibers of the aponeurosis supports both the longitudinal and the transverse arches.

The **long plantar ligament** (**8, 9**) braces the lateral row of the tarsals. It arises from the plantar side of the calcaneus, becomes wider distally and extends as a *long, superficial fibrous layer* (**8**) over the peroneus longus tendon to the bases of the metatarsals. Short fibers reach the tuberosity of the cuboid as the *plantar calcaneocuboid ligament* (**9**).

The **plantar calcaneonavicular ligament** (**10**) and the **short plantar ligaments** together form the deepest layer of ligaments. *It increases the size of the socket for the head of the talus.* On the inner surface it is covered by fibrocartilage, which sometimes may be calcified. This ligament may be up to 5 mm thick.

Plantar muscles. These also resist the effect of the weight of the body in spreading the foot, and they surround the arches like a clamp. *They are subject to fatigue and are weaker than the ligaments.* However, muscle tension can be regulated according to stress, and recent investigations have shown that it is brought into play under conditions of great stress. The action of the medial abductors is superior to that of the lateral abductors.

The plantar muscles are divided into the **intrinsic muscles of the foot** (**11**), which stretch between the tarsals and the metatarsals and phalanges, and the **tendons of the extrinsic muscles of the foot,** which descend from the leg and are inserted on the various tarsals, metatarsals, and phalanges. The intrinsic muscles of the foot permit movements of the digits with respect to the metatarsals and tarsals. In the standing or static position, the digits and metatarsals are pressed against the ground, and the intrinsic muscles of the foot function as tensor muscles of the plantar arch, as they counteract the sagging tendency of the metatarsals.

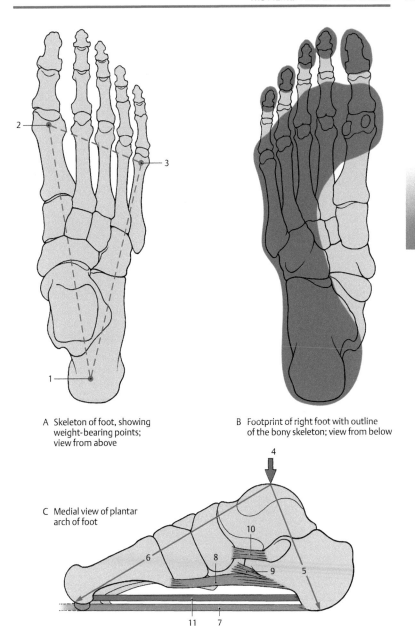

A Skeleton of foot, showing weight-bearing points; view from above

B Footprint of right foot with outline of the bony skeleton; view from below

C Medial view of plantar arch of foot

Foot Types (A–J)

The normal posture of the foot in the living can be determined by taking a footprint. In the **healthy foot**, **pes rectus** (**A**) the *print should show impressions of five digits, the anterior and posterior parts of the sole, and a strip joining them.* The *main load* on the healthy foot (**E**) *lies medially on the calcaneus* (**1**) *and the head of the first metatarsal* (**2**).

> **Clinical tip:** If the print shows a *wide, flattened impression* (**B**) of the entire sole, then the subject has a **flat foot**, **pes planus**. Flat feet are caused by a deficiency of the intrinsic plantar muscles, which leads to overstretching of the ligaments and thus to collapse of the plantar arch. When this occurs, there is a pronation of the talus, which may then slide medially over the calcaneus (**F**). The end result is a remodeling of all the involved tarsals (calcaneus, talus, navicular, and cuboid).
>
> The development of flat foot is associated with severe pain in the foot and leg due to overstretching of the long plantar muscles.
>
> A *footprint in two parts* (**C**) indicates a **high longitudinal arch, pes cavus** (**C**). In this case the calcaneus is supinated while the other bones of the foot are pronated.
>
> A **pes planovalgus** has a *footprint that bulges medially* (**D**). It represents a combination of flat foot and **pes valgus** (**H**); the calcaneus is pronated.

In the **healthy foot** (**G**), *the mechanical axis of the lower limb* (see also p. 214) *runs through the center of the calcaneus to its undersurface.*

> **Clinical tip:** In **pes valgus** (**H**), *the vertical axis through the talus and calcaneus is sharply angulated with respect to the longitudinal axis of the lower limb, thus forming an obtuse angle, open laterally.* The foot is everted (pronated). This posture of the foot may be caused by paralysis of the muscles of supination—triceps surae, tibialis posterior, flexor hallucis longus, flexor digitorum longus, and tibialis anterior.

> **Clubfoot, pes varus** (**J**), shows the exact opposite. *Here the long axis through the talus and calcaneus and the axis of the lower limb form an angle that is open medially.* This may be caused, for instance, by paralysis of the pronators, the peroneal muscles, extensor digitorum longus, and extensor hallucis longus, resulting in supination.

> In **pes rectus** (**G**) the lateral malleolus is lower than the medial malleolus. In **pes valgus** (**H**) this difference in height is increased, while in **clubfoot** (**J**) the difference is absent or may even be reversed.

> Other abnormal postures of the foot include **pes equinus** and **pes calcaneus**. Pes equinus is the result of a paralysis of the extensors, and pes calcaneus is caused by paralysis of the flexor muscles.

> A combination of pes varus and pes equinus is represented by **pes equinovarus**, which occurs after paralysis of the peroneal nerve and injury to the tibialis anterior.

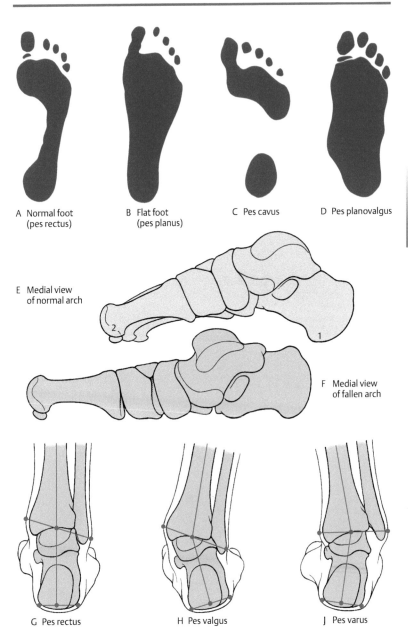

A Normal foot (pes rectus)

B Flat foot (pes planus)

C Pes cavus

D Pes planovalgus

E Medial view of normal arch

F Medial view of fallen arch

G Pes rectus

H Pes valgus

J Pes varus

Muscles, Fasciae, and Special Features

Muscles of the Hip and Thigh

Classification of the Muscles (A–C)

The hip muscles may be classified in several ways. Like the muscles of the shoulder girdle, they may be subdivided according to their locations or innervation from the ventral and dorsal divisions of the plexus layers (see Vol. 3). Further, they may also be grouped according to their development on the basis of their points of insertion. In this classification we distinguish between dorsal muscles with an anterior and posterior group, and ventral hip muscles. It is also possible to classify the muscles of the hip joint according to their function.

Thigh muscles may also be classified according to their location, function, or innervation. According to their location, we distinguish anterior and posterior thigh muscles and adductors. With the exception of the gracilis, all the adductors act solely on the hip joint and therefore insert on the femur. The true thigh muscles act primarily on the knee joint and are inserted into the leg. Here the extensors must be distinguished from the flexors. The extensors of the knee joint lie on the anterior surface of the femur and the flexors are on its posterior surface. Ontogenetically the sartorius is considered an extensor, since it has only been displaced secondarily and now flexes at the knee joint.

Discussion of the hip muscles will take into consideration their sites of insertion as well as their functions. The thigh muscles will be discussed first in terms of their location and then according to their function.

Dorsal Hip Muscles (see p. 234)

The anterior group, which is inserted in the region of the lesser trochanter, includes

Psoas major and iliacus, together forming the iliopsoas (**1**)
Psoas minor

The posterior group, which is inserted in the region of the greater trochanter and its continuation, includes

Piriformis (**2**)
Gluteus minimus (**3**)
Gluteus medius (**4**)
Tensor fasciae latae (**5**)
Gluteus maximus (**6**)

Ventral Hip Muscles and Adductors of the Thigh (see p. 238)

Obturator internus (**7**)
Gemelli (**8**)
Quadratus femoris (**9**)
Obturator externus (**10**)
Pectineus (**11**)
Gracilis (**12**)
Adductor brevis (**13**)
Adductor longus (**14**)
Adductor magnus (**15**)
Adductor minimus (**16**)

Anterior Thigh Muscles (see p. 248)

Quadriceps femoris, consisting of
– Rectus femoris (**17**)
– Vastus intermedius (**18**)
– Vastus medialis (**19**)
– Vastus lateralis (**20**)
Sartorius (**21**)

Posterior Thigh Muscles (see p. 250)

Biceps femoris (**22**)
Semitendinosus (**23**)
Semimembranosus (**24**)
Popliteus (see p. 264)

25 Fascia lata
26 Anteromedial intermuscular septum
27 Lateral femoral intermuscular septum
28 Neck of the femur
29 Femoral artery
30 Femoral vein
31 Saphenous nerve
32 Great saphenous vein
33 Sciatic nerve
34 Deep femoral artery
35 Femoral nerve

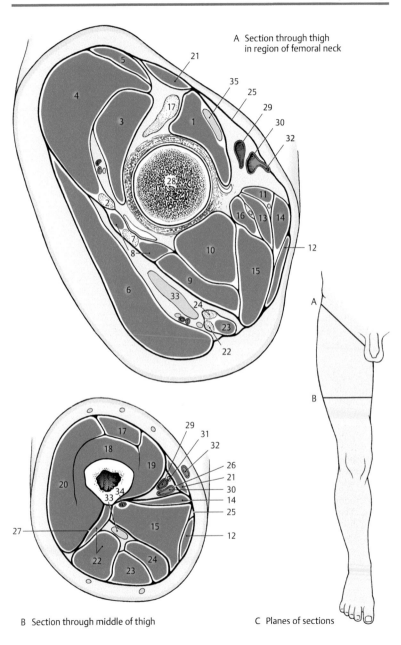

A Section through thigh in region of femoral neck

B Section through middle of thigh

C Plans of sections

Dorsal Hip Muscles

Anterior Group Inserted in the Region of the Lesser Trochanter (A, B)

The **psoas major** (**1**) is divided into a superficial and a deep part. The **superficial part** *arises from the lateral surfaces of the 12th thoracic vertebra and first to fourth lumbar vertebrae (**2**) as well as from their intervertebral disks.* The **deep part** *arises from the costal processes of the first to fifth lumbar vertebrae (**3**).*

The psoas major joins the iliacus (**4**) and, surrounded by the iliac fascia, proceeds as the **iliopsoas** (**5**) across the iliopubic eminence through the muscular lacuna to be inserted on the *lesser trochanter* (**6**). In the region of the iliopubic eminence, the iliopectineal bursa lies between the muscle and the bone and extends as far as the anterior surface of the capsule of the hip joint with which it communicates. Between the lesser trochanter and the attachment of the iliopsoas lies the iliac subtendinous bursa. The lumbar plexus lies between the two layers of the psoas major (see also p. 404).

The **iliacus** (**4**) *arises in the iliac fossa* (**7**) *and also from the region of the anterior inferior iliac spine.* It joins the psoas major (**1**) to form the **iliopsoas** (**5**). *The fibers of the iliacus are regularly inserted in front of the fibers of the psoas major and extend distally over the lesser trochanter.*

The iliopsoas is the most important muscle for lifting (flexing) the leg forward and makes walking possible. It also serves to bend the trunk forward and to lift the trunk when lying down. The iliopsoas is also a lateral rotator of the hip joint. In contrast to the iliacus, the psoas major acts on a number of joints, since it crosses vertebral and sacroiliac joints. It is therefore also involved in lateral bending.
Nerve supply: lumbar plexus and femoral nerve. Psoas major (L1–L3); iliac muscle (L2–L4).

■■ **Variants: The psoas minor** is present in less than 50% of the population. *It arises from the 12th thoracic and first lumbar vertebrae and projects into the iliac fascia.* It is either inserted on the iliopubic eminence or radiates into the iliopectineal arch.
Nerve supply: lumbar plexus (L1–L3).

The psoas major may also arise from the head of the 12th rib and the iliacus may arise from the capsule of the hip joint and from the sacrum.

Clinical tip: Migratory (hypostatic) abscesses, see page 94.

8 Pectineus
9 Adductor minimus
10 Adductor longus
11 Iliopectineal arch
12 Inguinal ligament

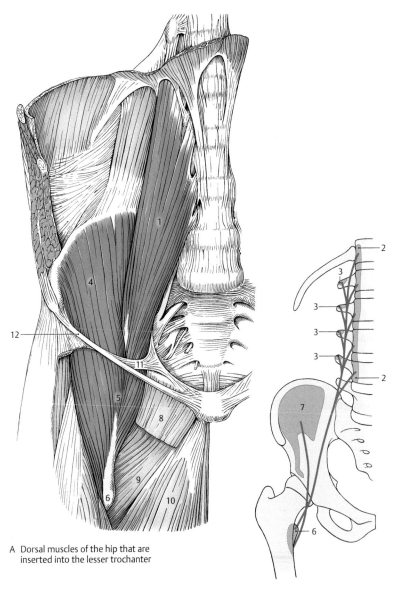

A Dorsal muscles of the hip that are
inserted into the lesser trochanter

B Diagram of origin, course,
and insertion of muscles

Dorsal Hip Muscles, continued

Posterior Group Inserted in the Region of the Greater Trochanter (A–D)

The **tensor fasciae latae** (**1**) *arises in the region of the anterior superior iliac spine* (**2**) *and extends distal to the greater trochanter into the iliotibial tract* (**3**), *which is inserted on the lateral tibial condyle.* It presses the head of the femur into the acetabulum. It is also a flexor, medial rotator, and abductor, and assists the anterior bundles of the gluteus medius and minimus.
Nerve supply: superior gluteal nerve (L4–L5).

The powerful **gluteus maximus** (**4**) has a **superficial** and a **deep origin**. The **superficial fibers** *arise from the iliac crest* (**5**), *the posterior superior iliac spine* (**6**), *the thoracolumbar fascia, the sacrum* (**7**), *and the coccyx* (**8**). The **deep fibers** *arise from the ala of the ilium* (**9**) behind the posterior gluteal line, *from the sacrotuberal ligament* (**10**) *and the fascia of the gluteus medius (Aponeurosis glutealis).* The **proximal part** *radiates into the iliotibial tract* (**3**) and the **distal part** *inserts into the gluteal tuberosity* (**11**). Between the latter and the greater trochanter lies the large trochanteric bursa of the gluteus maximus (**12**). Its relationship to the ischial tuberosity is dependent on the posture of the body. In the upright posture the muscle covers the ischial tuberosity but leaves it free in the seated position.

It is primarily an extensor and lateral rotator at the hip joint and represents a muscular defense against excessive forward tilting of the pelvis. It comes into action when climbing stairs and when changing from the sitting to the upright posture. With its different sites of insertion it is able to act as an adductor as well as an abductor. That part which tenses the fascia lata abducts, while the part inserted on the gluteal tuberosity adducts. Both glutei maximi may assist in contraction of the external sphincter ani.
Nerve supply: inferior gluteal nerve (L5–S2).

The **gluteus medius** (**13**) *arises from the gluteal surface of the ala of the ilium* (**14**), between the anterior and posterior gluteal lines, and *from the iliac crest* (**15**) *and its fascia (gluteal aponeurosis). It is inserted on the greater trochanter* (**16**) *like a cap.* Between the tendon of attachment and the greater trochanter lies the trochanteric bursa of the gluteus medius. The anterior fibers of the gluteus medius act as a medial rotator and flexor, and the posterior part as a lateral rotator and extensor of the hip, while the entire muscle can function as an abductor (for instance in dancing).
Nerve supply: superior gluteal nerve (L4–L5).

The **gluteus minimus** (**17**) *arises from the gluteal area on the iliac wing* (**18**) between the anterior and inferior gluteal lines *and is inserted into the greater trochanter* (**19**). It has a trochanteric bursa at its insertion. It corresponds in function to the gluteus medius, although it is a weaker abductor.
Nerve supply: superior gluteal nerve (L4–S1).

The **piriformis** (**20**) *originates as several slips from the pelvic surface of the sacrum,* lateral to the pelvic sacral foramina (**21**), *and from the margin of the greater sciatic notch.* It passes through the greater sciatic foramen *and is inserted on the anteromedial aspect of the tip of the greater trochanter* (**22**). In the upright posture it functions as a lateral rotator and abductor, and it also aids in extension of the thigh.
Nerve supply: sacral plexus (L5–S2).

■ **Variants:** The muscle may be divided into several parts by the sciatic nerve or other branches of the sacral plexus. Sometimes it may be partly or completely absent.

23 Obturator internus
24 Quadratus femoris

Lower Limb

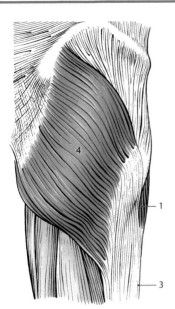

A Posterior group of hip muscles:
tensor of fascia lata and
gluteus maximus

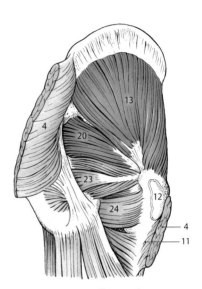

B Posterior group of hip muscles:
piriformis and gluteus medius

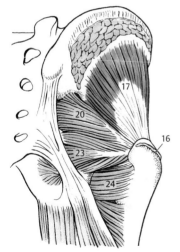

C Posterior group of hip muscles:
piriformis and gluteus minimus

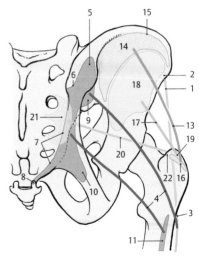

D Diagram of origin, course,
and insertion of muscles

Ventral Hip Muscles (A–D)

The ventral muscles, which are innervated by the ventral branches of the nerve plexus layer, function as lateral rotators. They are important in the control of the body's balance. Basically the lateral rotators are stronger than the medial rotators, and therefore, in the normal position of the limb, the apex of the foot points slightly outward to achieve better support for the body.

The **obturator internus** (**1**) *arises from the inner surface of the hip bone around the obturator foramen and from the obturator membrane*. It passes through the lesser sciatic foramen, almost filling it, and *is inserted into the trochanteric fossa* (**2**). The sciatic bursa of the obturator internus is found near the lesser sciatic notch. The bone acts as a fulcrum for this muscle. With the gluteus maximus and quadratus femoris it forms the strongest lateral rotator of the hip joint. In the sitting position, with the limb flexed in front, it acts as an abductor.

The two **gemelli** represent, as it were, marginal heads of the obturator internus. According to *Lanz* all three muscles together may be termed the **triceps coxae**. The **superior gemellus** (**3**) *arises from the ischial spine* (**4**), and the **inferior gemellus** (**5**) *from the ischial tuberosity* (**6**). *Both reach the trochanteric fossa* (**2**). Their function is to assist the obturator internus.
Nerve supply: inferior gluteal nerve, sacral plexus (L5–S2).

▬ Variants: It is quite common for one or the other gemellus, and sometimes both, to be absent. Occasionally the obturator internus receives extra bundles of muscle fibers arising from nearby ligaments.

The **quadratus femoris** (**7**) *arises from the ischial tuberosity* (**6**) and runs as a four-sided flattened muscle *to the intertrochanteric crest* (**8**). It acts as a strong lateral rotator and adductor of the thigh.
Nerve supply: inferior gluteal nerve, sacral plexus (L5–S2).

▬ Variants: It may be absent or it may fuse with the adductor magnus.

The **obturator externus** (**9**) *arises from the external surface of the medial bony margin of the obturator foramen and the obturator membrane. It extends to the trochanteric fossa* (**2**) *and (rarely) to the capsule of the hip joint*. This muscle lies deep and it only becomes visible when the adjacent muscles have been removed. At its origin it is covered by the adductors and in the thigh by the quadratus femoris. It is an external rotator and a weak adductor.
Nerve supply: obturator nerve (L1–L4).

10 Piriformis
11 Sacrum

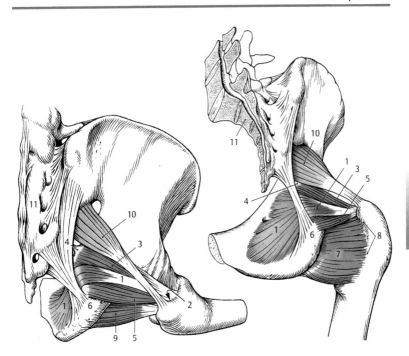

A Dorsal view of ventral muscles
of hip with thigh flexed

B Dorsal view of ventral muscles
of hip with thigh extended

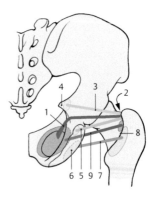

C Distal view of obturator
externus muscle

D Diagram of origin, course,
and insertion of muscles

Adductors of the Thigh (A–D)

The **functional** adductors of the thigh include

- The obturator externus (see p. 238)
- The gracilis
- The pectineus
- The adductor brevis
- The adductor longus (see p. 242)
- The adductor magnus (see p. 242)
- The adductor minimus (see p. 242)

All the adductors are innervated by the obturator nerve, but some receive additional fibers from the femoral nerve (pectineus) and tibial nerve (adductor magnus).

The **gracilis** (**1**) *arises near the symphysis from the inferior ramus of the pubis* (**2**), and, as the only muscle of the adductor group to act on two joints, it *extends as far as the medial surface of the tibia* (**3**), onto which it is inserted together with the semitendinosus and sartorius *as the pes anserinus superficialis* (**4**). It is the most medial muscle directly beneath the surface, and when the thigh is abducted, its origin can clearly be seen arching beneath the skin.

When the knee is extended, it acts as an adductor of the thigh and a flexor of the hip joint. It also flexes at the knee joint. In the region of the pes anserinus, between the three tendons of insertion of the muscles mentioned and the tibia, there is always a bursa, the anserine bursa.
Nerve supply: anterior branch of the obturator nerve (L2–L4).

The **pectineus** (**5**) *arises from the iliopubic eminence, along the pecten of the pubis* (**6**), as far as the pubic tubercle (**7**). It extends obliquely distalward and has an elongated rectangular shape. The proximal fibers run immediately behind the lesser trochanter. *It is inserted into the pectineal line* (**8**) *and into the proximal part of the linea aspera* (**9**). The pectineus and iliopsoas (see p. 234) together form the floor of the iliopectineal fossa. The pectineus flexes at the hip joint (anteversion), adducts the thigh,

and according to electromyographic investigations acts as a weak medial rotator.
Nerve supply: femoral nerve (L2–L3) and the anterior branch of the obturator nerve (L2–L4).

The **adductor brevis** (**10**) *arises from the inferior ramus of the pubis* (**11**) near the symphysis and *reaches the upper third of the medial lip of the linea aspera* (**9**). It lies very close to the adductor longus. In addition to its function as an adductor, it also acts as an external rotator and weak flexor at the hip joint.
Nerve supply: anterior branch of the obturator nerve (L2–L4).

12 Adductor longus
13 Adductor magnus
14 Adductor minimus
15 Obturator externus
16 Quadratus femoris
17 Semitendinosus
18 Sartorius
19 Iliopsoas

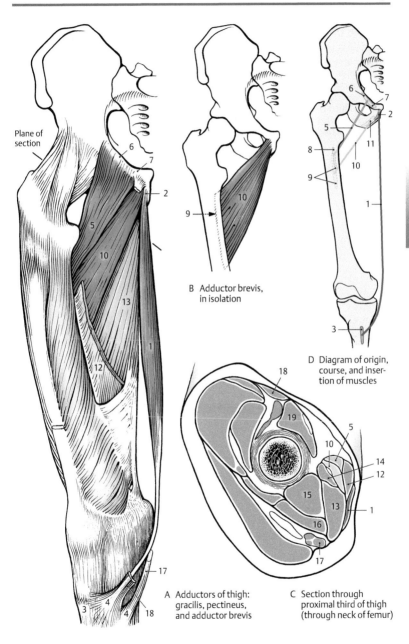

Plane of section

B Adductor brevis, in isolation

D Diagram of origin, course, and insertion of muscles

Lower Limb

A Adductors of thigh: gracilis, pectineus, and adductor brevis

C Section through proximal third of thigh (through neck of femur)

Adductors of the Thigh (A–D), continued

The **adductor longus** (**1**) *arises from the superior ramus of the pubis* (**2**) *and is inserted into the middle third of the medial lip of the linea aspera* (**3**). The adductor longus lies ventrally on the adductor magnus (**4**). Proximally and close to the femur, the adductor brevis (**5**) is interposed between them. The fibers of the adductor longus extend distally into the adductor canal (see below). It is primarily an adductor and a lateral rotator, but may also produce some degree of flexion (anteversion).

Nerve supply: anterior branch of the obturator nerve (L2–L4).

The **adductor magnus** (**4**) *arises from the anterior surface of the inferior ramus of the pubis* (**6**) *and the inferior ramus of the ischium* (**7**) *as far as the ischial tuberosity* (**8**). The large muscle belly passes downward on the medial side of the thigh and divides into **two parts**. **One part** (**9**) *is attached directly by its muscle fibers to the medial lip of the linea aspera* (**10**) and **the other part** (**11**) *is attached by a tendon to the adductor tubercle* (**12**) *of the medial epicondyle*. The tendinous part forms an intermuscular septum and on the medial side it separates the flexors from the extensors.

Between these insertions of the adductor magnus, there is a slitlike opening, the **adductor hiatus** (**13**). The tendinous portion may be palpated through the skin behind the vastus medialis and in front of the medial dimple of the knee.

The adductor magnus is a powerful adductor, which is particularly active when crossing the legs. The part attached to the linea aspera acts as a lateral rotator. Only the part that reaches the medial epicondyle acts as a medial rotator of the outwardly rotated and flexed leg, as well as an extensor of the hip joint.

The **adductor minimus** (**14**) is an incompletely separated division of the adductor magnus. *Its fibers arise from the inferior ramus of the pubis* (**6**) *as the most anterior part of the adductor magnus and run to the medial lip of the linea aspera* (**10**), crossing over the upper part of the fibers of the true adductor magnus. It adducts and externally rotates the femur.

Nerve supply: is common to both muscles. The obturator nerve supplies the part that is attached to the linea aspera, and the tibial nerve supplies the part inserted on the adductor tubercle (L3–L5).

Aponeurotic tendon fibers split off from the muscular part (**9**) of the adductor magnus (**4**) and pass over onto the tendinous surface of the vastus medialis (**15**; see p. 248). This is known as the **anteromedial intermuscular septum = subsartorial fascia = vastoadductor membrane** (**16**). Some fibers of the adductor longus (**1**) may radiate into this membrane. *Between the vastoadductor membrane and the adductor magnus, adductor longus, and vastus medialis, there is a tunnel*, the **adductor canal**, which opens through the **adductor hiatus** (see above) into the popliteal fossa.

17 Gracilis
18 Sartorius
19 Femur

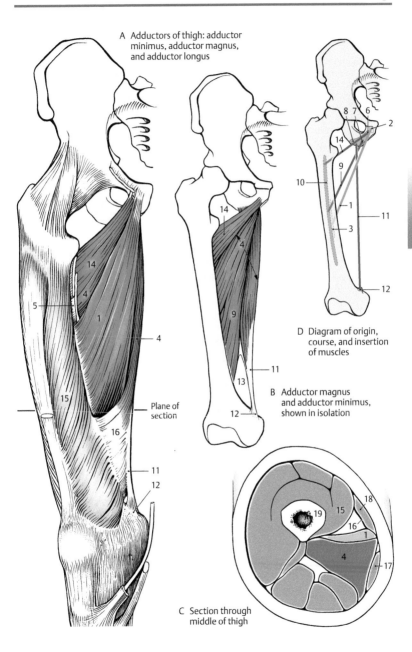

A Adductors of thigh: adductor minimus, adductor magnus, and adductor longus

D Diagram of origin, course, and insertion of muscles

B Adductor magnus and adductor minimus, shown in isolation

C Section through middle of thigh

Lower Limb

Function of the Hip Muscles and Adductors of the Thigh (A, B)

As some hip muscles have extensive areas of origin and insertion, the various parts of the muscle may produce very different movements. It must also be noted that some of the muscles span not only the hip joint but also vertebral joints and the knee joint.

Further influence on the vertebral joints by
– Psoas major

Further influence on the knee joint by
– Gracilis
– Tensor fasciae latae
– Sartorius
– Rectus femoris
– Semimembranosus
– Semitendinosus
– Long head of biceps femoris

Muscles of the thigh also act upon the hip joint, in addition to the muscles of the hip.

We distinguish **external** and **internal rotation** movements about the **longitudinal axis of the limb**. With the hip extended, internal rotation is more extensive than external rotation. With the hip flexed, the restrictive ligaments are tensed, so that the range of external rotation is then greater than that of internal rotation.

The movements about the **transverse axis** are **extension** (dorsiflexion, retroversion) and **flexion** (anteflexion, anteversion).

Abduction and **adduction** occur about a **sagittal axis**.

External rotation (A) is produced by

– Gluteus maximus (red, inferior gluteal nerve)
– Quadratus femoris (blue, inferior gluteal nerve, sacral plexus)
– Obturator internus (yellow, inferior gluteal nerve, sacral plexus)
– Gluteus medius and gluteus minimus with their dorsal fibers (orange, superior gluteal nerve)
– Iliopsoas (green, lumbar plexus, femoral nerve)
– All the functional adductors except the pectineus muscle and the gracilis (violet, obturator nerve, tibial nerve, p. 242)
– Piriformis (gray, sacral plexus)
– Sartorius (see p. 248; not shown)

Internal rotation (B) is produced by
– Anterior fibers of the gluteus medius and the gluteus minimus (red, superior gluteal nerve)
– Tensor fasciae latae (blue, superior gluteal nerve)
– The part of the adductor magnus inserted into the adductor tubercle (yellow, tibial nerve)

In the same way, the pectineus muscle (not shown) acts as an internal rotator with the leg abducted.

The color of the arrows represents the order of importance of the muscles in each movement:

red
blue
yellow
orange
green
brown
violet
gray

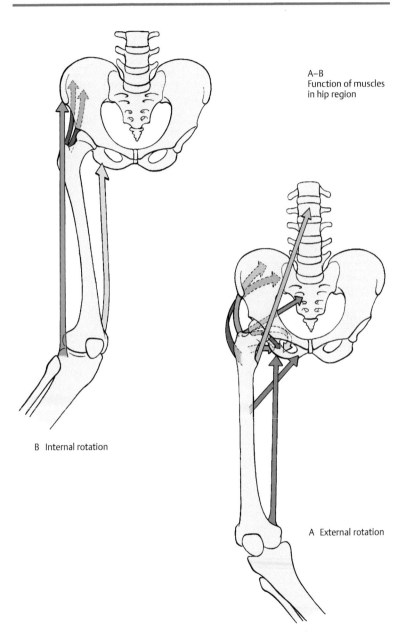

A–B
Function of muscles
in hip region

B Internal rotation

A External rotation

Lower Limb

Function of the Hip Muscles and Adductors of the Thigh, continued (A–D)

The **extensors** (A) of the hip joint are
- Gluteus maximus (red, inferior gluteal nerve)
- Dorsal fibers of the gluteus medius and gluteus minimus (blue, superior gluteal nerve)
- Adductor magnus (green, obturator nerve and tibial nerve, see p. 242)
- Piriformis (brown, sacral plexus)

The following thigh muscles (hamstring muscles) also function as extensors of the hip:
- Semimembranosus (yellow, tibial nerve, see p. 250)
- Semitendinosus (orange, tibial and common peroneal nerve, see p. 250)
- Long head of the biceps femoris (violet, tibial nerve, see p. 250)

Clinical tip: If the most important extensor, the gluteus maximus, is paralyzed, active standing from a sitting position is no longer possible, although standing and walking on a level plane can be done.

The **flexors** (B) of the hip joint are
- Iliopsoas (red, lumbar plexus, femoral nerve)
- Tensor fasciae latae (orange, superior gluteal nerve)
- Pectineus (green, femoral and obturator nerve)
- Adductor longus (brown, obturator nerve)
- Adductor brevis (brown, obturator nerve)
- Gracilis (brown, obturator nerve)

The following thigh muscles are flexors at the hip joint:
- Rectus femoris (blue, femoral nerve, see p. 248)
- Sartorius (yellow, femoral nerve, see p. 248)

Clinical tip: If the iliopsoas is paralyzed, flexion is no longer possible across the horizontal plane when in the sitting position.

Abduction (C) is carried out by
- Gluteus medius (red, superior gluteal nerve)
- Tensor fasciae latae (blue, superior gluteal nerve)
- Gluteus maximus with its attachment at the fascia lata (yellow, inferior gluteal nerve)
- Gluteus minimus (orange, superior gluteal nerve)
- Piriformis (green, sacral plexus)
- Obturator internus (brown, inferior gluteal nerve)

Clinical tip: If the abductors are paralyzed, the pelvis cannot be fixed on the unaffected side when standing on the affected leg. The pelvis falls toward the healthy side (unilateral positive **Trendelenburg test**). When abductor function is impaired on both sides (as in congenital dislocation of the hip), the patient develops a waddling gait (bilateral positive **Trendelenburg sign**).

Adduction (D) is produced by
- Adductor magnus with the adductor minimus (red, obturator and tibial nerve)
- Adductor longus (blue, obturator nerve)
- Adductor brevis (blue, obturator nerve)
- Gluteus maximus with its attachment at the gluteal tuberosity (yellow, inferior gluteal nerve)
- Gracilis (orange, obturator nerve)
- Pectineus (brown, obturator nerve)
- Quadratus femoris (violet, inferior gluteal nerve and sacral plexus)
- Obturator externus (not illustrated)

Of the thigh muscles, especially involved is the
- Semitendinosus (green, tibial nerve)

The color of the arrows in the following series indicates the importance of the muscles in the individual movements:

red
blue
yellow
orange
green
brown
violet

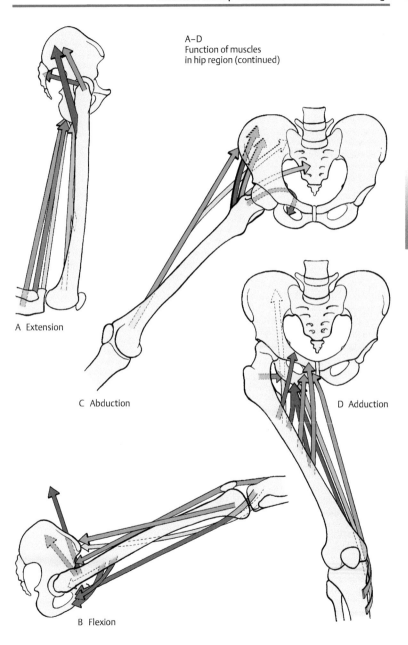

A–D
Function of muscles
in hip region (continued)

A Extension

B Flexion

C Abduction

D Adduction

Anterior Thigh Muscles (A–D)

The **quadriceps femoris** consists of **four parts**, of which the straight part, the rectus femoris, acting on two joints, runs in a channel formed by the other three single joint muscles.

The straight head of the **rectus femoris** (**1**) arises from the *anterior inferior iliac spine* (**2**) and the reflected head from the *upper rim of the acetabulum in the supra-acetabular groove.*

The **vastus intermedius** (**3**) *arises from the anterior and lateral surface of the femur* (**4**). It is easily distinguished from the vastus lateralis but is more difficult to separate from the vastus medialis. It covers the *articularis genu muscle*, which arises distal to it and radiates into the capsule of the knee joint.

The **vastus medialis** (**5**) *arises from the medial lip of the linea aspera* (**6**).

The **vastus lateralis** (**7**) *arises* (**8**) *from the lateral surface of the greater trochanter, the intertrochanteric line, the gluteal tuberosity, and the lateral lip of the linea aspera.*

The four muscles join to form a *common tendon that is inserted into the patella* (**9**). Distal to the patella, the tendon is continued as the *patellar ligament* (**10**) and is *inserted into the tibial tuberosity* (**11**). Superficial fibers run across the patella, while the deep tendon fibers insert into its upper and lateral margins.

Mainly fibers of the vastus medialis and a few fibers of the rectus femoris form the *medial patellar retinaculum*, and fibers of the vastus lateralis and rectus femoris form the *lateral patellar retinaculum*. Fibers from the iliotibial tract also radiate into the lateral patellar retinaculum. The retinacula extend distally around the patella to the tibial condyles.

The quadriceps femoris is *the* extensor of the knee joint. The rectus femoris also flexes the hip joint. The articularis genu muscle protects the capsule of the knee joint from being entrapped during extension.

Nerve supply: femoral nerve (L2–L4).

▬ **Variants:** The part of the rectus femoris that normally takes its origin from the superior rim of the acetabulum may be missing, and the articularis genu muscle may also be absent.

The **sartorius** (**12**) *arises from the anterior superior iliac spine* (**13**) and runs obliquely over the thigh in its fascial investment *to the pes anserinus superficialis* (**14**), by which it *is attached to the crural fascia* (**15**) and *is medial to the tibial tuberosity*. The sartorius acts on two joints, as a flexor at the knee joint and, if the knee is flexed, together with the other muscles of the pes anserinus, it functions as an internal rotator of the leg. In addition, it brings about flexion at the hip joint. Due to its course it also functions as an external rotator of the hip.

Nerve supply: femoral nerve (L2–L3).

16 Gracilis
17 Adductor longus
18 Adductor brevis
19 Pectineus
20 Iliopsoas
21 Tensor fasciae latae
22 Cut edge of fascia lata
23 Vastoadductor membrane = anteromedial intermuscular septum = subsartorial fascia

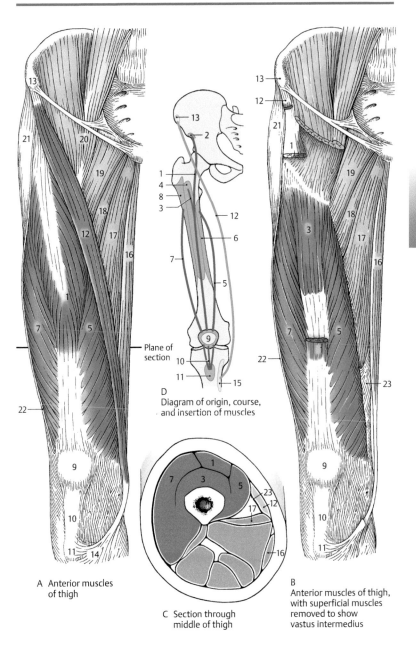

Lower Limb

D Diagram of origin, course, and insertion of muscles

Plane of section

A Anterior muscles of thigh

C Section through middle of thigh

B Anterior muscles of thigh, with superficial muscles removed to show vastus intermedius

Posterior Thigh Muscles (A–D)

The **biceps femoris** (**1**) has a **long head** and a **short head**. The **long head** (**2**), which acts over two joints, *arises from the ischial tuberosity* (**3**) in common with the semitendinosus (**4**). The **short head** (**5**), acting only over one joint, *originates from the middle third of the lateral lip of the linea aspera* (**6**) *and the lateral intermuscular septum*. The heads unite to form the biceps femoris (**1**), which is *inserted into the head of the fibula* (**7**). Between the muscle and the lateral collateral ligament of the knee is the inferior subtendinous bursa of the biceps femoris. The long head produces extension (retroversion) of the hip joint. The biceps femoris flexes at the knee joint and externally rotates the flexed leg. It is the only external rotator of the knee joint and thus opposes all the internal rotators.

Nerve supply: long head, tibial nerve (L5–S2); short head, common peroneal nerve (S1–S2).

▬▬ **Variants:** The short head may be absent; there may also be additional bundles of muscle fibers.

The **semitendinosus** (**4**) *arises* by a common head (see above) *from the ischial tuberosity* (**3**) *and runs toward the medial surface of the tibia* together with the gracilis (**9**) and sartorius (**10**) *to join the pes anserinus superficialis* (**8**). There is a large tibial intertendinous bursa (anserine bursa) between the surface of the tibia and the attachment to the pes anserinus. The muscle acts on two joints, being involved in extension of the hip joint, flexion of the knee joint and internal rotation of the leg.

Nerve supply: tibial nerve (L5–S2).

▬▬ **Variant:** Within its muscle belly there may be an oblique tendinous intersection.

The **semimembranosus** (**11**) *arises from the ischial tuberosity* (**3**). It is closely related to the semitendinosus. Below the medial collateral ligament, *its tendon divides into* **three parts**; the **first** runs anteriorly to the *medial tibial condyle*, the **second** goes into *the fascia of the popliteus*, and the **third** part *continues into the posterior wall of the capsule as the oblique popliteal ligament*. This tripartite division may also be called the "**deep**" **pes anserinus** ("goose's foot").

The muscle acts on two joints and has a function similar to the semitendinosus. It produces extension of the hip joint and flexion with simultaneous internal rotation of the knee joint. Between its tendon (before the division) and the medial head of the gastrocnemius lies the semimembranosus bursa, which is sometimes continuous with the medial subtendinous bursa of the gastrocnemius (see p. 210).

Nerve supply: tibial nerve (L5–S2).

▬▬ **Variants:** The muscle may sometimes be absent or may be completely fused with the semitendinosus. The oblique popliteal ligament need not always be present.

12 Adductor magnus
13 Adductor longus
14 Vastus medialis
15 Vastoadductor membrane = anteromedial intermuscular septum = subsartorial fascia

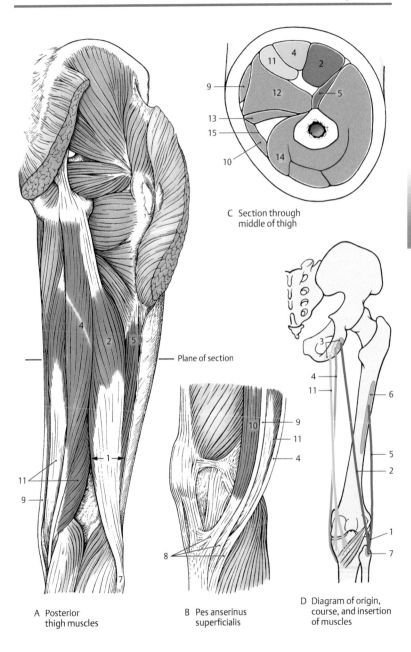

C Section through
middle of thigh

A Posterior
thigh muscles

B Pes anserinus
superficialis

D Diagram of origin,
course, and insertion
of muscles

Plane of section

Lower Limb

Function of the Knee Joint Muscles (A–D)

Only a few muscles act exclusively on the knee joint; the majority act also on the ankle and subtalar joint.

Extension and **flexion take place** around **transverse axes** that run through the femoral condyles (see p. 194). The rotary movements of **internal** and **external rotation take place around the long axis of the leg**. Rotation is possible only when the collateral ligaments are not tense (see p. 212); that is, in the extended position active rotation is impossible. Passively, in maximal extension, there is some external rotation of the leg on the nonweight-bearing side and internal rotation of the thigh of the weight-bearing limb of about 5°, known as terminal rotation or the "screw-home" mechanism (see p. 212). This rotation is produced by the anterior cruciate ligament aided by the shape of the medial femoral condyle and the iliotibial tract (see p. 254).

Extension (A) is carried out almost exclusively by the quadriceps femoris, with minimal assistance from the tensor fasciae latae. The action of the quadriceps is better when the hip joint is extended since then the rectus femoris (red) and the vasti muscles (blue) become fully active.

Clinical tip: The strength of the quadriceps femoris greatly exceeds that of all the other flexors. When this muscle is paralyzed, rising from the sitting position is not possible. Standing is only possible when the line of gravity of the body is in front of the transverse axis of movement.

Flexion (B) is produced by
- Semimembranosus (red, tibial nerve)
- Semitendinosus (blue, tibial nerve)
- Biceps femoris (yellow, tibial and common peroneal nerves)
- Gracilis (orange, obturator nerve)
- Sartorius (green, femoral nerve)
- Popliteus (brown, tibial nerve)
- Gastrocnemius (violet, tibial nerve)

Clinical tip: The gastrocnemius is only slightly active during flexion. Nevertheless, when there is a supracondylar fracture of the femoral shaft, it pulls the distal fragment posteriorly and distally.

The **internal rotators (C)** are
- Semimembranosus (red, tibial nerve)
- Semitendinosus (blue, tibial nerve)
- Gracilis (yellow, obturator nerve)
- Sartorius (orange, femoral nerve)
- Popliteus (green, tibital nerve)

External rotation (D) is carried out by
- biceps femoris (red, tibial and common peroneal nerve)

The biceps femoris is almost the only external rotator of the thigh and counterbalances all muscles acting as internal rotators. When the leg is nonweight-bearing, it can receive minimal support (in terminal rotation) from the tensor fasciae latae (not illustrated).

The color of the arrows in the following series indicates the importance of the muscles in the individual movements:

red
blue
yellow
orange
green
brown
violet

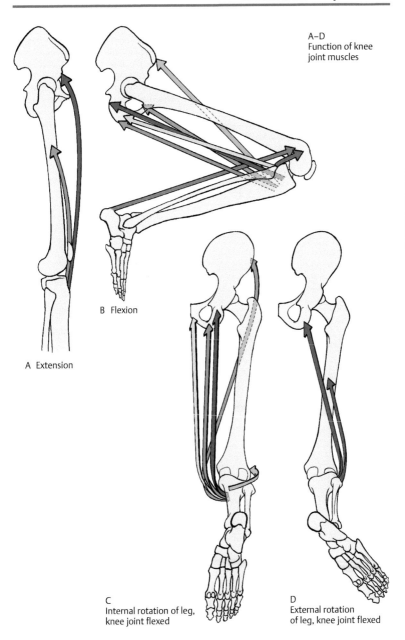

A–D
Function of knee
joint muscles

B Flexion

A Extension

C
Internal rotation of leg,
knee joint flexed

D
External rotation
of leg, knee joint flexed

Fasciae of the Hip and Thigh (A–C)

The muscles of the hip region are invested by various fasciae. For example, the iliopsoas muscle is covered by the **iliopsoas fascia**, which begins with the *psoas fascia* at the medial arcuate ligament as a sturdy fascial tube covering the psoas major and continues together with *iliac fascia* as far as the inguinal ligament. It forms the *iliopectineal arch*, which separates the muscular lacuna (see p. 100) from the vascular lacuna.

On the anterior surface, below the inguinal ligament, the pectineus is enclosed in a strong **pectineal fascia**, which is the pubic portion of the fascia lata (called also the Cowper ligament). The pectineal fascia combines with the iliac fascia to form the connective tissue lining of the iliopectineal fossa. The latter is bounded proximally by the inguinal ligament.

The gluteal region contains the delicate **gluteal fascia** (**1**) which covers the gluteus maximus and gives rise to septa that penetrate deeply between the individual muscle bundles. Between gluteus maximus and the underlying gluteus medius lies the firm, strong **gluteal aponeurosis** (see p. 236) from which portions of the gluteus maximus take their origin. In the region of the gluteal sulcus, the superficial gluteal fascia merges with the fascia lata (**2**), the fascia of the thigh.

On the lateral side of the thigh, the **fascia lata** forms a dense connective tissue layer of parallel fibers that becomes weaker medially. A band of fibers, the **iliotibial tract** (**3**; see pp. 236 and 422), is a conspicuous feature on the lateral side. The gluteus maximus and tensor fasciae latae radiate into this iliotibial tract. The iliotibial tract is several centimeters wide and extends distally on the lateral side to the lateral tibial condyle. In this region the lateral patellar retinaculum is intimately blended with it.

On the anterior surface of the thigh, the sartorius (**4**) possesses its own fascial covering. It overlies the *vastoadductor membrane* (**5**). Similarly the gracilis (**6**) is enclosed in its own fascial sheath, which can be separated from the other fasciae. All the thigh muscles have their own loose, delicate coverings that enable them to move relative to one other. From the fascia lata deep intermuscular septa project laterally and medially in the direction of the linea aspera. The *lateral intermuscular septum* (**7**) is relatively broad and provides an origin for several muscles. It divides the vastus lateralis (**8**) from the short head of the biceps femoris (**9**). The *medial intermuscular septum* (**10**) separates the vastus medialis (**11**) from the adductor canal (**12**).

On the anterior surface of the thigh below the inguinal ligament, in the region of the iliopectineal fossa that is covered superficially by the fascia lata, there is in the latter a porous area occupied by the **cribriform fascia**. This is pierced by vessels and nerves. Removal of this loose fascia reveals the **saphenous hiatus** (**13**), whose lateral edge, *the falciform margin*, called also *Hey's* or *Burns' ligament* (**14**), forms a sharply defined border. The falciform margin extends medially with a *superior horn* (**15**, called also the *Scarpa ligament*) and an *inferior horn* (**16**).

The femoral canal and femoral hernias are described on page 100.

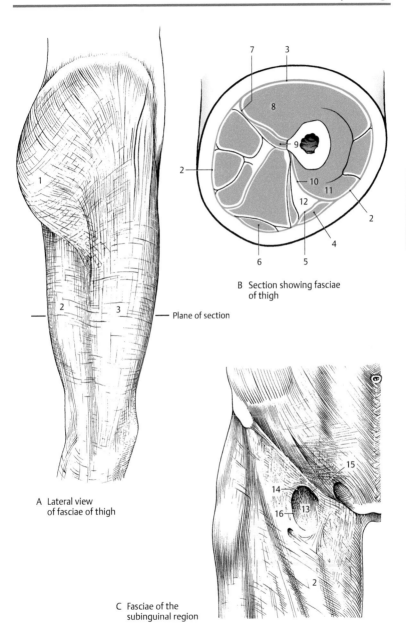

Lower Limb

A Lateral view
of fasciae of thigh

B Section showing fasciae
of thigh

C Fasciae of the
subinguinal region

Plane of section

Long Muscles of the Lower Leg and Foot

Classification of the Muscles (A–D)

All but one of the muscles that arise in the lower leg are attached to the bones of the foot. The only exception is the popliteus, which inserts on the tibia and must be classified with the thigh muscles. The muscles of the lower leg can be classified according to their location, principally into anterior and posterior groups. They are separated by the tibia and fibula and the interosseous membrane.

The two main groups are divided in turn into subgroups or layers. The anterior muscle group consists of the anterior extensors and the lateral subdivision of the peroneal group. The flexors on the back of the lower leg are subdivided into the superficial or calf muscles and the deep muscles.

Functionally the lower leg muscles can be subdivided into the extensors, lying on the anterior surface and responsible for dorsiflexion of the foot, and the flexors, which lie posteriorly and produce plantar flexion of the foot.

On the basis of their innervation, however, the muscles may be divided into those which receive nerves from the dorsal division of the plexus and those which are supplied by the ventral division.

For practical purposes the muscles of the lower leg, like those of the forearm, are best described according to their location.

Anterior Muscles of the Lower Leg

Extensor Group (see p. 258)

– Tibialis anterior (**1**)
– Extensor digitorum longus (**2**)
– Extensor hallucis longus (**3**)

Fibular (Peroneal) Group (see p. 260)

– Peroneus longus (**4**)
– Peroneus brevis (**5**)

Posterior Muscles of the Lower Leg

Superficial Layer (see p. 262)

– Triceps surae (**6**; with Achilles tendon) consisting of
– Soleus (**7**)
– Gastrocnemius (**8**)
– Plantaris (**9**)

Deep Layer (see p. 264)

– Tibialis posterior (**10**)
– Flexor hallucis longus (**11**)
– Flexor digitorum longus (**12**)

13 Popliteus
14 Semimembranosus
15 Sartorius
16 Gracilis
17 Semitendinosus
18 Popliteal artery and vein
19 Tibial nerve
20 Common peroneal (fibular) nerve
21 Long saphenous vein
22 Short saphenous vein
23 Saphenous nerve
24 Superficial peroneal (fibular) nerve
25 Deep peroneal (fibular) nerve
26 Lateral sural cutaneous nerve
27 Sural nerve
28 Peroneal (fibular) artery and vein
29 Anterior tibial artery and vein
30 Posterior tibial artery and vein
31 Tibia
32 Fibula

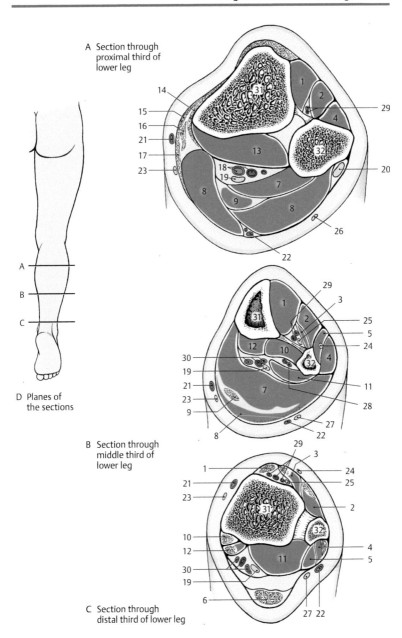

A Section through proximal third of lower leg

D Planes of the sections

B Section through middle third of lower leg

C Section through distal third of lower leg

Lower Limb

Anterior Lower Leg Muscles

Extensor Group (A–C)

The **tibialis anterior** (1) *arises from a wide area* (2) *of the lateral surface of the tibia, the interosseous membrane, and the crural fascia.* Its three-sided belly ends in a tendon that extends beneath the superior extensor retinaculum (3) and the inferior extensor retinaculum (4) surrounded by a synovial sheath. *It is inserted in the plantar surface of the medial cuneiform bone* (5) *and the first metatarsal* (6). The subtendinous bursa of the tibialis anterior lies between its tendon and the medial cuneiform.

When the leg is nonweight-bearing, the tibialis anterior dorsiflexes the foot while also raising the medial edge of the foot (supination). When the leg is bearing weight, the tibialis anterior moves the lower leg closer to the dorsum of the foot as, for example, in rapid walking, or in skiing. A slight participation in pronation has also been described.
Nerve supply: deep peroneal (fibular) nerve (L4–L5).

> **Clinical tip:** Under great stress the tibialis anterior may become fatigued, resulting in pain along the muscle.

The **extensor digitorum longus** (7) *arises from a large area* (8), namely from *the lateral condyle of the tibia, the head and anterior crest of the fibula, the deep fascia of the leg, and the interosseous membrane.* At the level of the ankle, the tendon in which the muscle ends is divided into four parts and extends to the second to fifth digits.

These tendons are enclosed in a common synovial sheath and run under the superior extensor retinaculum (3) and the inferior extensor retinaculum (4), lateral to the tendon of the tibialis anterior. They extend over the dorsum of the foot *into the dorsal aponeuroses of the second to fifth digits.*

In the nonweight-bearing leg, the muscle produces dorsiflexion of the digits and the foot. In the weight-bearing leg its function is the same as that of the tibialis anterior.
Nerve supply: deep peroneal (fibular) nerve (L5–S1).

Variants: The extensor digitorum longus may have an additional tendon that extends to the base of the fifth metatarsal and sometimes also to the base of the fourth metatarsal. This additional tendon is called the **peroneus tertius** (9), and as part of the extensor digitorum longus it may have a separate origin from the distal third of the anterior edge of the fibula. It acts as a pronator and abductor of the subtalar and talocalcaneonavicular joints.

The **extensor hallucis longus** (10) *arises from the medial surface of the fibula and the interosseous membrane* (11). It continues as a tendon that runs in its own synovial sheath between the sheath for the tendon of the tibialis anterior and that for the extensor digitorum longus, beneath the superior extensor retinaculum (3) and inferior extensor retinaculum (4). It stretches across the first metatarsal to the dorsal aponeurosis of the great digit and *is inserted into the distal phalanx* (12). The extensor hallucis longus dorsiflexes the great toe and in the nonweight-bearing leg it aids dorsiflexion of the foot. In the weight-bearing leg its function resembles that of the tibialis anterior, since it moves the leg closer to the dorsum of the foot. To a small extent it also aids in pronation and supination of the foot.
Nerve supply: deep peroneal (fibular) nerve (L4–S1).

Variants: An independent muscle or tendon bundle can frequently be split off and attached to the first metatarsal or in the region of the metatarsophalangeal joint as the **extensor hallucis accessorius** (13). This muscle is found primarily on the medial side of the main tendon.

14 Tibia
15 Fibula

Lower Limb

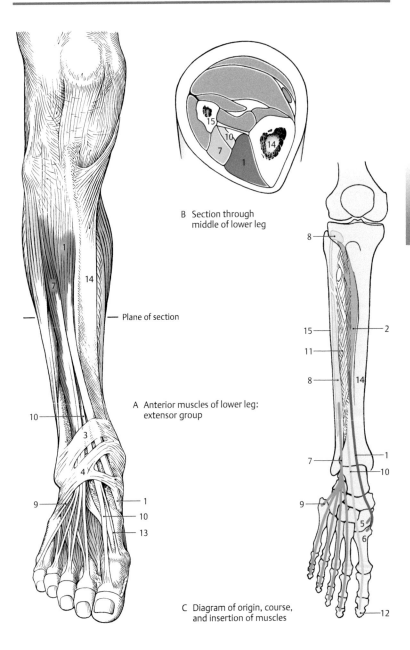

B Section through middle of lower leg

Plane of section

A Anterior muscles of lower leg: extensor group

C Diagram of origin, course, and insertion of muscles

Anterior Lower Leg Muscles, continued

Fibular (Peroneal) Group (A–D)

The peroneal muscles act as plantar flexors, a function they attained only secondarily, due to their displacement behind the lateral malleolus. Originally they lay in front of the malleolus, as can still be seen in predators.

The **fibularis (peroneus) longus** (**1**) *arises* (**2**) *from the capsule of the tibiofibular joint, the head of the fibula, and the proximal fibula.*

It ends in a long tendon that runs in the malleolar groove behind the lateral malleolus in a common synovial sheath with the *peroneus brevis tendon* (**3**) under the *superior peroneal retinaculum* (**4**). The peroneus longus tendon *extends distally* from the peroneal trochlea of the calcaneus in an evagination of the common synovial sheath (held in place by the *inferior peroneal retinaculum* [**5**]), *across the plantar surface to the tuberosity of the first metatarsal* (**6**) *and the medial cuneiform* (**7**). Its tendon reaches its site of insertion by coursing through the tendon groove of the *cuboid* (**8**) in a special fibrous canal, which runs from the lateral side behind the tuberosity of the fifth metatarsal obliquely to the medial border of the foot. Within this canal, on the sole of the foot, another synovial sheath encloses the tendon.

Due to this course its function is similar to that of a bow string (*Kummer*) and it braces the transverse arch of the foot. It depresses the medial border of the foot and, together with the peroneus brevis, it is the strongest pronator. It also aids in plantar flexion.
Nerve supply: superficial peroneal (fibular) nerve (L5–S1).

The **fibularis (peroneus) brevis** (**3**) *arises from the lateral surface of the fibula* (**9**). Its tendon, together with that of the peroneus longus, runs in a synovial sheath in the groove for the peroneus longus tendon, beneath the superior peroneal retinaculum (**4**). On the lateral surface of the calcaneus, the tendon becomes fixed proximally, that is, above the peroneal trochlea of the calcaneus, by the inferior peroneal retinaculum (**5**) where an evagination of the common synovial sheath surrounds the tendon. *This is attached to the tuberosity of the fifth metatarsal* (**10**). The muscle has the same action as the peroneus longus.
Nerve supply: superficial peroneal (fibular) nerve (L5–S1).

■■ **Variants:** The **peroneus quartus** is rarely present. It arises from the fibula and is attached to the lateral surface of the calcaneus or to the cuboid. It is closely associated with the tendons of extensor digitorum longus. It may also send a small tendon to the fifth digit.

11 Tibia
12 Fibula
13 Soleus
14 Gastrocnemius
15 Interosseous membrane

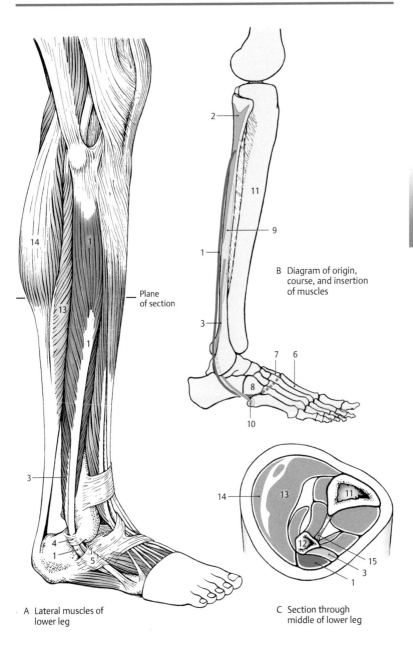

Plane
of section

B Diagram of origin,
course, and insertion
of muscles

A Lateral muscles of
lower leg

C Section through
middle of lower leg

Posterior Lower Leg Muscles

Superficial Layer (A–D)

The superficial layer of muscles is formed by the **triceps surae**, consisting of the **soleus** (**1**) and the **gastrocnemius** (**2**) with its medial and lateral heads. The *plantaris* (**3**) is also part of the superficial layer of muscles.

The **soleus** *arises from the head and upper third of the posterior surface of the fibula* (**4**), *from the line of the soleus muscle on the tibia* (**5**), and from the tendinous arch between the head of the fibula and the tibia, that is, the *tendinous arch of soleus* that lies distal to the popliteus (**6**). The large terminal tendon of the muscle joins the terminal tendon of the gastrocnemius and *is inserted into the tuber calcanei* (**8**) *as the Achilles (calcaneal) tendon* (**7**). Between the proximal surface of the tuber calcanei and this tendon lies the bursa of the Achilles tendon.

The **gastrocnemius** (**2**) *arises proximally to the medial femoral condyle* (**10**) with a **medial head** (**9**) and with a **lateral head** (**11**) *proximal to the lateral femoral condyle* (**12**). Some of the fibers from both heads also arise from the capsule of the knee joint. The two heads run distalward, forming the inferior borders of the popliteal fossa, and join the tendon of the soleus; they are *inserted into the tuber calcanei* (**8**).

The **plantaris** (**3**) is a slight, delicate muscle with a very long terminal tendon. *It arises* in the region of the lateral head of the gastrocnemius *proximally to the lateral femoral condyle and from the capsule of the knee joint*. Its tendon runs distally between the gastrocnemius and soleus and *adjoins the medial border of the Achilles tendon*.

Nerve supply: the tibial nerve (S1–S2) supplies all the muscles.

■■ Variant: The plantaris may be absent in 5 to 10% of cases.

The **triceps surae** is the cardinal muscle of plantar flexion. It can lift the weight of the body both in standing and walking. Its strength is most obvious in ballet dancing, which requires maximal plantar flexion. Full activity of the triceps surae is only possible with the knee extended, because when the knee is flexed the gastrocnemius is already shortened. Therefore, the gastrocnemius is particularly important in walking as it is involved not only in lifting the heel but also in flexing the knee. In this action it receives some assistance from the plantaris.

The triceps surae is also considered to be the strongest supinator of the subtalar joint.

> **Clinical tip: Achilles tendon tears** may occur in response to a high transistent stress. The most vulnerable people are those who are athletically unconditioned and who suddenly put stress on the tendon without any preliminary training. Most patients have a prior history of tendon injury, however.
> In approximately 10% of the population (mostly women), a small bean-shaped sesamoid bone called the **fabella** is embedded in the lateral head of the gastrocnemius. This bone may articulate with the lateral femoral condyle. It is projected behind and proximal to the joint space of the knee in lateral radiographs.
> A **painful fabella** causing localized pain or tenderness should be surgically removed.

13 Flexor digitorum longus
14 Flexor hallucis longus
15 Tibialis posterior
16 Interosseous membrane
17 Tibia
18 Fibula

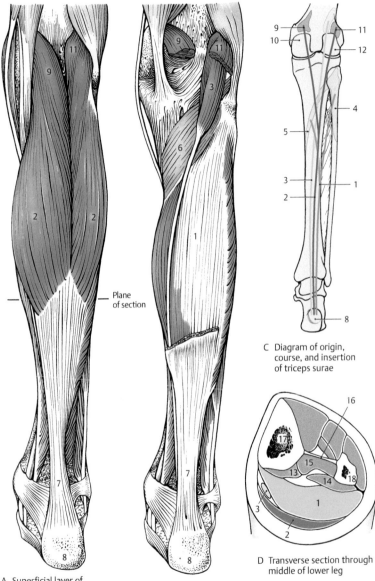

Plane of section

C Diagram of origin, course, and insertion of triceps surae

D Transverse section through middle of lower leg

A Superficial layer of posterior lower leg muscles: triceps surae muscle

B Soleus (with gastrocnemius removed)

Lower Limb

Posterior Lower Leg Muscles, continued

Deep Layer (A–E)

The **tibialis posterior** (**1**) *arises from the interosseous membrane* (**2**) *and the adjoining surfaces of the tibia* (**3**) *and fibula* (**4**). The tendon (**5**) runs downward in the malleolar groove behind the medial malleolus (**6**) in a synovial sheath between the sustentaculum tail and the navicular tuberosity and reaches the sole of the foot. *It divides into* **two parts**. The **thicker, medial part** (**7**) *is attached to the navicular tuberosity, while* the **lateral**, somewhat **weaker part** (**8**) *is inserted into the three cuneiform bones.* In the nonweight-bearing leg the tibialis posterior produces plantar flexion and simultaneous supination. In the weight-bearing leg it approximates the heel to the calf of the leg.
Nerve supply: tibial nerve (L4–L5).

■ **Variants:** The insertion of the muscle often extends also to the base of the second, third, and fourth metatarsals and the cuboid bone. Occasionally the muscle is absent.

The **flexor hallucis longus** (**9**) *arises from the distal two-thirds of the posterior surface of the fibula* (**10**), *the interosseous membrane* (**11**), *and the posterior crural intermuscular septum* (**12**). Its relatively thick muscle belly extends far distally and terminates in a tendon that lies in the groove for the tendon of the flexor hallucis longus in the talus and calcaneus, where it is invested by a synovial sheath. It extends beneath the flexor retinaculum (**13**) to the sole of the foot where *it inserted into the base of the distal phalanx of the first digit* (**14**). Distal to the sustentaculum tali it is crossed superficially by the tendon of the flexor digitorum longus. The flexor digitorum longus opposes development of a pes planovalgus by supporting the arch of the foot. It produces plantar flexion of the first digit and in some cases also of the others. It assists in supination.
Nerve supply: tibial nerve (S1–S3).

■ **Variants:** It may also give off terminal tendons to the second and third digits.

The **flexor digitorum longus** (**15**) *arises from the posterior surface of the tibia* (**16**), and its tendon (**17**) runs in a synovial sheath beneath the flexor retinaculum (**13**) to the sole of the foot. In the leg it posteriorly crosses the tibialis posterior and on the sole of the foot it superficially crosses the tendon of the flexor hallucis longus. In the sole of the foot the tendon divides into *four terminal tendons that extend to the distal phalanges* (**18**) of the lateral four digits. Distal to this division, the quadratus plantae radiates into it (see p. 274). In the region of the middle phalanges its terminal tendons penetrate the tendons of the flexor digitorum brevis. In the nonweight-bearing leg it plantar flexes the digits and then the foot. It also acts as a supinator. In the weight-bearing limb it assists in the support of the plantar arch.
Nerve supply: tibial nerve (S1–S3).

The **popliteus** (**19**; see also p. 232) *arises from the lateral femoral epicondyle* (**20**). *It inserts on the posterior tibial surface* (**21**). Between the muscle and the knee joint is the subpopliteal recess, which always communicates with the joint. The popliteus flexes the knee joint and rotates the leg internally.
Nerve supply: tibial nerve (L4–S1).

22 Gastrocnemius
23 Soleus
24 Plantaris

Figure **A**: Arrow is in the canal formed by the tendinous arch of the soleus muscle, which is traversed by the tibial nerve and the posterior tibial vessels.

In Figure **B** the flexor digitorum muscle and parts of the soleus origin have been removed.

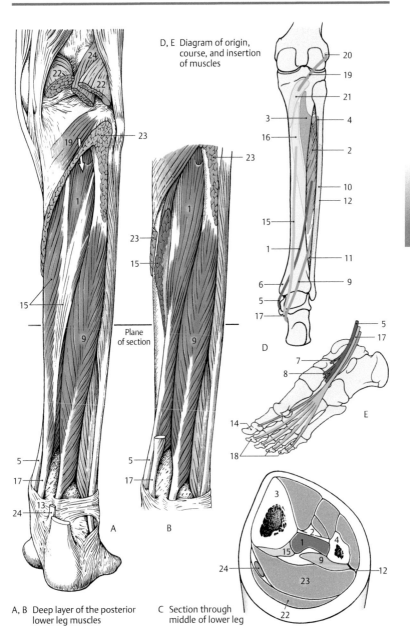

D, E Diagram of origin, course, and insertion of muscles

Plane of section

Lower Limb

D

E

A, B Deep layer of the posterior lower leg muscles

C Section through middle of lower leg

Function of the Ankle, Subtalar, and Talocalcaneonavicular Joint Muscles (A–D)

All these muscles act on multiple joints, but only their actions on the talocrural, subtalar, and talocalcaneonavicular joints will be described.

Dorsiflexion (extension) and **plantar flexion** (flexion) occur around the **transverse axis** of the talocrural (ankle) joint (see p. 222), which runs through the tip of the medial malleolus and the lateral malleolus.

Pronation, or eversion (elevation of the lateral border of the foot), and **supination**, or inversion (elevation of the medial border of the foot), occur around the **oblique axis** of the subtalar joint. This axis runs upward extending outward from the back and below and inward toward the front.

Dorsiflexion (A) is produced by
 – Tibialis anterior (red, deep peroneal nerve)
 – Extensor digitorum longus (blue, deep peroneal nerve)
 – Extensor hallucis longus (yellow, deep peroneal nerve)

Plantar flexion (B) is produced by
 – Triceps surae (red, tibial nerve)
 – Peroneus longus (blue, superficial peroneal nerve)
 – Peroneus brevis (yellow, superficial peroneal nerve)
 – Flexor digitorum longus (green, tibial nerve)
 – Tibialis posterior (brown, tibial nerve)

The triceps surae is the most important muscle in plantar flexion, whereas the remaining muscles contribute only a very slight action.

Pronation (C) is produced by
 – Peroneus longus (red, superficial peroneal nerve)
 – Peroneus brevis (blue, superficial peroneal nerve)
 – Extensor digitorum longus (yellow, deep peroneal nerve)
 – Peroneus tertius (orange, deep peroneal nerve)

Supination (D) is brought about by
 – Triceps surae (red, tibial nerve)
 – Tibialis posterior (blue, tibial nerve)
 – Flexor hallucis longus (yellow, tibial nerve)
 – Flexor digitorum longus (orange, tibial nerve)
 – Tibialis anterior (green, deep peroneal nerve)

The color of the arrows show the order of importance of the muscles in each movement:

red
blue
yellow
orange
green
brown

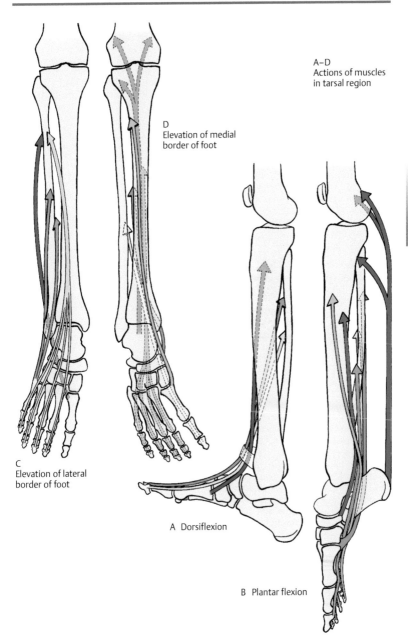

A–D
Actions of muscles
in tarsal region

D
Elevation of medial
border of foot

C
Elevation of lateral
border of foot

A Dorsiflexion

B Plantar flexion

Intrinsic Muscles of the Foot

As in the hand, only the tendons of the extrinsic muscles of the foot extend into the foot; the muscle bellies of these tendons are in the lower leg. In addition to these tendons there are the intrinsic muscles of the foot, which lie either on the dorsum or the sole of the foot. Apart from this topographical classification, the intrinsic muscles may be classified according to their innervation, the muscles of the dorsum of the foot being innervated by the dorsal division of the plexus and those of the sole of the foot by the ventral division. Like the muscles of the hand, the muscles of the sole of the foot may be divided into three groups: those of the middle plantar eminence and those which form the medial plantar eminence.

Muscles of the Dorsum of the Foot (A–C)

The tendons of **extensor digitorum longus** (**1**; see p. 258) and **extensor hallucis longus** (**2**; see p. 258) lie superficial to the intrinsic muscles of the dorsum of the foot. They are held in place by the *superior extensor retinaculum* (**3**; see p. 276) and the *inferior extensor retinaculum* (**4**; see p. 276). The tendons of the long extensors form the **dorsal aponeurosis** of the toes, into which the short extensors of the digits and the *plantar and dorsal interossei* also radiate (**5**; see p. 274).
Nerve supply: deep peroneal nerve (L5–S1).

The **extensor digitorum brevis** (**6**) *arises from the calcaneus* (**7**), near the entrance to the sinus tarsi, and *from one side of the inferior extensor retinaculum* (**4**). *It extends with three tendons to the dorsal aponeurosis* (**8**) *of the second to fourth digits.* It is responsible for dorsiflexion of these digits.
Nerve supply: deep peroneal (fibular) nerve (S1–S2).

■■ **Variants:** Individual tendons may be absent. The tendon for the fifth toe is only occasionally present.

The **extensor hallucis brevis** (**9**), *which extends into the dorsal aponeurosis of the first digit, splits off from the extensor digitorum brevis, with which it has a common origin from the calcaneus.* Like the latter muscle it serves to dorsiflex the first digit.
Nerve supply: deep peroneal (fibular) nerve (S1–S2).

10 Tibialis anterior
11 Peroneus tertius

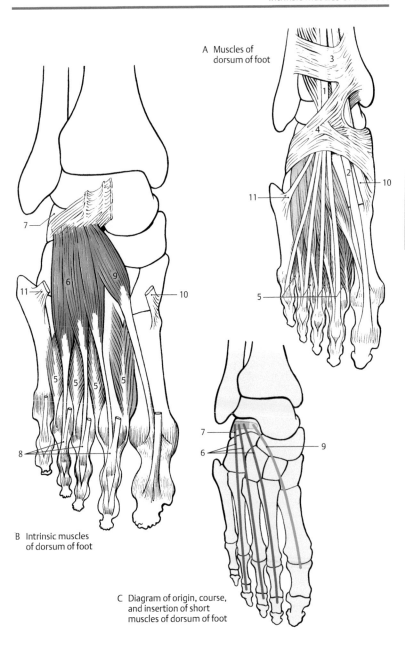

A Muscles of dorsum of foot

B Intrinsic muscles of dorsum of foot

C Diagram of origin, course, and insertion of short muscles of dorsum of foot

Muscles of the Sole of the Foot (A–C)

Three muscles groups may be distinguished in the sole of the foot—the muscles in the regions of
- **The great toe**
- **The little toe**
- **The middle region**

The abductor hallucis and the flexor hallucis brevis belong to the region of the big toe. In a broader sense it also includes the adductor hallucis, which originally formed a separate system.

The abductor digiti minimi, flexor digiti minimi brevis, and opponens digiti minimi belong to the region of the little toe.

The middle muscle group consists of the lumbricales, quadratus plantae, interossei, and flexor digitorum brevis.

All the plantar muscles of the foot are covered by the dense and strong **plantar aponeurosis** (**1**), which is derived from the superficial fascia. The plantar aponeurosis consists of **longitudinal fiber bundles** (**2**), which arise from the tuber calcanei and radiate into the digits. **Transverse fascicles** (**3**) interconnect these longitudinal fiber bundles. On the medial and lateral borders of the foot, the plantar aponeurosis merges into the thin dorsal fascia of the foot. Two tough septa extend deeply from the surfaces as the *medial* and *lateral plantar septa* (**4**). The former is attached to the first metatarsal, the medial cuneiform, and the navicular, and the latter to the fifth metatarsal and the long plantar ligament.

The three connective tissue spaces formed by these septa and the plantar aponeurosis each contain the three muscle groups referred to above, and fatty tissue. These cushions, formed by the muscles and fat, transmit the weight of the body to the underlying substrate. *The plantar aponeurosis, septa, muscles, fatty tissues, and skeleton of the foot form a functional unit.* Thus, the plantar aponeurosis makes an important contribution to maintenance of the longitudinal arch (see p. 226). In addition, the plantar aponeurosis acts to protect the vessels and nerves from pressure injuries.

Muscles of the Great Toe

The **abductor hallucis** (**5**) *arises from the medial process of the tuber calcanei* (**6**), *from the flexor retinaculum, and from the plantar aponeurosis* (**7**). Its origin makes a tendon arch beneath which the tendons of the long flexors of the digits run in the tarsal canal. *The muscle is inserted into the medial sesamoid bone* (**8**) *and the base of the proximal phalanx* (**9**). There is usually a synovial bursa between its tendon of insertion and the metatarsophalangeal joint. It acts as an abductor and a weak flexor and helps to maintain the arch of the foot.
Nerve supply: medial plantar nerve (L5–S1).

The **flexor hallucis brevis** (**10**) *arises from the medial cuneiform* (**11**), *the long plantar ligament, and the tibialis posterior tendon.* It has **two heads**; the **medial head** (**12**) is combined with the abductor hallucis and *extends to the medial sesamoid bone* (**13**) *and the proximal phalanx* (**14**), while the **lateral head** (**15**) joins the adductor hallucis and *is inserted into the lateral sesamoid bone* (**16**) *and the proximal phalanx* (**17**). It is an important plantar flexor and is needed particularly in ballet dancing.
Nerve supply: medial plantar nerve (L5–S1).

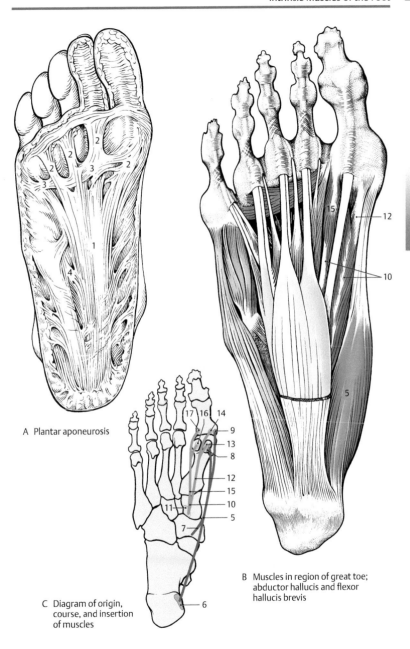

A Plantar aponeurosis

C Diagram of origin,
   course, and insertion
   of muscles

B Muscles in region of great toe;
   abductor hallucis and flexor
   hallucis brevis

Muscles of the Sole of the Foot, continued

Muscles of the Great Toe, continued (A–C)

The **adductor hallucis** (**1**) has **two heads**. It is visible only after the flexor digitorum longus and flexor digitorum brevis (**2**) have been removed (**A**). The strong **oblique head** (**3**) *arises from the cuboid* (**4**) *and lateral cuneiform* (**5**) *bones, and from the bases of the second and third metatarsals* (**6**). Other surfaces of origin may include the fourth metatarsal, the plantar calcaneocuboidal ligament, the long plantar ligament (**7**), and the plantar tendon sheath (**8**) of the peroneus longus. The **transverse head** (**9**) *arises from the capsular ligaments of the metatarsophalangeal joints of the third to fifth digits* (**10**) *and also from the deep transverse metatarsal ligament.* **Both heads** *are inserted into the lateral sesamoid bone* (**11**) *of the great toe.* The muscle acts especially as a tensor of the plantar arches. In addition it adducts the great toe and may then plantar flex the proximal phalanx.
Nerve supply: deep branch of the lateral plantar nerve (S1–S2).

Intrinsic Muscles of the Little Toe (A–C)

The **opponens digiti minimi** (**12**) *arises from the long plantar ligament* (**7**) *and from the plantar tendon sheath of the peroneus longus* (**13**). *It is inserted into the fifth metatarsal* (**14**). Its functions are to plantar flex the fifth metatarsal and support the plantar arch. It is quite often absent.
Nerve supply: lateral plantar nerve (S1–S2).

The **flexor digiti minimi** (**15**) *arises from the base of the fifth metatarsal* (**16**), *from the long plantar ligament* (**7**), *and from the plantar tendon sheath of the peroneus longus. It extends to the base of the proximal phalanx* (**17**) *of the fifth digit and usually merges with the abductor digiti minimi.* It acts as a plantar flexor of the little toe.
Nerve supply: lateral plantar nerve (S1–S2).

The **abductor digiti minimi** (**18**) is the largest and longest of the muscles of the little toe. In the main it actually forms the lateral border of the foot. *It arises from the lateral process of the tuber calcanei* (**19**), *the inferior surface of the calcaneus* (**20**), *the tuberosity of the fifth metatarsal* (**21**), *and the plantar aponeurosis and extends to the proximal phalanx* (**22**) *of the fifth digit.* Like the other muscles it supports the arch of the foot. In addition it plantar flexes the fifth digit and, to a small extent, it acts also as an abductor.
Nerve supply: lateral plantar nerve (S1–S2).

23 Quadratus plantae

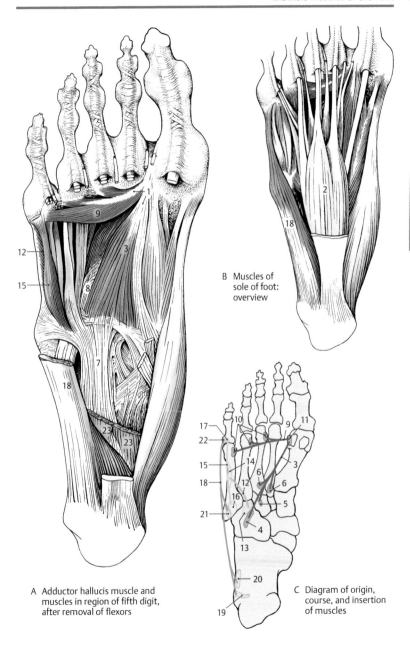

Lower Limb

B Muscles of
sole of foot:
overview

A Adductor hallucis muscle and
muscles in region of fifth digit,
after removal of flexors

C Diagram of origin,
course, and insertion
of muscles

Muscles of the Sole of the Foot, continued

Intrinsic Muscles at the Center of the Sole of the Foot (A–C)

The **four lumbricales** (**1**) *arise from the medial surfaces of the individual tendons* (**2**) *of the flexor digitorum longus. They extend to the medial margin of the proximal phalanges of the second to fifth digits and blend with the extensor (dorsal) aponeurosis.* The muscles are involved in plantar flexion and moving the four lateral digits toward the great toe. They also help to reinforce the plantar arch.

Nerve supply: medial plantar nerve to the first, second, and third lumbricales, and lateral plantar nerve to the fourth lumbricalis (L5–S2).

◾ **Variants:** In contrast to the lumbricales of the hand, those of the foot are quite variable. They may be absent or there may be more than four. They are inserted on the metatarsophalangeal joint capsules and on the proximal phalanges.

The **quadratus plantae** (**3**) is also known as the plantar head of the flexor digitorum longus (flexor accessorius). *It arises by two slips from the medial and lateral margins of the plantar surface of the calcaneus and projects into the lateral margin of the flexor digitorum longus tendon* (**4**).

Nerve supply: lateral plantar nerve (S1–S2).

◾ **Variants:** It may extend into the common tendon of the flexor digitorum longus or into the four divisions of that tendon, in which case it only extends to the two lateral tendons.

The **interossei** may be divided into **plantar** (**5**; blue) and **dorsal** (**6**; red) parts. They are arranged with respect to the second digit as the longitudinal axis of the foot.

The **three plantar interossei** each arise by a **single head** *from the medial side of the third to fifth metatarsals* (**7**) *and may receive additional fibers from the long plantar ligament. They extend to the medial side of the*

base of the proximal phalanx of the third to fifth digits (**8**).

The **four dorsal interossei** arise by **two heads** *from the opposing surfaces of all the metatarsals* (**9**) *and from the long plantar ligament. They are attached to the bases of the proximal phalanges of the second through fourth digits* (**10**).

The plantar interossei act as adductors and pull the third, fourth, and fifth digits toward the second digit. The dorsal interossei are abductors. The first and second are inserted into the proximal phalanx of the second digit, and the third and fourth are inserted into the proximal phalanx of the third and fourth digits.

In contrast to the interossei of the hand they usually do not reach the extensor aponeurosis. In addition to their functions as abductors and adductors they work together as plantar flexors of the metatarsophalangeal joints.

Nerve supply: deep branch of the lateral plantar nerve (S1–S2).

The **flexor digitorum brevis** (**11**) *arises from the undersurface of the tuber calcanei and from the proximal part of the plantar aponeurosis. Its tendons, which are inserted into the middle phalanx of the second through fourth digits, are divided near their termini* (**12**). The tendons of the flexor digitorum longus (**2**) run between these divided tendons. Thus, the flexor digitorum brevis is also called the perforatus. In this region the tendons together with the tendons of the flexor digitorum longus are surrounded by a tendon sheath. This muscle plantar flexes the middle phalanges.

Nerve supply: medial plantar nerve (L5–S1).

◾ **Variants:** The tendon to the fifth digit (little toe) is often absent. In some cases the entire muscle may be absent.

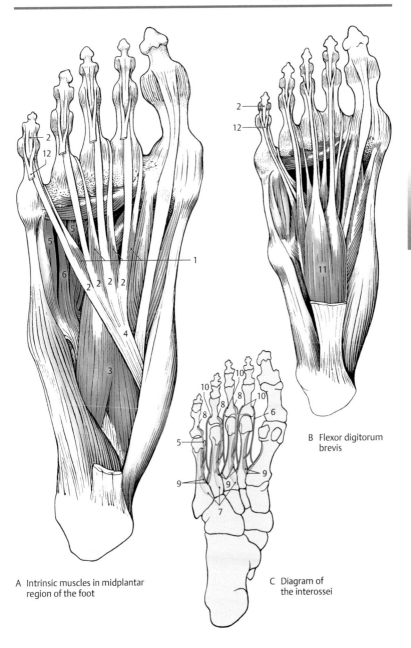

Lower Limb

A Intrinsic muscles in midplantar region of the foot

B Flexor digitorum brevis

C Diagram of the interossei

Fasciae of the Lower Leg and the Foot (A–D)

The superficial fascia of the lower leg, the **crural fascia** (**1**), is the continuation of the fascia lata and its special popliteal fascia. It encloses the superficial muscle layers of the leg. Strengthening fibers are interwoven into the crural fascia and delineate certain particular features. Thus over the extensors in the distal anterior part of the leg there are transverse strengthening fibers, forming the *superior extensor retinaculum* (**2**), and in the tarsal region on the dorsum of the foot as the *inferior extensor retinaculum* (**3**), which are visible due to reinforcing fibers within the fascia. The retinacula can be demonstrated with care in the fascia.

On the lateral side there is an intermuscular septum, both in front of and behind the peroneal muscles, which extends from the crural fascia deeply to the fibula. These are the *anterior* (**4**) and *posterior* (**5**) *intermuscular septa of the leg*. At the distal end, in the region of the lateral malleolus, strong fiber tracts are woven into the fascia and form the *superior* and *inferior peroneal retinacula* (**6**). Both can only be demonstrated by dissection.

The fascia over the dorsal crural muscles is thin. It is only strengthened distally, so that between the medial malleolus and the calcaneus there is a dense fibrous structure, the *flexor retinaculum* (**7**), the superficial layer of which serves as the boundary of the tendons of the deep muscles of the tibia.

The musculature of the calf may be divided into a superficial and a deep layer of muscles. Between the two groups lies the **deep fascia of the leg** (**8**), which arises proximal to the *tendinous arch of the soleus*. Part of the soleus also arises from it. At the distal end it has thicker fibers, and these form the *deep layer of the flexor retinaculum* on the medial side, and on the lateral side they contribute to the *superior peroneal retinaculum*. The four different muscle groups in the leg are separated in this way by these connective tissue layers and the interosseous membrane.

On the dorsum of the foot, the superficial **dorsal fascia of the foot** (**9**) lies distal to the *inferior extensor retinaculum* (**3**). It is very delicate and thin. It forms the immediate continuation of the crural fascia and extends distalward into the extensor aponeurosis of the digits. Laterally it is attached to the sides of the foot. Proximally, at the attachments of the superior extensor retinaculum, it forms the cross-shaped *inferior extensor retinaculum*, which, however, can be demonstrated only by careful dissection, and in which laterally the proximal crus is often absent. In this case these reinforcing fiber bundles within the fascia appear γ-shaped.

Deep to the tendons of the extensor digitorum longus is a connective tissue layer, the **deep dorsal fascia of foot**, which is dense and tight and is also attached to the borders of the foot.

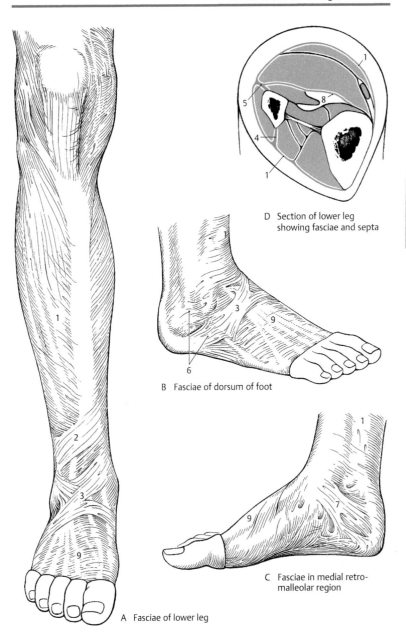

D Section of lower leg showing fasciae and septa

B Fasciae of dorsum of foot

C Fasciae in medial retro-malleolar region

A Fasciae of lower leg

Tendon Sheaths in the Foot (A–C)

As in the hand, various **tendon sheaths** are distinguished in the foot.

On the **dorsum of the foot**, **tendon sheaths** are found for the tendons of
- Tibialis anterior (**1**)
- Extensor hallucis longus (**2**)
- Extensor digitorum longus (**3**)
- Fibularis (peroneus) tertius (when present)

The tendons or tendon sheaths in this area are held in place by the *superior* (**4**) and *inferior* (**5**) *extensor retinacula*. On the lateral side of the tarsals in the region of the peroneal trochlea of the calcaneus is found the *common peroneal tendon sheath* of the *peroneus* (**6**). The peroneus longus tendon (**7**) leaves this common synovial sheath and continues across the plantar region within its own sheath, the *plantar tendon sheath* of the *peroneus longus*. The common tendon sheath for the peroneal muscles is held in place laterally by the *superior* (**8**) and *inferior* (**9**) *peroneal retinacula*.

The flexor tendons lie on the medial side of the foot directly behind the medial malleolus. Their tendon sheaths course below the *flexor retinaculum of the foot*, which comprises a *superficial layer* (**10**) of reinforced crural fascia, and a *deep layer* (**11**). Below this deep layer pass three tendons each of which is enclosed within its own synovial sheath—those of the tibialis posterior (tendon sheath of tibialis posterior, **12**), flexor digitorum longus (tendon sheath of flexor digitorum longus, **13**), and flexor hallucis longus (tendon sheath of flexor hallucis longus, **14**) muscles (see also p. 436).

On the **plantar aspect of the foot** are found **five tendon sheaths** corresponding to the individual toes (**15**). As a rule, these **synovial sheaths** do not communicate with each other and are strengthened by stout **fibrous sheaths** (**16**), each of which consists of an annular and a cruciform part. The *annular part* (**17**) consists of circular bundles of fibers and is located in the region of a joint. The *cruciform part* (**18**) is found between the joints and is composed of decussating connective tissue fibers. In contrast to the hand, no tendon sheaths are found in the middle compartment of the plantar surface of the foot. Only the two previously mentioned tendon sheaths, that of the flexor hallucis longus (**14**) and the flexor digitorum longus (**13**), extend into the metatarsal region.

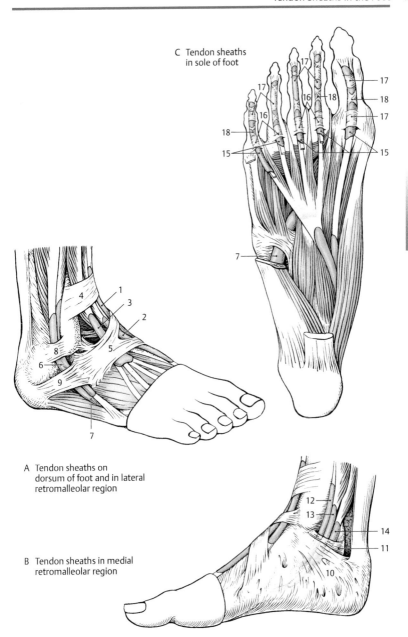

C Tendon sheaths
  in sole of foot

Lower Limb

A Tendon sheaths on
  dorsum of foot and in lateral
  retromalleolar region

B Tendon sheaths in medial
  retromalleolar region

Anatomical Terms and their Latin Equivalents

Lower Limb	Membrum inferius
Acetabulum labrum	Limbus acetabuli
Acetabular notch	Incisura acetabuli
Ankle joint	Articulatio talocruralis
Anteromedial intermuscular septum	Septum intermusculare vastoadductorium
Bones of the toes	Ossa digitorum pedis
Free part of lower limb	Pars libera membri inferioris
Hip bone	Os coxae
Infrapatellar fat pad	Corpus adiposum infrapatellare
Intermediate zone	Linea intermedia
Knee joint	Articulatio genus
Lunate surface	Facies lunata
Obturator externus (internus)	M. obturatorius externus (internus)
Obturator groove	Sulcus obturatorius
Pecten pubis	Pecten ossis pubis
Pubic crest	Crista pubica
Sciatic nerve	N. ischiadicus
Symphyseal surface	Facies symphysialis
Tendon sheath	Vagina tendinum
Wing of ilium	Ala ossis ilii

Systematic Anatomy of the Locomotor System

Head and Neck

Skull (A, B)

The bony part of the head, the skull or **cranium**, forms the upper end of the trunk. It acts as the capsule for the brain and sensory organs, forms the substructure of the face, and also contains the initial portions of the gastrointestinal and respiratory tracts. The structure of the skull is differentiated in accordance with the variety of its tasks.

Subdivisions of the Skull

The skull consists of two parts:
- The **neurocranium**, or cranial vault
- The **viscerocranium**, or facial skeleton

The boundary between the two is at the level of the nasal root and extends along the upper margin of the orbits to the external auditory canals.

The shape of the skull is partly determined by the muscles, which may produce certain changes due to their functions, and in part by the contents of the skull. Thus, there is a correlation between the neurocranium and the brain contained within it. The influence here is reciprocal, since excessive expansion of the brain may produce enlargement of the neurocranium, as in hydrocephalus (see p. 310). On the other hand, premature cessation of neurocranial growth may result in malformation of the brain. There is not only a reciprocal effect within the neurocranium but also a close relationship to the facial skeleton. Thus the development of the muscles and of the supporting system of the dura mater within the skull capsule are also interrelated.

Ossification of the Skull

Fundamentally there are two developmental processes in the skull, distinguishable by the type of bone formation. One is the **chondrocranium** and the other the **desmocranium**. In the chondrocranium there is replacement bone formation, while in the desmocranium the individual bones develop as membranous bones directly from condensations in the connective tissue. Both types of development occur in the two functional parts (the neurocranium and viscerocranium). However, portions of either desmal or chondral origin may fuse together to form a single bone, as illustrated by the temporal bone.

The neurocranium (A; orange) consists of the occipital bone (**1**), sphenoid bone (**2**), squamous (**3**) and mastoid portion of the petrous (**4**) parts of the temporal bone, the parietal bones (**5**), and the frontal bone (**6**).

The **viscerocranium** (A; violet) is composed of the ethmoid bone (**7**), the inferior nasal turbinates, the lacrimal bones (**8**), the nasal bones (**9**), the vomer, the maxillae (**10**) with the incisive bone, the palatine bones, the zygomas (**11**), the tympanic parts (**12**) and the styloid processes (**13**) of the temporal bones, the mandible (**14**), and the hyoid bone.

Bones preformed in cartilage (B; blue) include the occipital bone (**1**; with the exception of the upper part of its squama, **15**), the sphenoid bone (**2**; with the exception of the medial lamella of the pterygoid process), the temporal bone with its petrous part (**4**) and the auditory ossicles, the ethmoid bone (**7**), the inferior nasal turbinate, and the hyoid bone.

The following bones are formed by **ossification in connective tissue** (B; green): the upper part of the occipital squama (**15**), the sphenoidal concha, the medial lamella of the pterygoid process, the tympanic part (**12**), the squamous part of the temporal bone (**3**), the parietal bone (**5**), the frontal bone (**6**), the lacrimal bone (**8**), the nasal bone (**9**), the vomer, the maxilla (**10**), palatine bone, the zygoma (**11**), and the mandible (**14**).

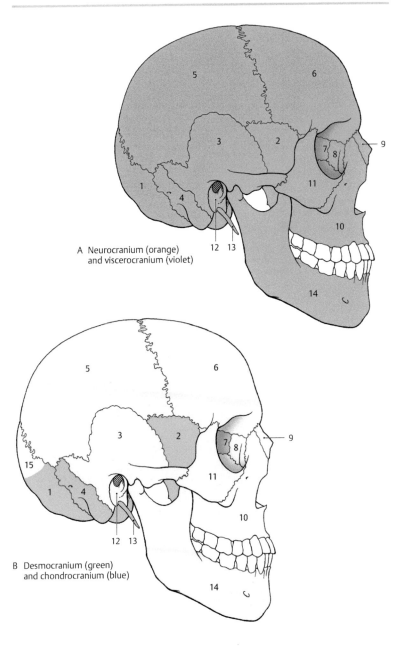

A Neurocranium (orange)
and viscerocranium (violet)

B Desmocranium (green)
and chondrocranium (blue)

Special Features of Intramembranous Ossification (A–D)

The calvaria develops in connective tissue and has several ossification centers from which bone formation radiates in all directions. In this way paired protuberances develop—*two frontal eminences* (**1**) and two *parietal eminences* (**2**). The bones develop from these eminences. At birth large connective tissue areas, the **fontanelles** or fonticuli, are still left between the individual bones.

The *anterior fontanelle* (**3**) is an unpaired opening sealed by connective tissue, which is almost square and at birth has a diagonal length of 2.5 to 3 cm. The smaller, unpaired *posterior fontanelle* (**4**) is also sealed by connective tissue and is triangular in shape. The anterior fontanelle lies between the two frontal bone anlages and both parietal anlages. The posterior fontanelle lies between the two parietal bone anlages and the anlage of the upper squama of the occipital bone. The paired fontanelles lie laterally, of which the *sphenoidal fontanelle* (**5**), closed by connective tissue, is the larger and should be distinguished from the small *mastoid fontanelle* (**6**), which is occluded by cartilage (corresponding to a synchondrosis). The sphenoidal fontanelle lies between the frontal, parietal, and sphenoid bones, and the mastoid fontanelle lies between the splenoid, temporal, and occipital bones.

The fontanelles close only after birth, the first being the posterior fontanelle in the 3rd month of life; the sphenoidal fontanelle follows in the 6th month, the mastoid fontanelle in the 18th month, and the anterior fontanelle in the 36th month.

> **Clinical tip:** In the newborn and in infants the anterior fontanelle can be used for taking blood samples from the dural sinuses. Venepuncture is also possible through the large anterior fontanelle.

Sutures and Synchondroses

The remnants of connective tissue between the cranial bones form cranial fibrous joints, cranial syndesmoses, the **sutures** (see p. 22), which permit continued growth of the bones. Only when the bones are completely fused as synostoses does growth cease.

Between some of the bones preformed in cartilage (chondrocranium) there are cranial cartilaginous joints, the **cranial synchondroses**. The *spheno-occipital synchondrosis*, which ossifies at around the 18th year of life, is of practical interest. In the region of the sphenoid body, the *intersphenoidal synchondrosis* is found, which ossifies early, while between the sphenoid and ethmoid bones is the *sphenoethmoidal synchondrosis*, which does not ossify until maturity. In addition, cartilage remnants are retained throughout life between the petrous part of the temporal bone and the adjacent bones, the *sphenopetrosal synchondrosis* and the *petro-occipital synchondrosis*.

Growth of the skull, as already stated, is dependent on the function and the contents of the skull. The neurocranium and viscerocranium do not grow at equal rates. Growth of the viscerocranium initially lags behind but becomes more rapid during the first years of life.

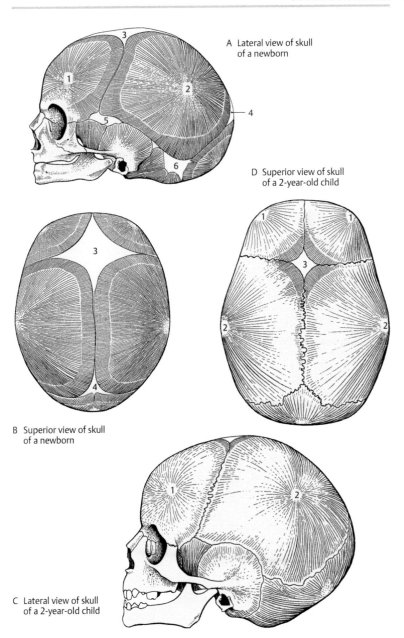

A Lateral view of skull of a newborn

D Superior view of skull of a 2-year-old child

B Superior view of skull of a newborn

C Lateral view of skull of a 2-year-old child

Structure of the Cranial Bones

Each of the flat bones of the skull consists of
- A compact **outer table** (lamina externa)
- A compact **inner table** (lamina interna)

and between the two lies
- The **diploë** (spongy layer), in which there are numerous veins within the diploic canals

Within other bones of the skull are certain air-filled spaces associated with the nasal cavities. The temporal bones contain the sensory organs of hearing and balance.

The outside of the skull is covered by the **pericranium**, and the inner surface of the skull is covered by the **endocranium** or **dura mater**.

It is useful first of all to take a unified view of the skull from its various aspects, in order to recognize the functional associations of the latter and to comprehend the special features of the individual cranial bones. The various cavities within the skull are also discussed below.

Calvaria (A–C)

The cranial vault, the calvaria, consists of a **frontal bone** (yellow), **parietal bones** (light brown), parts of the **temporal bones** (salmon), and the uppermost part of the **occipital bone** (orange). Examination of the outside of the skull will show first of all the sutures, for example the *coronal suture* (**1**), which separates the *squamous part* of the frontal bone (**2**) with the *frontal eminences* (**3**) from the parietal bones. Each parietal bone also has a *parietal eminence* (**4**). Between the parietal bones lies the *sagittal suture* (**5**), which runs from the coronal suture to the *lambdoid suture* (**6**), that is, the suture between the parietal bone and the *squamous part* of the occipital bone (**7**). Laterally, in the parietal region, are the *inferior* (**8**) and *superior* (**9**) *temporal lines*. In close relationship to the sagittal suture, immediately in front of the lambdoid suture, lie the *parietal foramina* (**10**). Special features are described on p. 290.

The sutures are also visible *on the inner surface of the adult calvaria*. On the cut surface the *outer table* (**11**), the *diploë* (**12**), and the *inner table* (**13**) are exposed. In the most anterior part of the squamous part of the frontal bone lies the *frontal crest* (**14**) which extends toward the parietal bones. In the region of the sagittal suture is the shallow *groove for the superior sagittal sinus* (**15**). The *arterial grooves* (**16**), which transmit the branches of the middle meningeal artery and its accompanying vein, ascend from the lateral toward the midline and posterior areas. Lateral to the groove for the superior sagittal sinus and lateral to the frontal crest there are a variable number of indentations of different size (*granular foveolae;* **17**) into which the arachnoidal granulations project.

On the internal and external surfaces of the parietal bone in the calvaria are the *frontal* (**18**) and *occipital* (**19**) *angles*, while the sphenoid and mastoid angles are found only at the skull base.

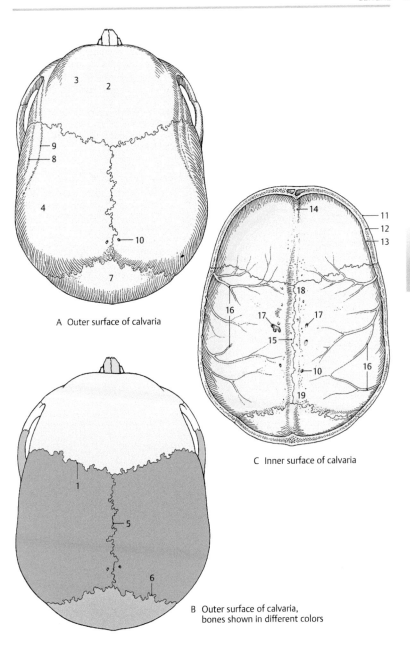

A Outer surface of calvaria

C Inner surface of calvaria

B Outer surface of calvaria,
bones shown in different colors

Lateral Aspect of the Skull (A–C)

In the orbitomeatal plane, which runs through the inferior margin of the orbit and the superior margin of the external acoustic meatus, the **neurocranium** shows the *temporal fossa* (**1**), which includes part of the **temporal bone** (salmon), the **parietal bone** (brown), portions of the **frontal bone** (yellow), and the **sphenoid bone** (brick red).

The temporal fossa is bounded above by the somewhat more prominent *inferior temporal line* (**2**) and the less obvious *superior temporal line* (**3**). From the *squamous part of the temporal bone* (**4**) the *zygomatic process* (**5**) extends anteriorly, and with the *temporal process* (**6**) of the *zygoma* (blue) it forms the *zygomatic arch* (**7**). Inferior to the root of the zygomatic process lies the *external acoustic meatus* (**8**), which is bordered mainly by the **tympanic part** (**9**; **C**, light blue), and to a lesser extent by the **squamous part** (**4**; **C**, salmon) of the **temporal bone** (**B**, salmon). Immediately above this there is often a small *suprameatal spine* (**10**) and a small cavity, the *foveola suprameatica = suprameatal triangle*. Posterior to the external meatus is the *mastoid process* (**11**), which originated as a muscular apophysis. Between this process and the tympanic part there is the variably developed *tympanomastoid fissure* (**12**). The *mastoid foramen* (**13**) lies at the root of the mastoid process. Below the tympanic part (**9**) there is the *styloid process* of variable size (**C**, light green).

On examining the **viscerocranium** we see the *supraciliary arch* (**14**) as a prominent ridge above the orbit. Below it is the *supraorbital margin* (**15**) with the *supraorbital notch*. The supraorbital margin is continued over the anterolateral margin of the orbital opening into the *infraorbital margin* (**16**). The latter is formed by the **zygoma** and the *frontal process of the maxilla* (**17**). Medially is a depression, the *fossa for the lacrimal sac* (**18**); (orbit, see p. 306).

There is one (or two) small foramen in the zygoma, the *zygomaticofacial foramen* (**19**). Below the infraorbital margin lies the *infraorbital foramen* (**20**). At the lowest point of the nasal opening the *anterior nasal spine* (**21**) is seen. The **maxilla** (light green) has an *alveolar process* (**22**) directed downward, which carries the maxillary teeth. The *maxillary tuberosity* (**23**) *bulges out posterior to this* (for details of the mandible, see p. 302).

Sutures

The *coronal suture* (**24**) separates the frontal and parietal bones. It meets the *sphenofrontal suture* (**25**), which lies between the *greater wing of the sphenoid bone* (**26**) and the frontal bone. The frontal bone and zygomas are joined by the *frontozygomatic suture* (**27**). The *zygomaticomaxillary suture* (**28**) lies between the zygoma and the maxilla, and the *temporozygomatic suture* (**29**) is found between the zygoma and temporal bone. The *frontomaxillary suture* (**30**) lies between the frontal bone and the maxilla, and the *nasomaxillary suture* (**31**) is between the maxilla and nasal bone (dark green). The *sphenosquamous suture* (**32**) forms the boundary between the greater wing of the sphenoid bone and the temporal squama. The temporal bone (light red) joins the parietal bone at the *squamous suture* (**33**). It may extend into the mastoid process as the *petrosquamous suture* (**34**) between its squamous (**C**, light red) and petrous (**C**, brown) parts.

The *lambdoid suture* (**35**) separates the parietal from the occipital bone (orange).

A small part of the greater wing of the sphenoid extends as far as the parietal bone, so that a *sphenoparietal suture* (**36**) can be described. Between the mastoid process and the parietal bone on the one hand and the occipital bone on the other lie the *parietomastoid* (**37**) and *occipitomastoid* (**38**) sutures.

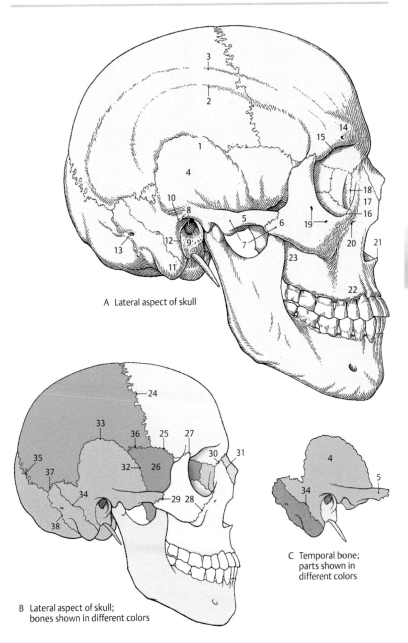

A Lateral aspect of skull

B Lateral aspect of skull;
bones shown in different colors

C Temporal bone;
parts shown in
different colors

Posterior Aspect of the Skull (A, B)

In the posterior view it is possible to see both **parietal bones** (brown, **1**), which are joined by the *sagittal suture* (**2**). The *lambdoid suture* (**3**) separates the two parietal bones from the **occipital bone** (orange, **4**).

The *external occipital protuberance* (**5**) is prominent on the occipital bone in the midline and is palpable through the skin. The *highest nuchal line* (**6**) extends upward and laterally from the external occipital protuberance. The line below is the *superior nuchal line* (**7**), which represents a transverse ridge lateral to the protuberance, and below it is the *inferior nuchal line* (**8**), which extends roughly in the center between the external occipital protuberance and the foramen magnum. The *inferior nuchal* line may begin at the *external occipital crest* (**9**), which shows variable development.

Lateral to the occipital bone lies the *mastoid process* (**11**), which is part of the temporal bone, but which is joined to the occipital bone by the *occipitomastoid suture* (**10**). A petrosquamous suture (**12**) may be present completely or in part in the mastoid process. This suture shows that the mastoid process is formed from both the squamous and the petrous parts of the temporal bone. In the region of the *occipitomastoid suture* (**10**) is the *mastoid foramen* (**13**), through which the mastoid emissary vein passes. On the medial side of the mastoid process lies the *mastoid notch* (**14**), medial to which is the *occipital groove* (**15**). *Parietal foramina* (**16**) are situated in the region of the parietal bones.

Variants: Sometimes the external occipital protuberance is particularly well developed. The upper squama may be present as a separate bone, the **incarial bone** (see p. 314).
The parietal foramina may be particularly large (**enlarged parietal foramina**) and may be a source of misinterpretation on radiographs (drill holes).

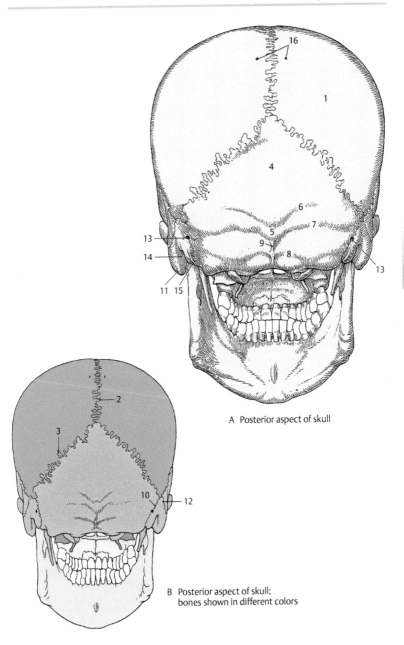

A Posterior aspect of skull

B Posterior aspect of skull;
bones shown in different colors

Anterior Aspect of the Skull (A, B)

From the front, the entire **viscerocranium** or facial skeleton is visible. The forehead region is formed by the **frontal bone** (yellow). In the region of the *squamous part* (**1**) the frontal bone is separated from the **parietal bones** (brown) by the *coronal suture* (**2**).

In the forehead, between the *supraciliary arches* (**3**), lies the *glabella* (**4**). The frontal bone marks the entrance to the orbits by forming the *supraorbital margin* (**5**), near the medial end of which is the variably sized, well-defined *supraorbital notch*. In some instances this notch is converted to a *supraorbital foramen* (**6**). Medial to it may lie a small *frontal notch* (**7**) or a frontal foramen.

Between the orbits the frontal bone is separated from the **nasal bones** (dark green) by the *frontonasal sutures* (**8**), and from the **maxillae** (light green) by the *frontomaxillary sutures* (**9**). The two nasal bones are joined by the *internasal suture* (**10**). Lateral to the orbital opening, the *frontozygomatic suture* (**11**) separates the frontal bone from the zygoma. The **zygoma** (blue) together with the maxilla forms a further part of the boundary of the orbital opening (for details of the orbital cavity, see p. 306).

The *infraorbital foramen* (**14**) is located in the **maxilla**, just below the *infraorbital margin* (**12**) and adjacent to the *zygomaticomaxillary suture* (**13**). It transmits a branch of the maxillary nerve, the infraorbital nerve, an artery, and a vein. Inferior to the orbit, in the region of the *maxillary body*, there is a deep depression, the *canine fossa* (**15**).

The *zygomatic process* (**16**) runs laterally from the maxillary body. The maxilla is attached to the frontal bone by the *frontal process* (**17**), which ascends from the maxillary body and connects with the nasal bone by the *nasomaxillary suture* (**18**). The *palatine process* (see p. 294) is directed medially and forms one of the foundations of the hard palate. Finally, in the tooth-bearing upper jaw, there is the downward-facing *alveolar process* (**19**).

The continuation of the infraorbital margin on the frontal process is the *anterior lacrimal crest* (**20**). The center of the maxilla is formed by the *body of the maxilla* mentioned above. The latter demarcates with its *nasal notch* (**21**) the *piriform aperture*, the entrance into the nasal cavities. At the lower margin of the aperture in the region of the *intermaxillary suture* (**22**), a spur, the *anterior nasal spine* (**23**), projects anteriorly. In the zygoma there are one or two *zygomaticofacial foramina* (**24**).

In the lower jaw, the **mandible** (light violet), the *body* (**25**), the *alveolar part* (**26**), and the *ramus* (**27**) are visible from the front. In the region of the body of the mandible, the *mental foramen* (**28**) lies vertically below the second premolar tooth. The *mental protuberance* (**29**) is found in the midline of the body of the mandible (see p. 302).

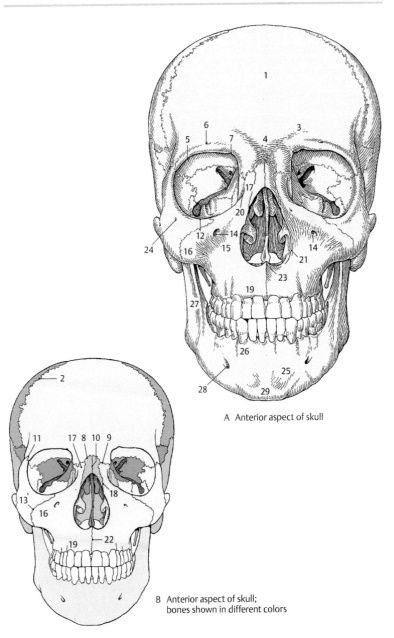

A Anterior aspect of skull

B Anterior aspect of skull;
   bones shown in different colors

Inferior Aspect of the Skull (A, B)

The external surface of the skull base consists of an anterior visceral part and a posterior neural part.

The **anterior part** is formed on each side by the *palatine process of the maxilla* (**1**, light green), the *horizontal plate of the palatine bone* (**2**, green), the *alveolar process* and *tuber of the maxilla* (**3**) and the **zygoma** (**4**, light blue). The **vomer** (dark blue) borders the *choanae* (**5**) medially. The two palatine processes are fused at the *median palatine suture* (**6**), the anterior end of which is indicated by the *incisive fossa* (**7**) housing the incisive canals. An *incisive suture* (**8**), which is often preserved, passes from the fossa up to the second incisor. The horizontal plate of the palatine bone contains the *greater* (**9**) and the *lesser* (**10**) *palatine foramina. Palatine grooves* pass anteriorly from the greater palatine foramen and are bordered by ridges, the *palatine spines*. The *transverse palatine suture* (**11**) is found between the **maxilla** (light green) and the **palatine bone** (medium green).

The **posterior part** of the skull base consists of the **sphenoid bone** (brick red), the **temporal bones** (salmon), and the **occipital bone** (orange). The pterygoid processes form the lateral borders of the choanae. We distinguish a *medial plate* (**12**), with its *hamulus*, and a *lateral plate* (**13**). Between them lies the *pterygoid fossa*. At the root of the medial plate is the *scaphoid fossa* (**14**) and next to it the *foramen lacerum* (**15**).

In the center lies the *body of the sphenoid bone* (**16**) and laterally its *greater wing* (**17**) with the *infratemporal crest* (**18**). The greater wing bears the *sphenoid spine* (**19**), whose base is pierced by the *foramen spinosum* (**20**). Between the foramen spinosum and the foramen lacerum is the *foramen ovale* (**21**), and between the sphenoid bone and the petrous part of the temporal bone we find the *sphenopetrosal fissure* (**22**). From the latter the groove of the *auditory tube* (**23**) extends posterolaterally. The

external opening of the cochlear canaliculus is found on the side of the *jugular fossa* (**25**) and adjacent to the *external opening of the carotid canal* (**24**). This is bounded laterally by the *jugular and occipital processes*. Between the jugular fossa and the external opening of the carotid canal is a small depression, the *petrosal fossula*, in which the canaliculus for the tympanic nerve opens. Next to this are the *tympanic part* (**26**) of the temporal bone and the *styloid process* (**27**) within its sheath. Immediately posterior to the process is the *stylomastoid foramen* (**28**). On the *mastoid process* (**29**) is the *mastoid notch* (**30**), and medial to it is the *occipitomastoid suture* (**31**) with the **occipital groove** for the occipital artery (**32**). Anterior to the mastoid process lies the opening of the *external acoustic meatus* (**34**), which is bounded by the *tympanic part* (**26**) and the *squamous part* (**33**).

The tympanic and squamous parts, as well as a small ridge of the petrous part, the *tegmental crest* bounded by the *petrotympanic* and *petrosquamous fissures*, form the *mandibular fossa* (**35**). This is bounded anteriorly by the *articular tubercle* (**36**). The *zygomatic process of the temporal bone* (**37**) extends anterolaterally.

The *basilar part* (**38**) of the occipital bone, which bears the *pharyngeal tubercle* (**39**), fuses with the *body of the sphenoid bone* (**16**). The *petro-occipital fissure* runs between the *petrous part* of the temporal bone and the occipital bone. The jugular fossa (**25**) is widened by the notch in the adjacent occipital bone to form the *jugular foramen*. The *foramen magnum* (**40**) is bordered laterally on each side by an *occipital condyle* (**41**), behind which lies the condylar fossa perforated by a *condylar canal* (**42**). Beginning directly behind the foramen magnum, the *external occipital crest* (**43**) passes upward to the *external occipital protuberance* (**44**).

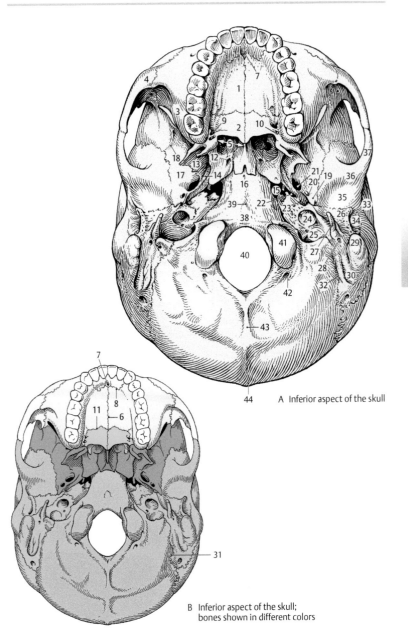

A Inferior aspect of the skull

B Inferior aspect of the skull;
bones shown in different colors

Internal Surface of Cranial Base (A, B)

The skull base is divided into three fossae:
- The **anterior cranial fossa**
- The **middle cranial fossa**
- The **posterior cranial fossa**

The following bones form the inner surface of the skull base: the **ethmoid bone** (blue-violet), the **frontal bone** (yellow), the **sphenoid bone** (brick red), the **temporal bones** (salmon), the **occipital bone** (orange), and the **parietal bones** (brown).

The anterior cranial fossa is separated from the middle fossa by the *lesser wings of the sphenoid* (**1**) and the *jugum sphenoidale* (**2**). The middle and posterior cranial fossae are separated from each other by the *superior borders* (**3**) of the petrous portions of the temporal bones and the *dorsum sellae* (**4**).

The anterior cranial fossa. The *cribriform plate* (**5**) formed by the ethmoid bone contains many small holes and bears in the midline the vertical *crista galli* (**6**) with its *ala of crista galli*. Anterior to the crista galli is the *foramen caecum* (**7**) and laterally lie the *orbital plates* (**8**) of the frontal bone with their *convolutional markings*. The cribriform plate is joined to the sphenoid bone by the *sphenoethmoidal suture* (**9**). In the middle, the *prechiasmatic groove* (**11**) lies between the *optic canals* (**10**). The *anterior clinoid processes* (**12**) border the optic canals.

At the center of the **middle cranial fossa** is the *sella turcica* with the *pituitary fossa* (**13**) and lateral to the sella the *carotid sulcus* (**14**), which is the prolongation of the carotid canal. The carotid canal, which lies on the anterior wall of the petrous part of the temporal bone, is split open in its medial portion near the *foramen lacerum* (**15**). The medial end of the canal is bounded by the *sphenoidal lingula* (**16**). Lateral to the carotid groove is the *foramen ovale* (**17**), in front the *foramen rotundum* (**18**), and lateral the *foramen spinosum* (**19**). The *groove for the middle meningeal artery* (**20**) runs laterally from the foramen spinosum.

Near the apex of the petrous part, the *trigeminal impression* (**21**) can be seen, and lateral and somewhat posterior to it is the *hiatus for the greater petrosal nerve* (**22**), which continues toward the sphenopetrosal fissure as the *groove for the greater petrosal nerve* (**23**). The *hiatus for the lesser petrosal nerve* (**24**) lies just antero-lateral to that of the greater petrosal nerve. The *superior border of the petrous part* (**3**) carries the more or less well-developed *groove of the superior petrosal sinus* (**25**). A prominent swelling, the *arcuate eminence* (**26**), is produced by the anterior semicircular canal. The squamous part of the temporal bone is joined to the sphenoid bone by the *sphenosquamous suture* (**27**).

The *foramen magnum* (**28**) lies in the middle of the **posterior cranial fossa**. The *clivus* (**29**) ascends anteriorly and ends in the *dorsum sellae* (**4**) and its *posterior clinoid processes* (**30**). Between the occipital bone and the petrous part of the temporal bone lies the *groove for the inferior petrosal sinus* (**31**) and also the petro-occipital synchondrosis, which may be seen in the macerated skull as the *petro-occipital fissure* (**32**). The groove for the inferior petrosal sinus ends in the *jugular foramen* (**33**). The *internal acoustic meatus* (**34**) opens onto the posterior surface of the petrous bone. Lateral to it, hidden under a small bony ridge, lies the *external opening of the vestibular aqueduct*.

The jugular foramen (**33**) is formed by the apposition of the jugular notches in the temporal and occipital bones. The *jugular notch in the occipital bone* is limited anteriorly by the projection of the *jugular tubercle*, and the jugular foramen is partly divided by the *intrajugular process of the temporal bone* (**35**). On its lateral side the jugular foramen is reached by the *groove of the sigmoid sinus* (**36**), which continues posteriorly into the *groove for the transverse sinus* (**37**). This extends to the *internal occipital protuberance* (**38**), from which the *internal occipital crest* (**39**) runs toward the foramen magnum (**28**). On either side of the anterior rim of the foramen magnum is the opening of the *hypoglossal canal* (**40**).

The clivus is formed by the body of the sphenoid bone and the basilar part of the occipital bone. During puberty they fuse (**os tribasilare**) but prior to this they are connected by the spheno-occipital synchondrosis.

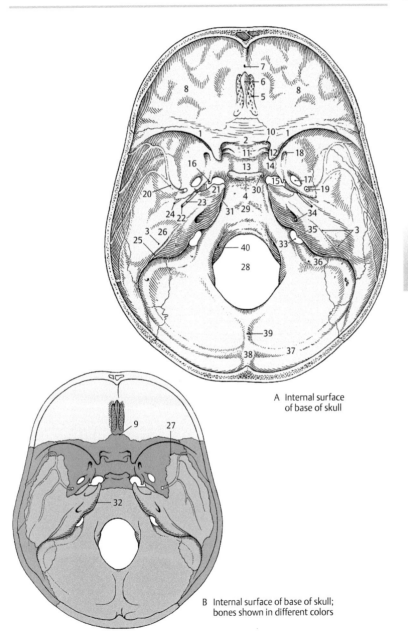

A Internal surface
of base of skull

B Internal surface of base of skull;
bones shown in different colors

Head and Neck

Variants of the Internal Surface of the Skull Base (A–E)

Imaging studies of the middle cranial fossa may demonstrate a number of variants in the region of the sella turcica.

In some cases the *sphenoidal lingula* (**1**), which is directed toward the temporal bone, may be fused with that bone. This distinctly demarcates the internal opening of the carotid canal.

Between the anterior and posterior clinoid processes there may be an additional process, the *middle clinoid process* (**2**). The latter may then fuse with the anterior clinoid process, when it forms a special opening, the *caroticoclinoid foramen* (**3**). As a result of this, the carotid notch, which lies medial to the anterior clinoid process, becomes an opening surrounded by bone on all sides.

Another variant is the presence of an *inter-clinoidal bridge* (**4**) between the anterior and posterior clinoid processes. This bony fusion of the two processes, when seen on radiographs, is termed the *sella bridge* (**4**). It may be present on one or both sides and may fuse (**5**) with the middle clinoid process if it is present.

A very rare variant is the presence of a *craniopharyngeal canal* (**6**) in the pituitary fossa.

Between the foramen ovale and the body of the sphenoid bone, there is sometimes an aperture, which serves as the exit point for a vein. This opening, the *foramen veno-sum* (**7**), is also called the sphenoidal emissarium or the foramen of Vesalius. It is not very uncommon and it establishes a communication between the cavernous sinus and extracranial veins. The foramen of Vesalius may be present on one or both sides.

In some cases the dorsum sellae may be so eroded laterally by extensive looping of the internal carotid artery that it no longer has any bony connection with the clivus. In that case, the dorsum sellae will be absent from the macerated skull (**D**).

Sometimes the internal occipital crest is divided into two and between the parts is the well-developed *groove of the occipital sinus*. This may extend into a *marginal groove* (**8**), running lateral to the *foramen magnum* (**9**), to the *jugular foramen* (**10**). The *condylar canal* (**11**) may empty with a particularly large opening into the sigmoid sinus.

The jugular foramina may be unequal in size, more often the left being smaller than the right. Rarely is the *groove for the inferior petrosal sinus* (**12**) very deep. *The hypoglossal canal may be divided into two* (**13**).

The apex of the petrous part of the temporal bone may have a bony connection with the dorsum sellae. This bony bridge is also known as the *abducent bridge* (**14**), since the abducent nerve runs beneath it.

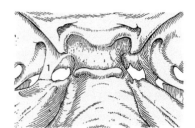

A   Sella turcica; right sphenoidal lingula fused with temporal bone

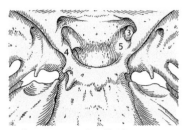

C   Sella turcica; interclinoid bridge, right caroticoclinoid foramen

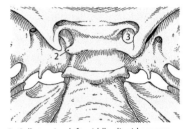

B   Sella turcica; left middle clinoid process, right caroticoclinoid foramen

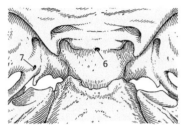

D   Sella turcica; craniopharyngeal canal, foramen venosum, absence of dorsum sellae

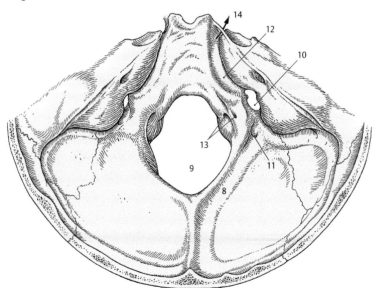

E   Groove of right occipital sinus, divided canal for hypoglossal nerve

Head and Neck

Head and Neck

Sites for Passage of Vessels and Nerves (A, B)

The openings in the skull base transmit vessels and nerves.

In the region of the anterior cranial fossa the *olfactory nerves* (**1**) and the *anterior ethmoidal artery* (**2**) pass through the **cribriform plate** to the nasal cavity.

The *optic nerve* (**3**) and the *ophthalmic artery* (**4**) run through the **optic canal**. Apart from the optic canal, the **superior orbital fissure** also forms a communication between the skull and the orbit. The *superior ophthalmic vein* (**5**), the *lacrimal nerve* (**6**), the *frontal nerve* (**7**), and the *trochlear nerve* (**8**) run in its lateral part. The *abducent nerve* (**9**), *oculomotor nerve* (**10**), and *nasociliary nerve* (**11**) pass through it more medially.

The *maxillary nerve* (**12**) passes through the **foramen rotundum**, while the *mandibular nerve* (**13**), together with the *venous plexus of the foramen ovale* that joins the cavernous sinus to the pterygoid plexus, runs through the **foramen ovale**. A recurrent branch of the mandibular nerve, the *meningeal branch* (**14**), together with the *middle meningeal artery* (**15**), reaches the cranial cavity through the **foramen spinosum**. The largest structure in the middle cranial fossa, the *internal carotid artery* (**16**), passes through the **carotid canal** into the cranial cavity. The internal carotid artery is surrounded by the *sympathetic carotid plexus* (**17**) *and internal carotid venous plexus*. The *greater petrosal nerve* (**18**) becomes visible at the **hiatus for the greater petrosal nerve**, and the *lesser petrosal nerve* (**19**) runs through the **hiatus for the lesser petrosal nerve** together with the *superior tympanic artery* (**20**).

In the posterior cranial fossa, the *medulla oblongata* (**21**) passes through the **foramen magnum**, accompanied on each side by the *spinal part of the accessory nerve* (**22**). Two large *vertebral arteries* (**23**), the small *anterior spinal artery* (**24**), the paired small *posterior spinal arteries* (**25**), and the *spinal vein* (**26**) also pass through the foramen magnum.

The *hypoglossal nerve* (**27**) and the *venous plexus of the hypoglossal canal* (**28**) pass through the **hypoglossal canal**.

The *glossopharyngeal nerve* (**29**), the *vagus* (**30**) and the *accessory nerve* (**31**), as well as the *inferior petrosal sinus* (**32**), the *internal jugular vein* (**33**), and the *posterior meningeal artery* (**34**) all pass through the **jugular foramen**.

The **internal acoustic meatus** transmits the *labyrinthine artery and vein* (**35**), the *vestibulocochlear nerve* (**36**), and the *facial nerve* (**37**).

On the *outer surface of the skull base* the *facial nerve* (**37**) becomes visible as it emerges from the **stylomastoid foramen**, through which the *stylomastoid artery* (**38**) enters the skull.

The *anterior tympanic artery* (**39**) and the *chorda tympani* (**40**) traverse the **petrotympanic fissure**.

The *greater palatine artery* (**41**) and the *greater palatine nerve* (**42**) pass through the **greater palatine foramen** in the hard palate, and the *lesser palatine arteries and nerves* (**43**) run through the **lesser palatine foramina**. The *nasopalatine nerve* and an artery (**44**) run through the **incisive canal** toward the palate.

The *condylar emissary vein* (**45**) runs through the **condylar canal**.

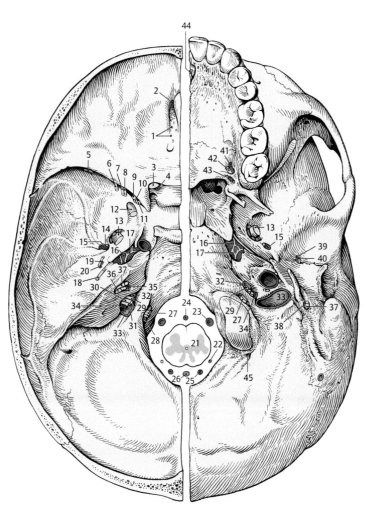

A Internal view of skull base,
   left half

B Inferior view of skull base,
   left half

Mandible (A–C)

The lower jaw (mandible) is only connected with the other bones of the skull by synovial joints. It is preformed in connective tissue and consists of the **body** (**1**) with an ascending **ramus** (**2**) on each side.

In the adult the **body of the mandible bears** the *alveolar part* (**3**), which is marked on its outer surface by the *alveolar eminences* (**4**). In old age, that is, after loss of the teeth, the alveolar part undergoes regression (see p. 304). On the front of the body of the mandible is the *mental protuberance* (**5**), which is elevated on each side to form the *mental tubercle*. On the outer surface, on a vertical line through the second premolar, there is an opening, the *mental foramen* (**6**). The inferior surface of the mandible is called the base of the mandible. The *oblique line* (**7**) ascends from the body to the ramus of the mandible. Posteriorly, the body of the mandible merges at the *mandibular angle* (**8**) with the ramus.

The **ramus of the mandible** has two processes, the anterior *coronoid process* (**9**) for insertion of a muscle, and the posterior *condylar process* (**10**) for the articular surface.

Between the processes lies the *mandibular notch* (**11**). The condylar process has a *neck* (**12**) and supports the *head of the mandible* with its *articular surface* (**13**). The mandibular head is also known as the *mandibular condyle* because of its cylindrical shape. On the inner aspect of the head of the mandible, below the articular surface, is a small pit, *the pterygoid fovea* (**14**), for the insertion of part of the lateral pterygoid muscle. Near the angle of the mandible there is sometimes a roughened area, the *masseteric tuberosity* (**15**), which gives attachment to the masseter muscle. On the inner surface of the mandible in the region of the ramus lies the *mandibular foramen* (**16**), which is the entrance to the *mandibular canal*. The opening is partly concealed

by a delicate spur of bone, the *lingula of the mandible* (**17**). The *mylohyoid groove* (**18**) begins directly at the mandibular foramen and runs obliquely downward. Below the mylohyoid groove, at the angle of the mandible, is the *pterygoid tuberosity* (**19**), which serves for the insertion of the medial pterygoid muscle.

The inner surface of the body of the mandible is divided by an oblique ridge, the *mylohyoid line* (**20**). Below this line, from which the mylohyoid muscle arises, we find the *submandibular fossa* (**21**), while above it and somewhat more anterior is the *sublingual fossa* (**22**).

The dental alveoli are separated by the *interalveolar septa* (**23**). Within the alveoli of the molars, *interradicular septa* may be seen. Posterior to the last molar is the *retromolar triangle*, which is of variable size.

Anteriorly, on the inner surface of the body, lies the *mental spine* (**24**), from which muscles arise (also called genial spines), and laterally, somewhat lower, are the *digastric fossae* (**25**), the points of insertion of the digastric muscles.

▬▬ **Variant:** Two mental spines are sometimes present, one situated above the other.

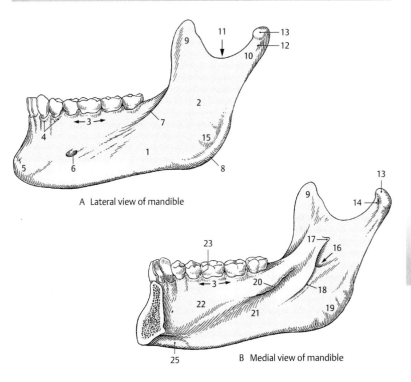

A Lateral view of mandible

B Medial view of mandible

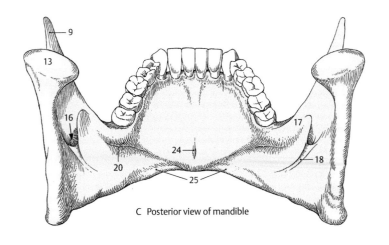

C Posterior view of mandible

Shape of Mandible (A–E)

The *angle of the mandible* differs at various stages of life. In the newborn (**A**) it is still relatively large, approximately 150°, while during childhood (**B**) it becomes smaller. In the adult (**C**) it is reduced to approximately 120 to 130°. In old age (**D**) it again increases to approximately 140°.

The change in the angle of the mandible is dependent on the presence of the alveolar part with its alveolar arch and the teeth. With eruption of the teeth there is an alteration in the mandibular angle of the infant, and it changes again in old age when the teeth are lost.

Apart from the change in the angle of the mandible at the various stages of life, the body of the mandible also shows variations. The body of the mandible bears the alveolar process, which undergoes resorption in old age as the teeth are lost. During this regression the size of the body of the mandible becomes reduced and sometimes flattened, which may push the chin forward.

> **Clinical tip:** Modern dentistry can reduce age-related changes in the shape of the jaw with **dental implants**. Metal posts are surgically implanted in the mandible or maxilla to serve as abutments for dental crowns. This eliminates the need for a denture, which always constitutes a foreign body. However, dental implantation requires a highly experienced surgeon.

The alveolar part of the mandible may vary in its orientation. In some instances, particularly among the primates, there may be an alveolar part protruding outward and the position of the teeth differs from that in modern humans.

Ossification: As noted on p. 282, the mandible is preformed in connective tissue. It appears on both sides in the first visceral arch as intermembranous bone, formed on *Meckel's* cartilage (*Meckel's jun. 1781–1833*). In the region of the symphysis, that is, anteriorly, parts of *Meckel's* cartilage form the basis of those parts of the mental ossicles that develop in cartilage. They fuse with the mandible. The first bone cells appear in the 6th intrauterine week. Together with the clavicle, it is the first bone in the body to develop. The synostosis of the two parts of the mandible begins in the 2nd month of life.

Hyoid Bone (F)

The hyoid bone, which is considered part of the cranial skeleton, is not directly connected but is joined to it by muscles and ligaments. It consists of a *body* (**1**), the anterior part, and two *greater horns* (**2**) located laterally. One can see an upward-directed *lesser horn* (**3**) and a larger, posteriorly directed *greater horn* (**2**).

Ossification: In the body and the greater horn of the hyoid bone, ossification centers develop in cartilage just before birth, while in the lesser horn the center develops much later in about the 20th year of life. The lesser horn need not ossify but may remain cartilaginous. Like the mandible, the hyoid bone develops from the skeleton of the visceral arches.

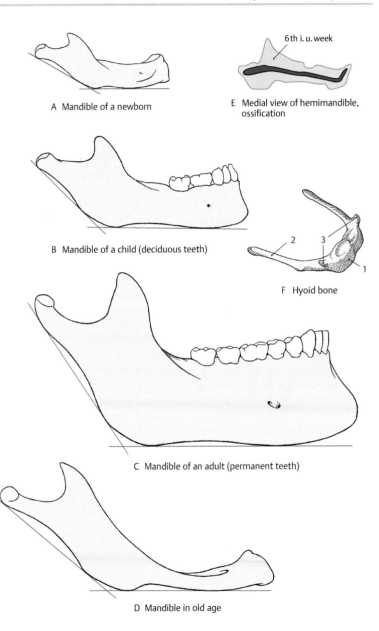

A Mandible of a newborn

6th i. u. week

E Medial view of hemimandible, ossification

B Mandible of a child (deciduous teeth)

2   3

1

F Hyoid bone

C Mandible of an adult (permanent teeth)

D Mandible in old age

Orbit (A, B)

Each **orbital cavity** is shaped like a four-sided pyramid, the apex lying deep inside and the base forming the orbital opening. It is bounded by various bones.

Roof: The roof of the orbit is formed anteriorly by the *orbital plate of the frontal bone* (**1**) and posteriorly by the *lesser wing of the sphenoid* (**2**).

Lateral wall: The lateral wall consists of the *zygoma* (**3**) and the *greater wing of the sphenoid* (**4**).

Floor: The anterior part of the floor is formed by the orbital surface of the *body of the maxilla* (**5**) and posteriorly by the *orbital process of the palatine bone* (**6**). Along the infraorbital margin, the floor is completed anteriorly by the zygoma (**3**).

Medial wall: The thin medial wall is formed by the *orbital plate of the ethmoid bone* (**7**), the *lacrimal bone* (**8**), and the *sphenoid* (**9**). In addition, the frontal bone (**1**) and the *maxilla* provide smaller contributions to this wall.

Orbital openings: The *supraorbital* and *infraorbital margins* of the entrance to the orbit have already been described (see p. 292). Medially and laterally they are joined together at the *medial* and *lateral margins*. Posteriorly there are two converging fissures, the *superior orbital fissure* (**10**), which opens into the cranial cavity, and the *inferior orbital fissure* (**11**) for communication with the pterygopalatine fossa. The fissures converge medially, and immediately above the junction lies the *optic canal* (**12**). From the inferior orbital fissure runs the *infraorbital groove* (**13**), which becomes the *infraorbital canal* to open below the infraorbital margin as the *infraorbital foramen* (**14**). On the lateral wall the zygomatic nerve passes through the *zygomatico-orbital foramen* (**15**). On the medial wall, where the ethmoid bone meets the frontal bone, are the *anterior* (**16**) and *posterior* (**17**) *ethmoidal foramina*. The homonymous nerves and arteries exit through these foramina.

The anterior ethmoidal foramen opens into the cranial cavity, while the posterior foramen leads into the ethmoid cells. Near the entrance to the orbit lies the *fossa for the lacrimal sac* (**18**), which is bounded anteriorly and posteriorly by the *anterior* (**19**) and *posterior* (**20**) *lacrimal crests*. It leads into the *nasolacrimal canal*, which opens into the nasal cavity (see p. 308).

In the immediate neighborhood of the orbits are the **paranasal sinuses**. The variably sized orbital recess of the *frontal sinus* (**21**) extends into the roof of the orbit. Medially lie the ethmoid cells and posteriorly the sphenoidal sinus. Inferiorly the orbit is separated from the *maxillary sinus* (**22**) by a thin plate of bone.

Pterygopalatine Fossa (B, C)

The **pterygopalatine fossa** may be approached from the lateral side through the *pterygomaxillary fissure* (**23**). Anterior to it lies the *maxilla* (**24**), posteriorly the *pterygoid process* (**25**), and medially the *perpendicular plate of the palatine bone* (**26**). It is an important junction area for vessels and nerves. It is connected to the cranial cavity by the *foramen rotundum* (**27**) and to the inferior surface of the skull base by the *pterygoid canal* (**28**). The *greater palatine canal* (**29**) and the lesser palatine canal lead to the palate, the *sphenopalatine foramen* (**30**) to the nasal cavity, and the inferior orbital fissure (**11**) into the orbital cavity.

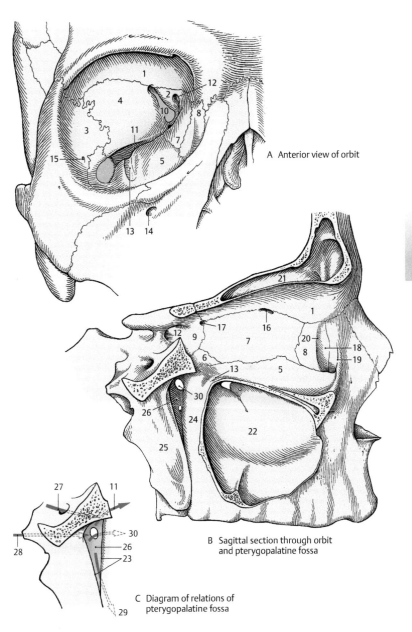

A Anterior view of orbit

B Sagittal section through orbit
and pterygopalatine fossa

C Diagram of relations of
pterygopalatine fossa

Nasal Cavity (A–C)

The **bony nasal cavity consists of right and left halves** separated medially by the **nasal septum**. The septum often deviates from the midline. The nasal cavities open anteriorly into the **piriform aperture** (see p. 292) and posteriorly each opens via the **choana**, the posterior nasal aperture, into the pharynx (see Vol. 2).

The **nasal septum** (**A**) consists of **cartilaginous** and **bony elements**. The **cartilaginous septum** (**1**) with its *posterior process* (**2**) completes the bony partition between the two nasal cavities. The **medial crus of the major alar cartilage** (**3**) is superimposed on each side on the septal cartilage as the medial border of the anterior opening of the nose. The bony partition, the **bony nasal septum**, is formed by the *perpendicular plate of the ethmoid* (**4**), the *sphenoid crest* (**5**), and the *vomer* (**6**).

The **floor** of the nasal cavity is formed by the *maxilla* (**7**) and *the palatine bone* (**8**).

The **roof** is formed anteriorly by the *nasal bone* (**9**) and posteriorly and superiorly by the *cribriform plate of the ethmoid* (**10**).

The **lateral wall** (**B**, **C**) of each nasal cavity is made irregular by the **three** turbinate bones or conchae and the underlying ethmoid cells. The *superior* (**11**) and *middle* (**12**) *turbinates* are part of the ethmoid bone, while the *inferior turbinate* (**13**) is a separate bone of the skull.

Behind the superior turbinate lies the *sphenoethmoidal recess* (**14**) into which the sphenoidal sinuses open. The *sphenopalatine foramen* (**15**) lies in the lateral wall of the recess. It connects it to the pterygopalatine fossa (see p. 306). After removal of the three turbinates, the *superior, medial,* and *inferior nasal meati* are revealed, and *the perpendicular plate of the palatine bone* (**16**) is fully exposed. The openings (**17**) of the posterior ethmoid cells can be seen in the superior nasal meatus.

In the middle nasal meatus, the *uncinate process* (**18**) partly covers the *maxillary hiatus* (**19**), which connects the maxillary sinus with the nasal cavity. Superior to this process is the *ethmoid bulla* (**20**), a particularly large anterior ethmoid cell. Above and below the bulla the middle and the anterior ethmoid cells open into the middle meatus of the nasal cavity.

Between the ethmoid bulla and the uncinate process is the *ethmoid infundibulum* (**21**), across which the *frontal sinus* (**22**), the *maxillary sinus* (**23**), and the anterior ethmoid cells are connected with the nasal cavity. The uncinate process also partly covers the *lacrimal bone* (**24**), which forms the lateral wall together with the maxilla (**7**) and the ethmoid bone.

The *nasal opening* (**25**) *of the nasolacrimal canal* lies in the inferior nasal meatus.

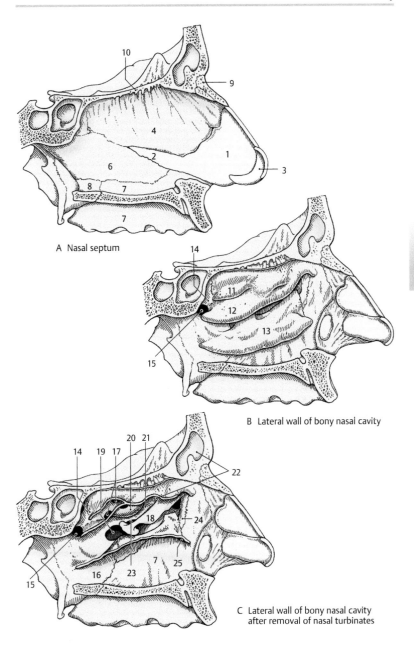

A Nasal septum

B Lateral wall of bony nasal cavity

C Lateral wall of bony nasal cavity
   after removal of nasal turbinates

Cranial Shapes (A–C)

Anatomy and anthropology recognize a number of craniometric points, lines, and angles that permit comparison of the various types of normal skull (**A**) and also permit recognition of abnormal forms (**B**, **C**).

Some of the important points for measurement include: the *glabella* (**1**) = the smooth area between the eyebrows; the *opisthocranion* = the most posterior protruding point of the occipital bone in the midline sagittal plane; the *basion* = the anterior margin of the foramen magnum; the *bregma* (**2**) = the point of contact between the sagittal suture and coronal suture; the *nasion* (**3**) = the crossing point of the nasofrontal suture with the median sagittal plane; the *gnathion* (**4**) = that point on the base of mandible in the median sagittal plane which protrudes furthest downward; and the *zygion* (**5**) = the most laterally protruding point of the zygomatic arch. Also of interest are the *gonion* (**6**) = the widest, downward, backward, and laterally directed point at the angle of the mandible; the *vertex* = the highest point of the skull in the midsagittal plane when oriented to the orbitomeatal plane; and the *inion* = the most prominent point (center) of the external occipital protuberance.

Other points of measurement, lines, and angles may be found in textbooks of anthropology.

The most important indices based on a comparison of the distances between the individual points of measurement are presented below.

Length–Width Index of the Neurocranium

$$\frac{\text{greatest width of the skull}}{\substack{\text{greatest length of the skull} \\ \text{(glabella–opisthocranion)}}} \times 100 \, (= I)$$

Long head (dolichocephaly) $I < 75$
Normal head (mesocephaly) $I = 75–80$
Short head (brachycephaly) $I > 80$

Length–Height Index of the Neurocranium

$$\frac{\substack{\text{height of the skull} \\ \text{(basion–bregma)}}}{\substack{\text{greatest length of the skull} \\ \text{(glabella–opisthocranion)}}} \times 100 \, (= I)$$

Wide head (platycephaly) $I < 70$
Normal head (orthocephaly) $I = 70–75$
Steeple head (hypsicephaly) $I > 75$

Facial Index

$$\frac{\substack{\text{height of the face} \\ \text{(nasion–gnathion)}}}{\substack{\text{width of the zygomatic} \\ \text{arch}}} \times 100 \, (= I)$$

Wide face (euryprosopy) $I < 85$
Medium face (mesoprosopy) $I = 85–90$
Narrow face (leptoprosopy) $I > 90$

A basic reciprocity exists between the growth of the brain and that of the skull. If there is an abnormal increase in the volume of the contents of the skull, this will result in significant enlargement of the bony skull. Abnormal enlargement of the brain is due to enlargement of the cerebral cavities that are filled with cerebrospinal fluid, and it may be associated with an excessive production of cerebrospinal fluid (see also Vol. 3).

Clinical tip: Malformations. In **hydrocephalus** (**B**), the cranial vault (neurocranium) is abnormally large in relation to the facial skeleton (viscerocranium). The cranial bones are thin. There is delayed closure of the enlarged fontanelles, and frontal and parietal bossing is present. The orbits are small and shallow.

Premature closure of the cranial sutures leads to **microcephaly** (**C**). The premature closure may result, for instance, from reduced brain growth. Microcephaly is characterized by deep orbits and thick zygomatic arches.

Other malformations include **scaphocephaly**, in which there is premature fusion of the sagittal suture, and **oxycephaly**, in which the coronal suture ossifies prematurely.

These various malformations must be distinguished from artificially deformed skulls.

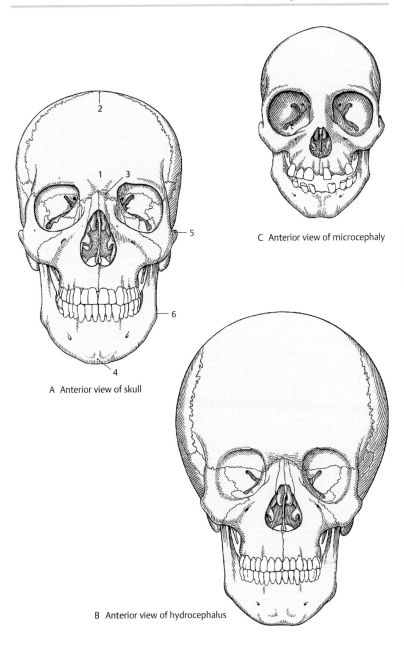

A Anterior view of skull

C Anterior view of microcephaly

B Anterior view of hydrocephalus

Special Cranial Shapes and Sutures (A–D)

The size and shape of the neurocranium depends on the growth of the brain, and the size of the viscerocranium will be substantially influenced by the activity of the masticatory apparatus. The influence of other elements, such as the supporting system of the dura mater, must also be taken into account. The various forms of the cranial sutures are also of interest in this regard.

In the skull, in the region of the intramembranous bones, there are three different types of sutures:
- **Sutura plana**
- **Sutura serrata**
- **Sutura squamosa** (see p. 22)

During development all the sutures are at first fairly straight and could be described as harmonious. It is only during the course of development that their shapes change. Occasionally subforms are seen, such as the *sutura limbosa*, which is a special form of the sutura squamosa. There are also more sutures in the newborn than in adults; for example, because of the paired anlages of the frontal bones there is *a frontal = metopic suture* (**1**), which usually closes between the 1st and 2nd years of life. If it persists (**A**) the skull is termed a **"crossed skull,"** *as there is a cruciform suture where the coronal* (**2**), *frontal* (**1**), *and sagittal* (**3**) *sutures meet. Remnants of the frontal suture may often be seen near the root of the nose* (**4**). If the frontal suture does persist, the forehead may become particularly prominent because of the more marked growth of both parts of the frontal bone.

Clinical tip: Atypical ossification centers may produce **additional bone.** An incarial bone (see p. 314) produces a transverse occipital suture. A *horizontal parietal suture* (**5**) is a special feature produced by the presence of a *superior parietal bone* (**6**) and an *inferior parietal bone* (**7**). The atypical sutures may lead to misinterpretations on radiographs (fractures).

By approximately 30 years of age, the individual sutures fuse and bone growth ceases. The first to fuse is usually the sagittal suture, but less frequently it is the coronal suture.

Clinical tip: If there is an early general fusion of sutures, microcephaly results (see p. 310). If only one suture fuses, the skull acquires an abnormal shape such as scaphocephaly or oxycephaly. If only one part of a suture fuses prematurely, as may happen in the coronal suture, **plagiocephaly** or "crooked skull" (**C, D**) results. A plagiocephalic skull should be distinguished from an artificially deformed skull.

8   Outline of a plagiocephalic skull
9   Outline of a normally developed skull

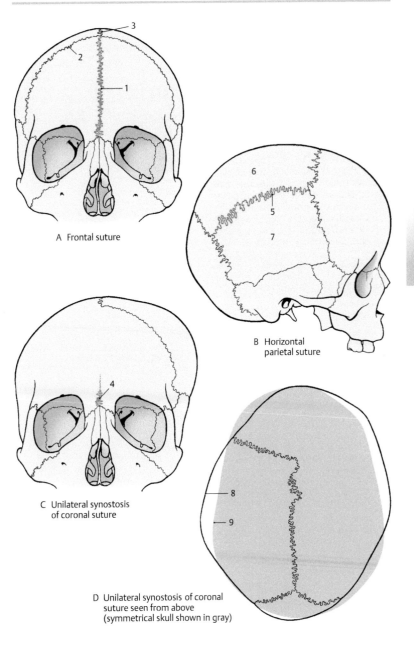

A Frontal suture

B Horizontal
   parietal suture

C Unilateral synostosis
   of coronal suture

D Unilateral synostosis of coronal
   suture seen from above
   (symmetrical skull shown in gray)

Accessory Bones of the Skull (A–E)

Quite often there are supernumerary independent bones between or within the other bones of the skull. They are either called **epactal bones** or, if they lie between the other bones of the skull, wormian or **sutural bones**. These supernumerary bones, the majority of which develop in connective tissue, can be divided into two groups.

One group consists of bones that arise at typical sites and occasionally may be symmetrical. These may be bones that have specific anlages during development but fail to unite with the other bones. They are of considerable practical interest, as the sutures between these bony elements may be mistaken for artificial fissures in radiographs. The second group of supernumerary bones are those that are completely irregular in number, shape, and location, and commonly show individual variations.

A prime example of the first group is the **incarial bone** (1). This term is derived from the word Inca, as the bone has frequently (20%) been found in old Peruvian skulls. *It corresponds to the superior part of the interparietal bone, which developed from connective tissue,* and forms the superior squama of the occipital bone.

The lower part of the interparietal bone (*triangular plate*) fuses as a connective tissue component with the part that develops by chondral ossification (*supraoccipital bone*) and forms the inferior squama. The incarial bone is bounded by both parietal bones (2) and by the inferior squama (3) of the occipital bone. The suture between the incarial bone and inferior occipital squama corresponds to the *sutura mendosa* of the fetus, and is called the *transverse occipital suture* (4). The incarial bone may also be divided into two or three parts.

Other bones that occur in a typical position are those in the fontanelle region. Immediately adjacent to the incarial bone, in the posterior fontanelle, is the **apical bone** (5), which may persist as an independent bone. In the region of the greater fontanelle the **bregmatic bone** (6), also called the frontoparietal bone, occurs less commonly. It is an epactal bone, either circular or rhomboidal in shape, and is uncommon.

Another typical epactal bone is the **epipteric bone** (7) or pterion ossicle, in which we distinguish *anterior* and *posterior parts*. It is found in the sphenoidal fontanelle, where it is bounded by the frontal bone (8), the parietal bone (2), the squamous part of the temporal bone (9), and the sphenoid bone (10). An anterior epipteric bone may not always extend to the parietal bone, and a posterior epipteric bone may not always reach the frontal bone. An undivided epipteric bone may occur, or both types mentioned above may be present, or only one of them. Lastly a separate ossification center (11) may be found in the region of the posterior lateral fontanelle.

The second group comprises specifically the wormian bones, which are particularly common. They occur in the region of the lambdoid, sagittal, and coronal (12) sutures. In addition, they may be found in the transverse occipital suture (see above).

Rarely an independent ossification center (13) may be found within a bone. Epactal bones appear occasionally in the parietal bone (2) and very rarely in the frontal bone.

> **Clinical tip:** Intercalated and wormian bones may extend through the full thickness of the skull; they may be seen only on the surface, or only in the interior of the vault.

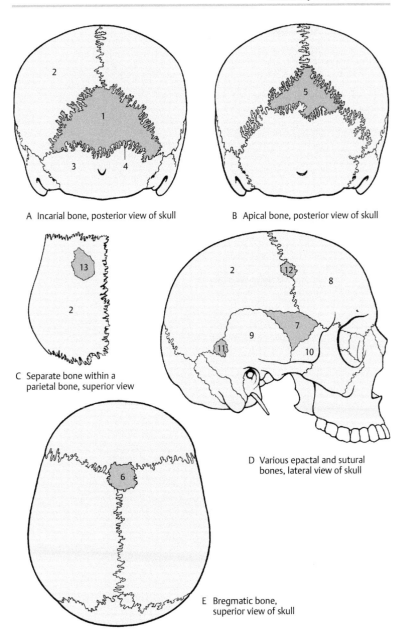

A Incarial bone, posterior view of skull

B Apical bone, posterior view of skull

C Separate bone within a
parietal bone, superior view

D Various epactal and sutural
bones, lateral view of skull

E Bregmatic bone,
superior view of skull

Temporomandibular Joint (A–C)

The **temporomandibular joint** is divided into two compartments by the **articular disc** (**1**). The joint is formed, on the one side, by the **head of the mandible** (**2**) and, on the other side, by the **mandibular fossa** (**3**) with the **articular tubercle** (**4**).

The approximately cylindrical head of the mandible is so positioned that its longitudinal axis forms an angle, in the median plane just in front of the foramen magnum, of approximately 160° with the longitudinal axis of the joint of the opposite side. The head is covered by fibrocartilage and the mandibular fossa likewise possesses a lining of fibrocartilage.

The **articular disc** (**1**) represents a movable socket for the head of the mandible. Its anterior portion consists of fibrous material with interspersed chondrocytes; its posterior part is bilaminar. The upper portion (**5**), which is attached to the posterior wall of the mandibular fossa, consists of loose fibroelastic tissue, whereas the lower portion (**6**), which is fixed to the posterior margin of the head of the mandible, is composed of very taut fibrous tissue. Between these parts lies a retroarticular venous plexus that serves as a malleable cushion (*Zenker*). Anteriorly the articular disc is firmly attached to the joint capsule and the *infratemporal head* of the *lateral pterygoid muscle* (**7**).

The **temporomandibular joint capsule** (**8**) is relatively thin and lax and is reinforced by the *lateral ligament* (**9**), particularly on its lateral side. This ligament extends from the zygomatic arch to the condylar process directly below the head of the mandible, where it very often exhibits an eminence, sometimes a ridgelike elevation or, more rarely, a pitlike depression. In the older literature this was considered a condylar tubercle, a condylar crest, or a condylar fossa, respectively.

The **stylomandibular** (**10**) and **sphenomandibular** (**11**) **ligaments** act as guiding ligaments, although neither has a direct connection with the capsule. The sphenomandibular ligament extends from the *spine of the sphenoid* (**12**) to the *lingula of the mandible* (**13**), whereas the stylomandibular ligament stretches from the *styloid process* (**14**) to the *angle of the mandible* (**15**) and is in connection with the *stylohyoid ligament* (**16**). In addition, fibrous tracts extend from the angle of the mandible to the hyoid bone and are designated as the *hyomandibular ligament* (**17**).

Functionally the temporomandibular joint represents a combination of two joints: an articulation between the articular disc and the head of the mandible and an articulation between the articular disc and the mandibular fossa. Active opening of the mouth always involves a **rotational movement** at the lower joint and an anterior **sliding movement** in the upper joint. The sliding movement is especially brought about by the lateral pterygoid muscle. Besides opening movements, lateral or **grinding movements** are also possible.

The temporomandibular joint and the shape of its articular surfaces are dependent on the development of the dentition and consequently on the age of the individual. In edentulous jaws (infants, elderly), the mandibular fossa is flat and the articular tubercle is inconspicuous.

The **external acoustic meatus** (**18**) lies directly behind the temporomandibular joint and the middle cranial fossa directly above it. The parotid gland (see Vol. 2) and various vessels and nerves are also closely related to this joint.

19 Hyoid bone

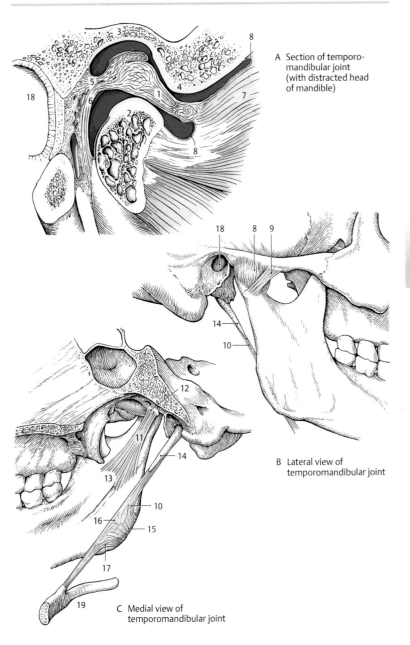

A Section of temporomandibular joint (with distracted head of mandible)

B Lateral view of temporomandibular joint

C Medial view of temporomandibular joint

Muscles and Fasciae

Muscles of the Head

Mimetic Muscles

The mimetic muscles radiate into the skin of the face and the head, and their contraction causes displacement of the skin. This displacement, which takes the form of folds and wrinkles, is the basis of facial expression.

The expression is dependent on racial characteristics, intellectual capacity and the age of the individual. In youthful elastic skin these changes are reversible after muscle contraction, while in old age, when skin elasticity is diminished, wrinkles may persist. In the following section the function of each muscle will be described.

> **Clinical tip:** Facial expressions depend on the state of health. Various diseases of the heart, thyroid gland, stomach, and liver may have an effect on facial expression. Facial expressions may be especially affected by paralysis of the facial nerve.

The mimetic muscles are divided into
- **Muscles of the calvaria**
- **Muscles about the palpebral fissure**
- **Muscles about the nose**
- **Muscles about the mouth**

Mimetic Muscles of the Calvaria (A, B)

The muscles of the calvaria are known collectively as the **epicranius muscle**. This muscle is attached only loosely to the periosteum but very firmly to the scalp. Between the paired anterior and posterior bellies stretches a taut tendon, the **galea aponeurotica** (**1**), from which the fibers of the temporoparietal muscles also arise.

The **occipitofrontalis** consists of an *occipital belly* (**2**) and a *frontal belly* (**3**) on each side. *The former arises from the lateral two-thirds of the highest nuchal line and the latter* lacks a bony origin but instead *arises from the skin and the subcutaneous tissue of the eyebrow and glabellar region.* The frontal belly is also closely related to the orbicularis oculi (**4**).

The **temporoparietalis** (**5**) *arises in the region of the galea aponeurotica and reaches the auricular cartilage.* The most posterior part of the muscle is also known as the *superior auricular muscle.*

The epicranius, particularly its anterior bellies, produces wrinkles in the forehead. In addition, contractions of both frontal bellies may lift the eyebrows and the upper eyelids. This produces the facial expression of astonishment.

Nerve supply: facial nerve.

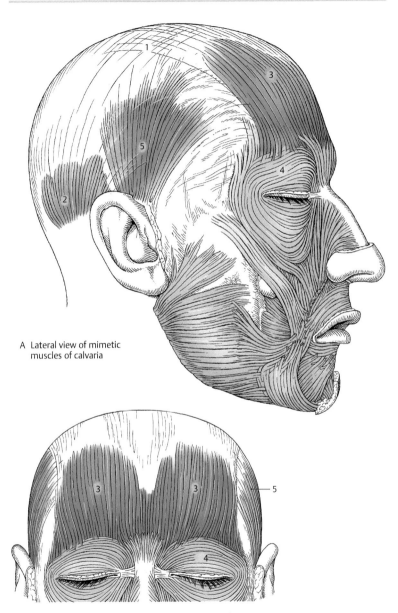

A Lateral view of mimetic
    muscles of calvaria

B Anterior view of mimetic muscles of forehead

Head and Neck

Muscles of the Head, continued

Mimetic Muscles about the Palpebral Fissure (A–F)

The **orbicularis oculi** consists of three parts: **orbital** (**1**), **palpebral** (**2**), and **lacrimal** (**3**); the latter is also regarded as the deep part of the palpebral part. The thick **orbital part** (**1**) *is arranged circularly around the orbit and is attached to the palpebral ligament* (**4**), *the frontal process of the maxilla, and the anterior lacrimal crest.* In the upper lid the medial fibers of the orbital part fan out in the direction of the eyebrows. These fibers are also known as the *depressor supercilii.* The more delicate **palpebral part** (**2**) *lies directly on the eyelids and extends also to the palpebral ligament.* The fibers lie partly on the tarsal plates (**5**) and partly on the orbital septum. The **lacrimal part** (**3**; *Horner's muscle,* deep part of the palpebral part) *lies medial to the deep crus of the palpebral ligament and arises chiefly from the posterior lacrimal crest* (**6**).

The orbital part is concerned with tight closure of the lid, while the palpebral part is primarily concerned with the blink reflex. The function of the lacrimal part is not fully understood. It is thought to expand the lacrimal sac or to expel its contents.

Owing to the close relationship of muscle fibers to the skin, radial folds develop about the lateral canthus of the eye; these are called "crow's feet." The orbicularis oculi produces an expression of worry (**D**) or concern.

The **corrugator supercilii** (**7**) penetrates the orbicularis oculi and the frontal belly (**8**) of the epicranius. *It arises from the glabella and the supraorbital margin and radiates into the skin of the eyebrows.*

It pulls the skin of the eyebrows downward and medially and produces a vertical furrow. It has a protective action in bright sunlight and is called the muscle of pathetic pain. Its contraction produces the expression of a "thinker's brow" (**E**).

Mimetic Muscles about the Nose (A–G)

The **procerus** (**9**) *arises from the nasal dorsum and radiates into the skin of the forehead.* As a relatively thin muscle plate, it produces a transverse fold across the root of the nose.

From a mimetic standpoint, it produces a menacing expression. In old age these folds often become permanent.

The **nasalis** consists of **transverse** (**10**) and **alar** (**11**) **parts**. *It arises from the alveolar eminences of the canine tooth and the lateral incisor, and reaches the skin on the side of the nose.* The transverse part is a thin, broad plate, which is joined by a flattened tendon to the transverse part of the muscle on the opposite side, while the alar part radiates into the skin of the nasal ala.

Contraction of this muscle pulls the nasal ala downward and backward and reduces the size of the nostril. It produces a happy, astonished expression and gives the impression of desire, yearning, and sensuousness (**F**).

The **levator labii superioris alaeque nasi** (**12**) *arises from the infraorbital margin and extends down into the skin of the upper lip and nasal ala.* It elevates not only the skin of the nasal ala but also that of the upper lip. Simultaneous bilateral contraction slightly lifts the tip of the nose.

It elevates the nasal ala and widens the nostrils. Stronger contractions produce a fold in the skin. The resulting facial expression is one of displeasure and discontent (**G**).

In Figure **C**, the orbicularis oculi is reflected medially along with the tarsal plates. View of the posterior surface.

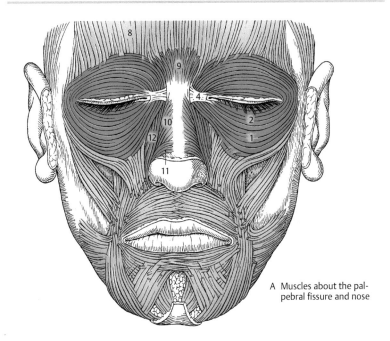

A Muscles about the palpebral fissure and nose

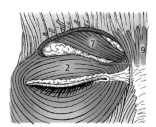

B Corrugator supercilii muscle

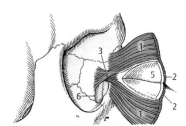

C Internal view of lacrimal part of orbicularis oculi

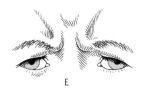

D–G Effects of muscles on facial expression (from *Rouillé*)

Muscles of the Head, continued

Mimetic Muscles about the Mouth (A–L)

The **orbicularis oris** (**1**) looks like a circular muscle, but in fact it consists of four parts (**A**). It also has an inner **labial** and an outer **marginal part**. The shape of the mouth is determined by its tone and the shape of the underlying bone and teeth.

In weak contraction the lips are in contact or closed, while in strong contraction they pout forward and protrude in a sucking shape. The primary function of this muscle is seen in eating and drinking. Its contraction gives a facial expression of reserve (**D**).

The quadrilateral **buccinator** (**2**) *arises from the mandible in the region of the first and second molars and from the pterygomandibular raphe* (**3**). It extends to the oral commissure and forms the lateral wall of the oral vestibule.

It enables air to be blown out of the mouth, pulls the oral commissure laterally, and keeps the mucous membrane of the cheeks free of folds. It is involved in laughing and crying, and, when contracted, produces a facial expression of satisfaction (**E**).

The **zygomaticus major** (**4**) *arises from the zygoma and extends toward the oral commissure.* Some of its fibers decussate with those of depressor anguli oris.

It lifts the corner of the mouth upward and laterally. It produces the facial expression of laughter or pleasure (**F**).

The **zygomaticus minor** (**5**) *extends from the outer surface of the zygoma to the nasolabial groove.*

The **risorius** (**6**) consists of superficial muscle bundles that *arise from the masseteric fascia and run to the oral commissure.* Together with the zygomaticus major it produces the nasolabial folds. They are called, therefore,

the laughing muscles. Contraction of the muscle produces an expression of action (**G**).

The **levator labii superioris** (**7**) is associated with the levator labii superioris alaeque nasi. *It arises from the infraorbital margin and extends into the skin of the upper lip.*

The **levator anguli oris** (**8**) *arises below the infraorbital foramen and runs to the oral commissure.*
It lifts the oral commissure and produces an expression of self-confidence (**H**).

The triangular **depressor anguli oris** (**9**) arises from the inferior border of the mandible and also extends to the oral commissure.
It pulls the oral commissure downward to produce an expression of sadness (**I**).

The **transversus menti** is present only when the depressor anguli oris is well developed. It runs across the mental region and may be associated with the formation of a double chin.

The **depressor labii inferioris** (**10**) *arises from the mandible below the mental foramen and radiates into the skin of the lower lip.*
It pulls the lower lip down and produces an expression of perseverance (**K**).

The **mentalis** (**11**) *arises from the mandible in the region of the alveolar eminence of the lateral incisor and radiates into the skin of the chin.*
It produces the labiomental crease and produces an expression of doubt and indecision (**L**).

The **platysma** (**12**) *radiates from the neck into the facial region* and is connected to the risorius and the depressors of the oral commissure and lower lip.

All mimetic muscles are innervated by the facial nerve.

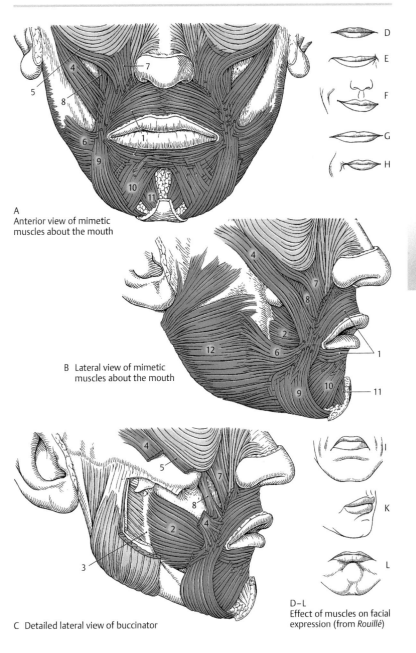

A
Anterior view of mimetic
muscles about the mouth

B Lateral view of mimetic
muscles about the mouth

C Detailed lateral view of buccinator

D–L
Effect of muscles on facial
expression (from *Rouillé*)

Muscles of the Head, continued

Muscles of Mastication (A–E)

The muscles of mastication are innervated by branches of the mandibular nerve. They develop phylogenetically from the first visceral arch.

In a strict sense they include the
- **Masseter** (**1**)
- **Temporalis** (**2**)
- **Lateral pterygoid** (**3**)
- **Medial pterygoid** (**4**)

The **masseter** (**1**) *arises from the zygomatic arch* (**5**) *and is inserted into the masseteric tuberosity* (**6**) on the angle of the mandible. The muscle is divided into a strong **superficial part** (**7**) with oblique fibers and a **deep part** (**8**) whose vertical fibers arise from the inner surface of the zygomatic process of the temporal bone and from the temporal fascia. The masseter, like the temporalis, powerfully closes the jaws by elevating the mandible.
Nerve supply: masseteric nerve.

The **temporalis** (**2**) is the strongest elevator of the lower jaw. *It arises from the temporal fossa* (**9**) *as far as the inferior temporal line and from the temporal fascia* (**10**). It *is inserted by a strong tendon into the coronoid process* of the mandible (**11**). Its insertion also extends downward on the interior and anterior side of the mandibular ramus.
Nerve supply: deep temporal nerves.

The **lateral pterygoid** (**3**) is involved in all movements of the mandible. It serves as the guiding muscle of the mandibular joint. It consists of two parts: an **inferior head** (**12**) *arising from the lateral surface of the lateral plate of the pterygoid process*, and a **superior head** (**13**) *arising from the infratemporal surface* (**14**) *and the infratemporal crest of the greater wing of the sphenoid. The latter part extends to the articular disk, while the former part is inserted into the pterygoid fovea* (**15**).
Nerve supply: lateral pterygoid nerve.

The **medial pterygoid** (**4**) runs almost at right angles to the muscle just described. *It arises in the pterygoid fossa, the* **larger part** *from the medial surface of the lateral pterygoid plate* and the **smaller part** *from the lateral surface of that plate as well as with a few fibers from the maxillary tuberosity. It extends to the* angle of the mandible *where it is inserted into the pterygoid tuberosity*, so that the angle of the mandible lies in a sling formed by the masseter and medial pterygoid. It elevates the mandible and also pushes it forward. It may also be involved in lateral displacement of the lower jaw and may contribute in rotary movements.
Nerve supply: medial pterygoid nerve.

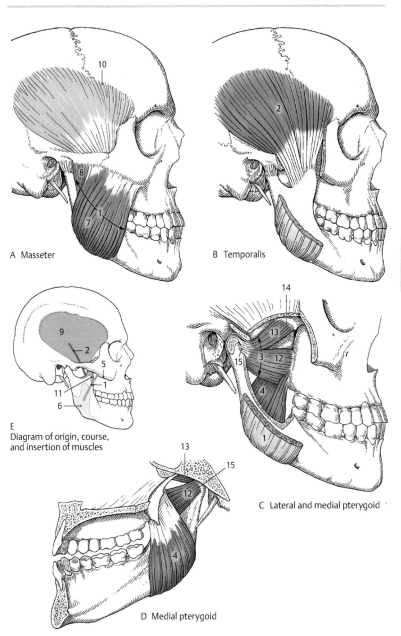

A Masseter

B Temporalis

E
Diagram of origin, course,
and insertion of muscles

C Lateral and medial pterygoid

D Medial pterygoid

Anterior Muscles of the Neck

Infrahyoid Muscles (A, B)

The infrahyoid muscles act on the hyoid bone and thus on the mandible, as well as on the cervical spine.

The infrahyoid muscles include the
- **Sternohyoid**
- **Omohyoid**
- **Sternothyroid**
- **Thyrohyoid**

Phylogenetically they belong to the great ventral longitudinal muscle system. The omohyoid is also included in the muscles of the shoulder girdle (see p. 146).

The **sternohyoid** (**1**) arises from *the posterior surface of the manubrium* (**2**), *from the sternoclavicular joint, and sometimes from the sternal end of the clavicle. It is inserted into the lateral inner surface of the body of the hyoid bone* (**3**).

The **omohyoid** (**4**) has two bellies, a **superior** and an **inferior**, which are connected by an intervening tendon. The **inferior belly** *arises from the superior margin of the scapula*, adjacent to the scapular notch (**5**), and ascends obliquely. In the lateral region of the neck it is closely connected with the middle cervical fascia and it ends in an intermediate tendon that crosses the neurovascular bundle of the neck. The **superior belly** arises from the intermediate tendon and ascends obliquely to the hyoid bone. *It is inserted*, usually without muscle fibers, *into the lateral third of the lower edge of the body of the hyoid* and with some fibers onto the inner surface of the body of the hyoid bone (**6**).

The **sternothyroid** (**7**) is wider than the sternohyoid, which lies superficial to it. *It arises from the posterior surface of the sternal manubrium* (**8**) *and reaches the oblique line of the thyroid cartilage* (**9**). It directly adjoins the thyroid capsule.

The **thyrohyoid** (**10**) is the continuation of the sternothyroid. *It arises from the oblique line of the thyroid cartilage* (**9**) and is *inserted onto the inner surface of the lateral third* (**11**) and *the inferior margin of the median surface of the greater horn* (Fischer).

All the infrahyoid muscles work together and specifically they may approximate the thyroid cartilage to the hyoid bone or, when the mouth is being opened, stabilize the laryngeal cartilages and the hyoid bone, or pull them downward. Because of its relationship to the neurovascular trunk and middle cervical fascia, the omohyoid has the additional function of preventing pressure on the large underlying vein. It keeps the internal jugular vein patent, thereby aiding the return of blood from the head to the superior vena cava.

The infrahyoid and suprahyoid muscles (see Vol. 2) can flex the head forward with the mouth closed. The omohyoid muscle is an accessory muscle in opening the mouth and in flexion, side bending, and rotation of the head (*Fischer and Ransmayr*).
Nerve supply: deep cervical ansa and thyrohyoid branch (C1, C2, and C3).

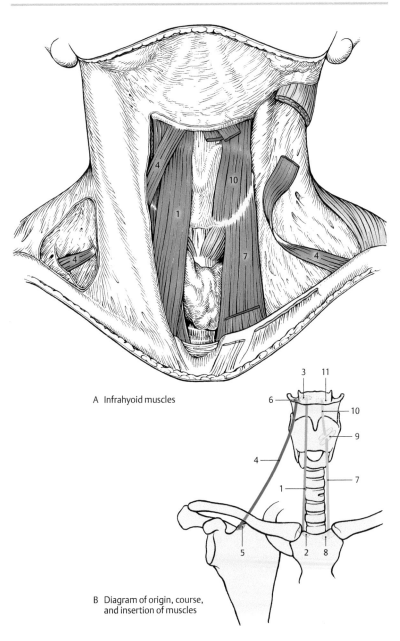

A Infrahyoid muscles

B Diagram of origin, course,
and insertion of muscles

Head and Neck *(sidebar)*

Head Muscles Inserted on the Shoulder Girdle (A–C)

The two muscles of the head that are inserted into the shoulder girdle are the trapezius and sternocleidomastoid.

The **trapezius** (**1**; see also p. 146) is divided into **descending** (**2**), **transverse** (**3**), and **ascending** (**4**) **parts**.

The **descending part** *arises from the superior nuchal line, the external occipital protuberance* (**5**), *and the ligamentum nuchae* (**6**; see p. 56) *and is inserted into the lateral third of the clavicle* (**7**). The **transverse part** *arises from the seventh cervical to the third thoracic vertebrae* (**8**; from the spinous processes and supraspinous ligaments) *and is inserted into the acromial end of the clavicle* (**9**), *the acromion* (**10**), *and part of scapular spine* (**11**). The **ascending part** *arises from the third through 12th thoracic vertebrae* (**12**; from the spinous processes and supraspinous ligaments) *and is inserted onto the spinal trigone and the adjacent part of the scapular spine* (**13**).

The primary function of the trapezius is a static one: it supports the scapula and thus stabilizes the shoulder girdle. Its contraction pulls the scapula and the clavicle backward and toward the vertebral column. The descending and ascending parts rotate the scapula. In addition to producing adduction, the descending part produces slight elevation of the shoulder, assisting the serratus anterior. If the latter muscle is paralyzed, the descending part is able to raise the arm slightly above the horizontal.
Nerve supply: accessory nerve and trapezius branch (C2–C4).

The **sternocleidomastoid** (**14**; see also p. 146) *arises by* **one head** *from the sternum* (**15**) *and by the* **other** *from the clavicle* (**16**). *It is inserted into the mastoid process and the superior nuchal line.* There it has a tendinous connection with the origin of the trapezius.

Unilateral action of the sternocleidomastoid turns the head to the opposite side and bends it to the ipsilateral side. **Bilateral contraction lifts the head.** This muscle is often incorrectly called the flexor of the head. Finally the sternocleidomastoid can be an accessory muscle of respiration if the head is fixed and the intercostal muscles are paralyzed. If the intercostal muscles are still functioning, however, the sternocleidomastoid does not become active.
Nerve supply: accessory nerve and fibers C1–C2 from the cervical plexus.

▬ **Variants:** Since the sternocleidomastoid and trapezius develop from the same material, they sometimes maintain a close relationship. The insertion of the trapezius on the clavicle may be considerably extended medially, and conversely the origin of the sternocleidomastoid may be shifted laterally. In this case the greater supraclavicular fossa, which is bordered by these two muscles and the clavicle, is reduced in size.

Clinical tip: Erb's point 2 (**17**) is located 2 to 3 cm above the clavicle and 1 to 2 cm past the posterior border of the sternocleidomastoid muscle. Stimulation applied at this point contracts various arm muscles by stimulating the upper part of the brachial plexus (see Vol. 3).

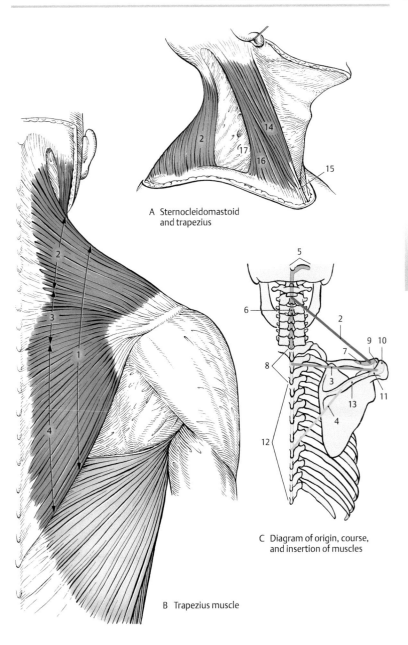

A Sternocleidomastoid and trapezius

B Trapezius muscle

C Diagram of origin, course, and insertion of muscles

Fasciae of the Neck (A, B)

There are three layers of muscular fasciae in the neck between the hyoid bone and the shoulder girdle.

The **superficial layer or investing layer** (**1**) **of the cervical fascia** (**1**) encloses all the structures of the neck except the platysma (**2**) and is continued posteriorly into the nuchal fascia. The sternocleidomastoid (**3**) and trapezius (**4**) are embedded within it. The fascia extends from the mandible to the manubrium sterni and the clavicles. The portion between the hyoid bone and mandible is called the cervical fascia (see below).

Just below the superficial layer is the **middle** or **pretracheal layer** (**5**) **of the cervical fascia** in which the infrahyoid musculature is embedded (see p. 326). This fascia has a firm consistency in the region of the infrahyoid muscles (**6**). It does not end at the lateral borders of the omohyoid muscles, however, but continues laterally as a thin sheet. It comes into contact with the deep or prevertebral layer of the cervical fascia (**7**) and blends with it. It is also connected to the connective tissue sheath around the neurovascular bundle (common carotid artery, internal jugular vein, vagus nerve) as the **carotid sheath** (fasciae cervicalis; **8**).

The pretracheal layer extends in a craniocaudal direction from the hyoid bone to the manubrium sterni and the clavicles. Cranial to the hyoid bone the pretracheal layer blends with the superficial layer of the cervical fascia.

Between the superficial (**1**) and pretracheal (**5**) layers of the cervical fascia is the *suprasternal interfascial* space (**9**; see p. 354) in the region of the middle compartment of the neck.

The **deep** or **prevertebral layer** (**7**) **of the cervical fascia** covers the vertebral column and the deep cervical muscles associated with it. The deep muscles of the neck include the longus capitis, the longus colli (**10**), and the scalene muscles (**11**). The prevertebral layer arises from the skull base and extends into the thoracic cavity, where it is continuous with the endothoracic fascia.

The contents of the neck, larynx, esophagus (**12**), trachea (**13**), and thyroid gland (**14**), with the parathyroid glands, lie between the pretracheal and prevertebral layers.

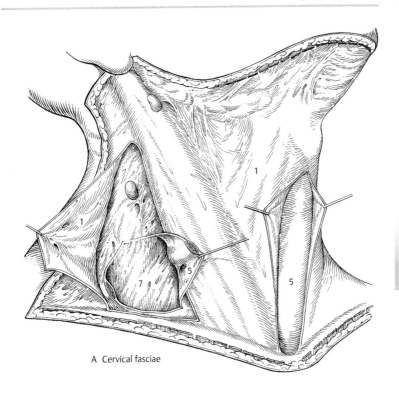

A Cervical fasciae

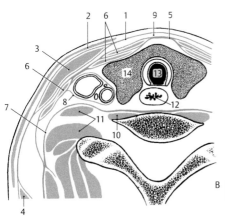

B Section through the neck
   to show cervical fasciae

Anatomical Terms and their Latin Equivalent

Head	Caput
Cribriform plate	Lamina cribrosa
Groove for the inferior petrosal sinus	Sulcus sinus petrosi inferioris
Groove for the lesser petrosal nerve	Sulcus nervi petrosi minoris
Hiatus for greater (lesser) petrosal nerve	
Highest nuchal line	Linea nuchalis suprema
Lesser (greater) wing of sphenoid	Ala minor (major) ossis sphenoidalis
Mastoid (frontal) notch	Incisura mastoidea (frontalis)
Occipital groove	Sulcus arteriae occipitalis
Prechiasmatic groove	Sulcus prechiasmaticus
Superior border of petrous part	Margo superior partis petrosae
Upper (lower) jaw	Maxilla (Mandibula)

Topography of Peripheral Nerves and Vessels

Head and Neck

Regions (A, B)

The head is separated from the neck by a line beginning at the chin and continuing over the body of the mandible, the mastoid process, and the superior nuchal line to the external occipital protuberance.

The jugular notch of the sternum and the clavicles mark the boundary between the neck and trunk. A precise posterior boundary line cannot be identified.

Regions of the Head

The **frontal region** (**1**) comprises the forehead up to the coronal suture. Adjacent to it, over the parietal bone on each side, is the **parietal region** (**2**), and over the squamous part of the temporal bone lies the **temporal region** (**3**). The **infratemporal region** (**4**) is covered by the zygomatic arch. Posteriorly the **occipital region** (**5**) lies over the occipital bone.

The various **anterior fascial regions** are the **nasal region** (**6**), the **oral region** (**7**), and the **chin or mental region** (**8**). The **orbital region** (**9**) lies around the eyes, the **infraorbital region** (**10**) is the area lateral to the nose, and the **buccal region** (**11**) is lateral to the oral region. The **zygomatic region** (**12**) lies about the zygomatic bone, and the **parotid region** (**13**) contains the masseter muscle and the parotid gland.

Regions of the Neck

The **neck** is divided into a **posterior cervical region** or **nuchal region** (**14**) and anterolateral region. The latter is subdivided by the **sternocleidomastoid region** (**15**) into an unpaired **anterior cervical triangle** and the paired lateral cervical regions. The anterior cervical triangle includes the area between the Morton neuroma and the anterior borders of both sternocleidomastoids. It can be further subdivided. In the center lies the **median cervical region** (**16**), which is bounded by the hyoid bone, the omohyoids and sternocleidomastoids, and inferiorly by the jugular notch of the sternum. The depressed part of the median cervical region, which lies just above the sternal jugular notch, is called the *suprasternal fossa* (**17**).

The **submental triangle** or **submental region** (**18**) extends between the hyoid bone and the mental region. Laterally it is separated from the **submandibular triangle** (**19**) by the anterior belly of the digastric muscle. This triangular area is bounded superiorly by the mandible. It may be helpful to use the angular tract of the cervical fascia to separate the submandibular triangle from its superoposterior part, the **retromandibular fossa** (**20**), which contains the cervical part of the parotid gland and the trunk of the facial nerve. The **carotid triangle** (**21**) is of great practical importance as it contains the bifurcation of the common carotid artery. It is bounded superiorly by the posterior belly of the digastric, anteriorly by the superior belly of the omohyoid, and posteriorly by the sternocleidomastoid.

The **lateral cervical region** (**22**), or **posterior cervical triangle**, ends anteriorly at the sternocleidomastoid, posteriorly at the trapezius, and inferiorly at the clavicle. The omoclavicular triangle, or *greater supraclavicular fossa* (**23**), merits special mention in this area. It is bounded by the sternocleidomastoid, the inferior belly of the omohyoid, and the clavicle. In thin individuals it may also be possible to see the *lesser supraclavicular fossa* (**24**) between the two origins of the sternocleidomastoid.

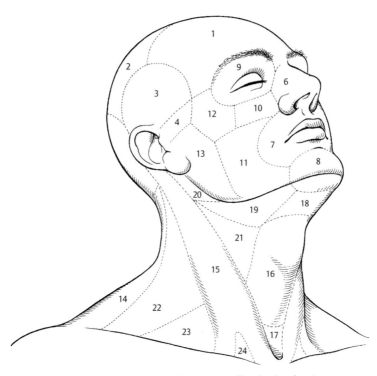

A Lateral view of head and neck regions

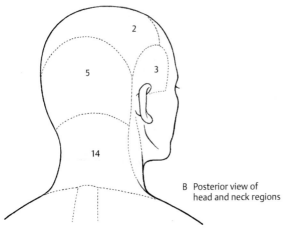

B Posterior view of
head and neck regions

Peripheral Pathways

Anterior Facial Regions (A, B)

The blood supply of the face comes primarily from branches of the external carotid artery and to a lesser extent from the internal carotid artery. On the anterior margin of the *masseter* (**1**), the *facial artery* (**2**) ascends and anastomoses via the *angular artery* (**3**) with the *dorsal nasal artery* (**4**), which springs from the ophthalmic artery. By way of larger branches in the facial region, the facial artery sends smaller branches to the lip region (see p. 340). The lateral facial region is supplied either by the facial artery or by the *transverse facial artery* (**5**), which is a branch of the *superficial temporal artery* (**6**). The deep layers of the anterior facial region receive their blood supply from the *infraorbital artery* (**7**), a terminal branch of the maxillary artery. The superficial temporal artery (**6**) supplies the temporal and parietal regions, and the forehead area proper is supplied by the *supratrochlear* (**8**) and *supraorbital* (**9**) arteries, both being terminal branches of the ophthalmic artery. Among the larger superficial veins of the facial region only the *facial vein* (**10**), which anastomoses via the *angular vein* (**11**) with the *dorsal nasal vein* and the *superficial temporal vein* (**12**), lies superficially.

The mimetic muscles are supplied by branches of the facial nerve. These are the *temporal* (**13**), *zygomatic* (**14**), and *buccal* (**15**) *branches* and the *marginal mandibular branch* (**16**).

The sensory innervation to the skin of the face is derived from branches of the **trigeminal nerve**, the ophthalmic, the maxillary, and the mandibular nerves.

Ophthalmic nerve: The skin of the forehead is supplied by the frontal nerve with its *supratrochlear nerve* (**17**) and the *supraorbital nerve* (**18**). Near the lateral canthus of the eye the *lacrimal nerve* (**19**) penetrates the orbicularis oculi (**20**) with a few of its branches and innervates the skin in that region. The *external nasal nerve* (**21**), a branch of the nasociliary nerve, supplies the dorsum and tip of the nose.

Maxillary nerve: The lower eyelid, the cheek area, the lateral nasal region, the upper lip, and the anterior temporal region are innervated by branches of the *infraorbital nerve* (**22**) and the *zygomaticofacial* and *zygomaticotemporal branches* of the zygomatic nerve.

Mandibular nerve: The skin of the lower lip, mandible (except its angle) and the mental area are supplied by the *mental nerve* (**23**), while the *auriculotemporal nerve* (**24**) innervates the skin over the mandibular ramus, the concha of the auricle, the largest portion of the external acoustic meatus, most of the external surface of the tympanic membrane, and the posterior temporal region. The mental nerve exits from the mental foramen, and the auriculotemporal nerve ascends in front of the external ear together with the superficial temporal artery and vein.

Clinical tip: The anastomosis between the facial vein (**10**) and dorsal nasal vein is important since it affords a direct connection to the cavernous sinus (see Vol. 2), through which an infection may spread from an extracranial site (e.g., a furuncle on the lip) into the interior of the skull.

Pressure Points of the Trigeminal Nerve (B) and Clinical Relevance

The sensitivity of the three main divisions of the trigeminal nerve can be tested at the branch points of these divisions. The *supraorbital notch* (**25**) serves as a pressure point for the **supraorbital nerve** (**18**), the *infraorbital foramen* (**26**) as a pressure point for the **infraorbital nerve** (**22**), and the **mental foramen** (**27**) for the **mental nerve** (**23**). All three pressure points lie roughly along a vertical line (**28**) running through the center of pupil approximately 2 to 3 cm lateral to the midline.

The broken blue lines in Figure **B** indicate the boundaries between the regions supplied by the three divisions of the trigeminal nerve.

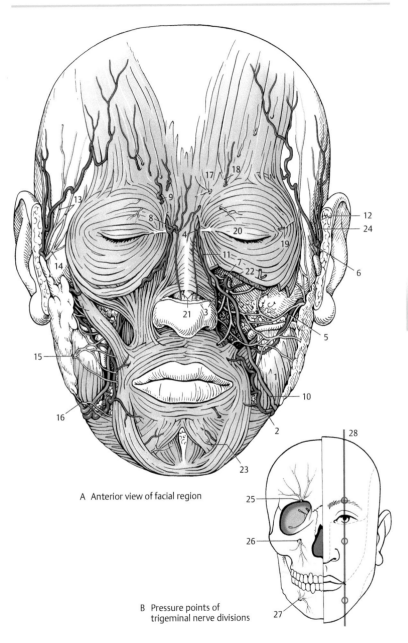

A Anterior view of facial region

B Pressure points of
trigeminal nerve divisions

Orbital Region (A, B)

In an anterior view the orbital region roughly corresponds to the region of the orbicularis oculi. In this area there are anastomoses between the facial vessels and vessels from the interior of the skull. These anastomoses are of practical importance, both as a source of collateral circulation and for the spread of bacteria from the facial skin through the veins to the interior of the skull.

In the **orbital region** (**A**) the *orbital septum* (**1**) separates the superficial structures from the contents of the orbital cavity. Superficially the vessels are a continuation of *the facial artery and vein* (**2**), namely the *angular artery and vein* (**3**). The *dorsal nasal artery and vein* (**5**) lie in front of the *palpebral ligament* (**4**). The dorsal nasal artery may branch from the *supratrochlear artery* (**6**) outside (see figure) or inside the orbit. Together with the dorsal nasal artery, the *infratrochlear nerve* (**7**) also pierces the orbital septum. It often anastomoses with the *supratrochlear nerve* (**8**), which is only separated from it by the *trochlea* (**B9**).

The supratrochlear nerve innervates the skin on the medial part of the forehead and nasal root and is accompanied by the *supratrochlear artery and veins* (**10**). Lateral to the supratrochlear nerve, the *medial branch* (**11**) of the supraorbital nerve pierces the septum and adjacent to it is the *lateral branch* (**12**) of the supraorbital nerve, accompanied by the *supraorbital artery* (**13**). This artery and nerve leave an indentation in the bone, the supraorbital notch, which is sometimes closed to form a supraorbital foramen (see p. 292).

In the lateral canthus of the eye, branches of the *lacrimal nerve* (**14**) pierce the orbital septum. The upper eyelid is innervated by these nerves and by branches of the frontal nerve. The lower eyelid is innervated by inferior palpebral branches of the *infraorbital nerve* (**15**), which emerges from the infraorbital foramen together with the *infraorbital artery* (**16**).

Within the **orbit** (**B**), after removal of the orbital septum, the *superior oblique muscle of the eye* (**17**) becomes visible as it bends around the trochlea (**9**). The *levator palpebrae superioris* (**18**) and the *superior tarsal muscle* (**19**) can also be seen. A lateral tendon slip from levator palpebrae superioris divides the lacrimal gland into an *orbital part* (**20**), also called the *Galen gland*, and a *palpebral part* (**21**), formerly called the *Rosenmüller* or *Cloquet gland*. Below the eyeball the *inferior oblique muscle of the eye* (**22**) arises from the infraorbital margin.

In the medial canthus of the eye, the outer limb of the (medial) palpebral ligament can be divided to expose the *lacrimal sac* (**23**) with the *lacrimal canaliculi* (**24**) which open into it.

25 Cut edge of the lateral part of the levator palpebrae superioris tendon
26 Outer limb of the (medial) palpebral ligament, divided and reflected

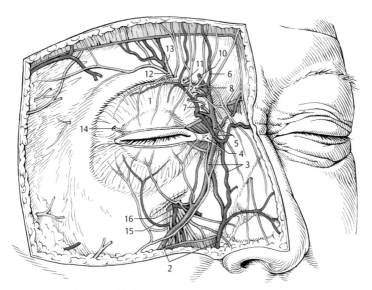

A Orbital region: orbital septum

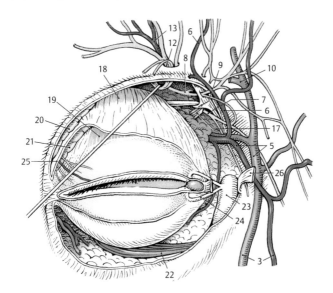

B Orbital region: lacrimal apparatus, intraorbital vessels and nerves

Lateral Facial Regions (A, B)

Parotidomasseteric Region (A)

The **parotidomasseteric region** is the most important of the lateral facial regions. In it lies the parotid gland (see Vol. 2), which is differentiated into a superficial and a deep part. Anteriorly the *parotid gland* (**1**) lies on the *masseter muscle* (**2**) and posteriorly it occupies the retromandibular fossa. At the anterior margin of the parotid gland the *parotid* or *Stensen's duct* (**3**) leaves the gland and runs deeply in front of the *buccal fat pad* (**4**). It is accompanied by the somewhat variable *transverse facial artery* (**5**), a branch of the *superficial temporal artery* (**6**). This vessel supplies blood to portions of the face.

Between the superficial and deep parts of the gland is the parotid plexus of the facial nerve, whose branches—the *temporal* (**7**), *zygomatic* (**8**), *buccal* (**9**), and *marginal mandibular* (**10**)—become visible on the superior and anterior borders of the gland and are distributed to the mimetic muscles. At the inferior border of the parotid gland, the *cervical branch of the facial nerve* (**11**) is seen, which sometimes runs for a distance together with the marginal mandibular branch and which forms the superficial ansa cervicalis with the transverse cervical nerve (see p. 358).

At the inferior margin of the parotid gland the *retromandibular vein* (**12**) runs with the cervical branch of the facial nerve or with the marginal mandibular branch. This vein is joined by the *facial vein* (**13**) as it runs along the anterior border of the masseter muscle (**2**). Usually the *facial artery* (**14**) passes in front of the facial vein around the mandible (**bony pressure point**). It continues as the angular artery (see p. 336) to the medial canthus of the eye and gives off the *inferior* (**15**) and *superior* (**16**) *labial arteries*.

The superficial temporal artery (**6**) lies at the superior margin of the parotid gland, directly in front of the external ear, where it gives off *anterior auricular branches* to the external ear, as well as the *zygomatico-orbital artery*. Finally, after providing a *middle temporal artery*, it divides into a *frontal* (**17**) and *parietal* (**18**) *branch*. It can take a very tortuous course and is accompanied by the *superficial temporal vein* (**19**). The *auriculotemporal nerve* (**20**), a twig from the mandibular nerve, follows the parietal branch (**18**) and innervates the skin of the posterior temporal region. *Superficial parotid lymph nodes* (**21**) are found in variable numbers usually directly in front of the external ear.

Intraparotid Plexus (B)

Upon removal of the superficial part of the parotid gland there is usually a *superior* (**22**) and an *inferior branch* (**23**) of the facial nerve. The superior branch sends out the temporal branches (**7**) and the zygomatic branches (**8**), while the inferior branch gives off the buccal branches (**9**), the marginal mandibular branch (**10**), and the cervical branch (**11**). Both branches and their ramifications usually connect with one another by anastomoses, thus creating the intraparotid plexus.

Parallel to the inferior branch runs the retromandibular vein (**12**). An accessory parotid gland, which is sometimes present (**24**), may be small and is then covered by the superficial part of the parotid gland. If it is larger, it adjoins the parotid duct anterior to the parotid gland.

Clinical tip: Malignant tumors of the parotid gland may cause damage to the facial nerve and its branches.
The superficial temporal artery **pulse** is palpable at the superior border of the parotid gland, just in front of the external auditory canal. The facial artery pulse is palpable at the anterior border of the masseter muscle at the base of the mandible.

25 Great auricular nerve
26 Platysma

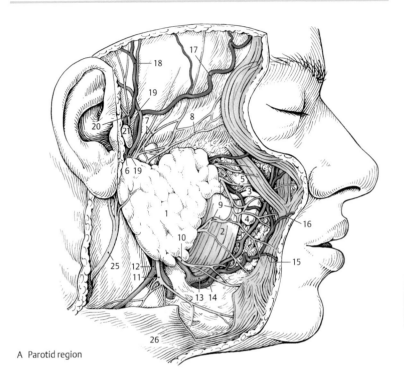

A Parotid region

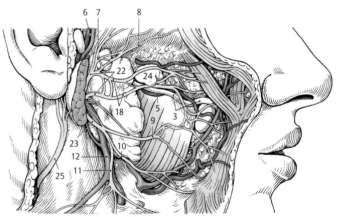

B Parotid plexus

Infratemporal Fossa (A–G)

First Layer (A)

Access to the infratemporal fossa is gained by removal of the zygomatic arch and the coronoid process of the mandible. The *lateral* (**1**) and *medial* (**2**) *pterygoid muscles* then become visible. The infratemporal fossa is bounded anteriorly by the *maxillary tuberosity* (**3**) and the *pterygomandibular raphe* (**4**).

The *maxillary artery* (**5**) may run between the two heads of the lateral pterygoid muscle. In this region it gives off the *buccal artery* (**6**) and the *superior posterior alveolar artery* (**7**) in addition to branches to the masticatory muscles, before descending into the pterygopalatine fossa. The maxillary artery is surrounded by a venous plexus, the *pterygoid plexus*, which is continuous with the *maxillary veins*. The *buccal nerve* (**8**) also runs between the two heads of the lateral pterygoid muscle. The *lingual* (**9**) and *inferior alveolar* (**10**) *nerves* are visible below the lateral pterygoid, and the *masseteric nerve* (**11**) above it.

Second Layer (B)

The vessels and nerves of the infratemporal fossa only become fully visible after removal of the lateral pterygoid muscle and the condylar process of the mandible. The maxillary artery (**5**) lies lateral to the *sphenomandibular ligament* (**12**) and to the large branches of the *mandibular nerve* (**13**) and can be traced throughout its entire length. Its mandibular part gives off the *anterior tympanic artery* (**14**), the *deep auricular artery* (**15**), and the *middle meningeal artery* (**16**), which reaches the interior of the skull through the foramen spinosum.

The middle meningeal artery passes between the two roots of the *auriculotemporal nerve* (**17**), which may often receive additional fibers (**18**) from the inferior alveolar nerve (**10**). The auriculotemporal nerve (**17**) anastomoses with *communicating branches* (**19**) from the *facial nerve* (**20**).

By means of this anastomosis, which can wind around the *superficial temporal artery* (**21**), parasympathetic fibers pass from the otic ganglion to the facial nerve, which then transmits them to the parotid gland (see Vol. 3).

Before it reaches the mandibular canal, the inferior alveolar nerve (**10**) gives off the *mylohyoid nerve* (**22**), which is accompanied by the *mylohyoid artery* (**23**), a branch of the *inferior alveolar artery* (**24**). The *chorda tympani* (**25**), which carries parasympathetic and sensory fibers, descends to join the lingual nerve. From the anterior part of the mandibular nerve (**13**), the buccal nerve (**8**) arises to innervate the buccal mucosa and supply parasympathetic fibers from the otic ganglion to the buccal glands. Purely motor branches, such as the masseteric nerve (**11**), the medial and lateral pterygoid nerves, and the *deep temporal nerves* (**26**) arise also from the anterior part.

Special Features (C–G)

The maxillary artery has a highly variable course because of its development. Thus the maxillary artery (**5**) often runs lateral to the lateral pterygoid muscle (**C**) and less often medial to it (**A, D**). When it does lie medially the artery usually runs to the pterygopalatine fossa, laterally (**E**) to the inferior alveolar nerve (**10**) and the lingual nerve (**9**), but medially to the buccal nerve (**8**). However, the artery may run between the branches (**F**) or, more rarely, medial to the trunk of the mandibular nerve (**G**).

> **Clinical tip:** The infratemporal fossa provides clinical access to the trigeminal ganglion. **Trigeminal neuralgia** can be treated with injections and other procedures on the trigeminal ganglion, which is accessible through the foramen ovale.

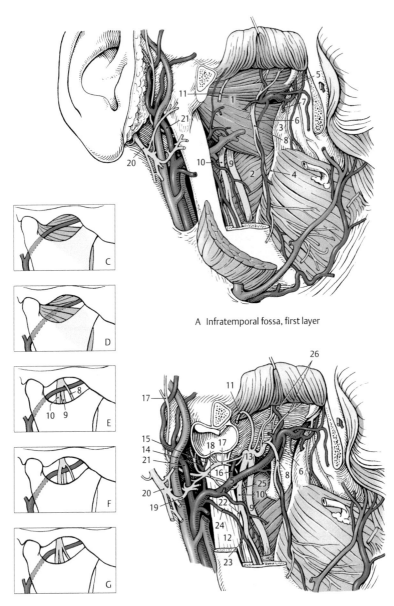

A Infratemporal fossa, first layer

C–G Variants of maxillary artery

B Infratemporal fossa, second layer

Superior View of the Orbit (A, B)

Only a few vessels and nerves of the orbit can be seen when viewed from the front, and a clear view of their relationships can be gained only by removing of the roof of the orbit.

First Layer (A)

After removal of the orbital roof and periorbita, it is possible to see the nerves that run through the lateral part of the superior orbital fissure. The most medial is the *trochlear nerve* (**1**), which innervates the *superior oblique muscle of the eyeball* (**2**). Alongside runs the relatively thick *frontal nerve* (**3**), which lies on the *levator palpebrae superioris* (**4**). The *supraorbital artery* (**5**) accompanies its lateral branch, the *supraorbital nerve* (**6**), while the medial branch, the *supratrochlear nerve* (**7**), runs along with the *supratrochlear artery* (**8**). The most lateral is the *lacrimal nerve* (**9**), which innervates the *lacrimal gland* (**10**) with the fibers received from the zygomatic nerve, and the skin at the lateral canthus of the eye.

The *superior ophthalmic vein* (**11**) also passes through the lateral part of the superior orbital fissure. One of its tributaries crosses below the *superior rectus muscle* (**12**) having anastomosed with the external facial veins (see p. 336) in the region of the *trochlea* (**13**); the other branch runs together with the *lacrimal artery* (**14**), which may give off small branches to muscles, and the *short posterior ciliary arteries* (**B 15**). Covered by the superior oblique (**2**) on the medial side lie the *anterior ethmoidal artery and nerve* (**16**), and superior to this muscle and more posteriorly run *the posterior ethmoidal artery and nerve* (**17**).

Second Layer (B)

After division and reflection of the levator palpebrae superioris (**4**) and the superior rectus (**12**), the *optic nerve* (**18**), the *ophthalmic artery* (**19**), and the nerves that pass through the medial part of the superior orbital fissure become visible.

The *abducens nerve* (**20**), which innervates the *lateral rectus* (**21**), is the most lateral of these. Just medial to it runs the *oculomotor nerve* (**22**), which divides into two branches. The *superior branch* (**23**) supplies the levator palpebrae superioris (**4**) and the superior rectus (**12**). The *inferior branch* (**24**) innervates the *medial rectus* (**25**), inferior rectus, and inferior oblique. In addition, the inferior branch sends the *oculomotor root* (**26**) to the *ciliary ganglion* (**27**), which adjoins the optic nerve (**18**). The ganglion is connected with the *nasociliary nerve* (**29**) via a *nasociliary root* (**28**). From the ganglion the *short ciliary nerves* (**30**), which contain postganglionic parasympathetic fibers for innervating the ciliary muscle and sphincter pupillae, run to the *eyeball* (**31**). The short ciliary nerves also carry sensory and sympathetic fibers; the latter reach the ganglion from a sympathetic network (not shown) around the ophthalmic artery as the sympathetic root of the ciliary ganglion. Sensory fibers from the nasociliary nerve also run to the eyeball via the *long ciliary nerves* (**32**). The nasociliary nerve, which gives off the ethmoidal nerves, is continued as the *infratrochlear nerve* (**33**).

> **Clinical tip:** The superior ophthalmic vein is important, as it anastomoses with the facial veins and opens into the cavernous sinus. It provides a route by which infection in the facial region can spread to the cavernous sinus.

> ▬ **Variant:** There is sometimes a *meningo-orbital artery* (**34**) that joins the middle meningeal with the lacrimal artery (anastomotic branch with the lacrimal artery).

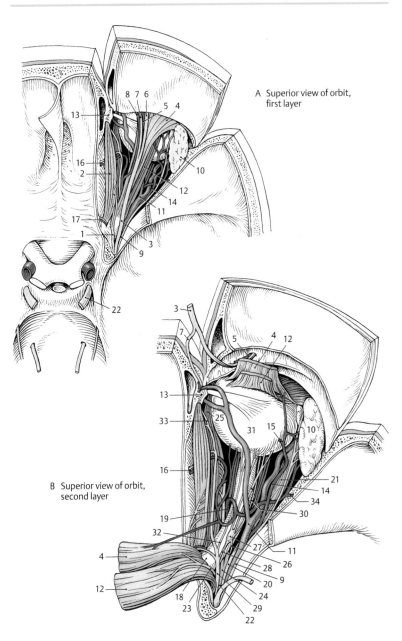

A Superior view of orbit,
first layer

B Superior view of orbit,
second layer

Occipital Region and Posterior Cervical (Nuchal) Region (A)

The vessels and nerves that supply the nuchal skin are subcutaneous. The *occipital artery* (**1**) penetrates the nuchal fascia above the tendinous arch (**2**) that extends between the attachment sites of the *sternocleidomastoid* (**3**) and *trapezius* (**4**). The occipital artery is accompanied by an *occipital vein* (**5**) of variable caliber, which is sometimes absent and may be completely replaced by a large median vessel, the "*nuchal azygos vein*" (**6**).

In close proximity to the occipital artery and vein, the *greater occipital nerve* (**7**) becomes subcutaneous. This nerve is the dorsal branch of the second cervical spinal nerve. Together with the *lesser occipital nerve* (**8**) from the cervical plexus, it innervates the skin on the back of the head. Anastomoses almost always exist between branches of the greater and lesser occipital nerves. Immediately behind the ear the skin is also supplied by the posterior branch of the *great auricular nerve* (**9**). In addition, segmental dorsal branches, of which the third *occipital nerve* (**10**) is the more strongly developed, contribute to the cutaneous innervation of this region. *Occipital lymph nodes* (**11**) are found at the points where the vessels and nerves pass through the nuchal fascia.

Suboccipital Triangle (B)

The suboccipital triangle, or "vertebral artery triangle," can be seen only after removal of all the superficial muscles (**A**; sternocleidomastoid [**3**], trapezius [**4**], *splenius capitis* [**12**], and *semispinalis capitis* [**13**]). The *vertebral artery* (**14**) is located in this region. It passes through the transverse foramina (cervical part) of the upper six cervical vertebrae; then its atlantic part lies in the groove for the vertebral artery on the *posterior arch of the atlas* (**15**) and enters the cranial cavity by piercing the posterior atlantooccipital membrane.

The triangle is bounded by the *rectus capitis posterior major* (**16**), *obliquus capitis superior* (**17**) and *obliquus capitis inferior* (**18**). In this area the vertebral artery gives off a branch (**19**) to the surrounding muscles. Between the artery and the posterior arch of the atlas lies the *suboccipital nerve* (**20**), which, as the dorsal branch of the first cervical spinal nerve, innervates the muscles mentioned above and the *rectus capitis posterior minor* (**21**).

Clinical tip: Suboccipital puncture is performed in these regions, using a needle to withdraw cerebrospinal fluid from the cerebellomedullary cistern (see Vol. 3). The needle is introduced in the midline (**24**) between the external occipital protuberance and the spinous process of the axis. The needle tip is aimed directly at the root of the nose and is advanced between the recti capitis posteriores minores, piercing the posterior atlantooccipital membrane and the dura mater. A firm resistance is felt as the needle pierces the dura. The cerebellomedullary cistern lies directly beneath the dura mater. The depth of needle insertion should **not** exceed **4 to 5 cm** in adults!
Note that if *contraindications* exist (e.g., papilledema, mass lesions), suboccipital puncture should be withheld in favor of lumbar puncture (see p. 42).

22 Parotid gland
23 Mastoid lymph node

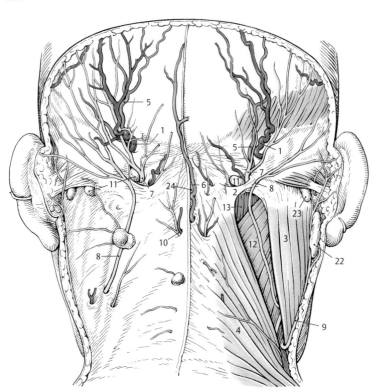

A Occipital and posterior cervical regions
   Left: subcutaneous layer
   Right: subfascial layer

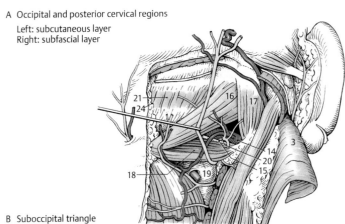

B Suboccipital triangle

Peripheral Pathways

Lateral Pharyngeal and Retropharyngeal Spaces (A)

The vessels and nerves between the head and the trunk run through the neck in a space that is lateral and posterior to the pharynx.

The most posterior structure is the *sympathetic trunk* (**1**), which divides at the *superior cervical ganglion* (**2**) into the *jugular nerve* (**3**) and the *internal carotid nerve* (**4**). While the carotid nerve follows the *internal carotid artery* (**5**), the jugular nerve turns toward the *inferior ganglion* (**6**) of the *vagus nerve* (**7**). In addition, there are connections to the *hypoglossal nerve* (**8**) and to the *carotid body* (**9**), which also receives fibers from the *nerve to the carotid sinus*, the *carotid branch* (**10**). Furthermore, the superior cervical ganglion sends out delicate descending branches, the *external carotid nerves* (not illustrated), to the external carotid plexus as well as laryngopharyngeal branches and the superior cervical cardiac nerve.

The vagus nerve (**7**) passes through the jugular foramen, has a superior and inferior ganglion (**6**), and descends between the internal carotid artery (**5**) and the *internal jugular vein* (**11**). In addition to small branches and anastomoses the vagus nerve running medial to the internal carotid artery gives off the *superior laryngeal nerve* (**12**), which divides into an *external* (**13**) and an *internal* (**14**) *branch*. Other branches include the auricular branch and the *pharyngeal rami* (**15**), which run along with the pharyngeal branches (**16**) of the *glossopharyngeal nerve* (**17**) to supply the muscles of the pharynx and the pharyngeal mucosa.

The glossopharyngeal nerve (**17**), separated from the vagus nerve (**7**) and from *the external branch of the accessory nerve* (**19**) by a bridge of dura (**18**), traverses the jugular foramen. After giving off pharyngeal branches and the nerve to the carotid sinus, the *carotid branch* (**10**), it runs downward and forward between the inter-

nal carotid artery (**5**) and *external carotid artery* (**20**).

The external branch of the accessory nerve (**19**) usually takes a course posterior to the *superior bulb* (**21**) of the internal jugular vein (**11**). Then it runs laterally and passes through the *sternocleidomastoid* (**22**) or medial to it in the lateral cervical region, also called the posterior cervical triangle (see p. 358).

The hypoglossal nerve (**8**) reaches the retropharyngeal space by passing through the hypoglossal nerve canal. It then enters the lateral pharyngeal space and runs anteriorly, passing lateral to both carotid arteries. Immediately below the skull base it receives fibers (**23**) from the first and second cervical segments. It gives off most of these fibers as the *superior* (*anterior) root of the "deep" cervical ansa* (**24**; see p. 362).

The external carotid artery gives off its posterior branch, the *ascending pharyngeal artery* (**25**), which ascends alongside the pharynx and reaches the interior of the skull base passing through the jugular foramen by its branch, the posterior meningeal artery.

26 Pharyngobasilar fascia
27 Pharyngeal raphe
28 Superior pharyngeal constrictor
29 Middle pharyngeal constrictor
30 Inferior pharyngeal constrictor
31 Stylopharyngeus
32 Facial nerve
33 Thyroid gland
34 Right superior parathyroid gland

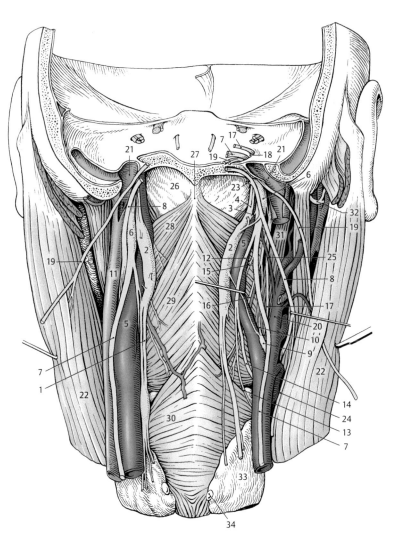

A Lateral pharyngeal and retropharyngeal spaces

Submandibular Triangle (A, B)

The submandibular triangle (**A**) is bounded by the *base of mandible* (**1**), the *anterior belly* (**2**) of the *digastric,* and the *angular tract of the cervical fascia* (**3**) with the interglandular septum. Deep down, starting from the tractus angularis, the interglandular septum separates the submandibular compartment from the parotid compartment. When it is removed, a communication can be established between the submandibular triangle and the retromandibular fossa (**B**).

Submandibular Triangle, Superficial Layer (A)

The *submandibular gland* (**4**) lies superficial to the *mylohyoid* (**5**), around the posterior margin of which winds the *submandibular duct* (**6**) accompanied by a more or less well-developed *uncinate process.*

Thus, the mylohyoid divides the submandibular triangle into a superficial and a deep compartment. The *facial artery and vein* (**7**) pass through the gland. While in the gland the facial artery gives off the *submental artery* (**8**), which runs to the chin superficial to the mylohyoid (**5**), accompanied by the homonymous vein. The *mylohyoid nerve* (**9**), which arises from the inferior alveolar nerve, lies in the same plane and innervates the mylohoid muscle and the anterior belly (**2**) of the digastric.

One or more *submental lymph nodes* (**10**) are on the outside of the mylohyoid and collect lymph from the chin and lower lip regions. Deep and medial to the mylohyoid, the *lingual nerve* (**11**) runs in an arch toward the tongue and is connected to the *submandibular ganglion* (**12**) by *ganglionic branches. Glandular branches* run from the ganglion to the submandibular gland. The submandibular duct (**6**) runs in the immediate vicinity of the ganglion together with the *hypoglossal nerve* (**13**) and an *accompanying vein.*

Submandibular Triangle, Deep Layer (B)

The *geniohyoid* (**14**) and *hyoglossus* (**15**) are exposed by dividing the anterior belly of the digastric (**2**) and the mylohyoid (**5**). The styloglossus radiates forward into the tongue. Inferior to the hypoglossal nerve (**13**), the fibers of the hyoglossus (**15**) can be separated to demonstrate the underlying *lingual artery* (**16**), which is sometimes accompanied by a small lingual vein. The area where the artery is identified is called the **lingual artery triangle.** It is formed by the hypoglossal nerve, the anterior belly of the digastric, and the posterior border of the mylohyoid muscle (see **A**).

Medial to the hyoglossus the *glossopharyngeal nerve* (**17**) descends from the retromandibular fossa and is crossed by the *ascending palatine artery* (**18**), a branch of the facial artery. The *stylohyoid ligament* (**19**) runs parallel to the glossopharyngeal nerve.

20 External carotid artery
21 Facial nerve
22 Masseter
23 Sternocleidomastoid
24 External jugular vein

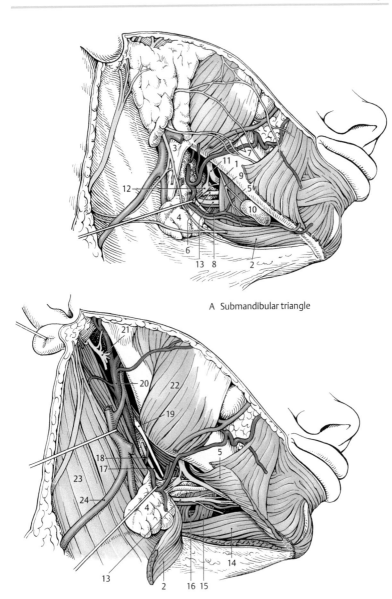

A Submandibular triangle

B Submandibular triangle (deep layer)
and retromandibular fossa

Retromandibular Fossa (A)

The retromandibular fossa is bounded by the *ramus of the mandible* (**1**), the posterior belly of the digastric, and the *angular tract of the cervical fascia* (**2**). It contains the deep portion of the parotid gland.

With the parotid gland removed, the *facial nerve* (**3**) can be seen emerging from the stylomastoid foramen and dividing into its branches. The first branch to be given off outside the skull is the *posterior auricular nerve* (**4**), which supplies the occipital belly of the occipitofrontal muscle with the occipital branch and the posterior muscles of the ear with the auricular branch. The next branches to leave the trunk of the facial nerve are the *digastric* (**5**) and *stylohyoid* (**6**) *branches.* The facial nerve then splits up into the *parotid plexus* (**7**), which lies between the superficial and deep parts of the parotid gland. This plexus also forms loops around the neighboring vessels and sends branches to the mimetic muscles, that is, the *temporal* (**8**), *zygomatic* (**9**), and *buccal* (**10**) *branches and the marginal mandibular branch* (**11**). The *cervical branch of the facial nerve* (**12**) also arises from the parotid plexus. It innervates the platysma and forms the "superficial ansa cervicalis" with the transverse cervical nerve.

Deep in the retromandibular fossa is the *external carotid artery* (**13**), which divides into the *maxillary artery* (**14**) and the *superficial temporal artery* (**15**). The first branch of the superficial temporal artery is usually the *transverse facial artery* (**16**), but that vessel may also spring directly from the external carotid artery (see figure). The external carotid artery is accompanied by the *retromandibular vein* (**17**), which is formed by the *superficial temporal* (**18**) and the *maxillary* (**19**) *veins.*

When the retromandibular vein runs superficially, it anastomoses with the *facial vein* (**20**) and continues into the *external jugular vein* (**21**). In this case we find deep accompanying veins (**22**) of the external carotid artery. The *posterior auricular artery* (**23**) ascends behind the retromandibular vein. At the superior margin of the retromandibular fossa, the superficial temporal artery and vein cross the *auriculotemporal nerve* (**24**), which emerges from the infratemporal fossa and innervates the skin of the posterior temporal region.

25 Great auricular nerve
26 Anastomosis with transverse cervical nerve ("superficial ansa cervicalis")
27 Parotid duct (cut)
28 Buccal nerve
29 Facial artery
30 Masseter
31 Buccinator

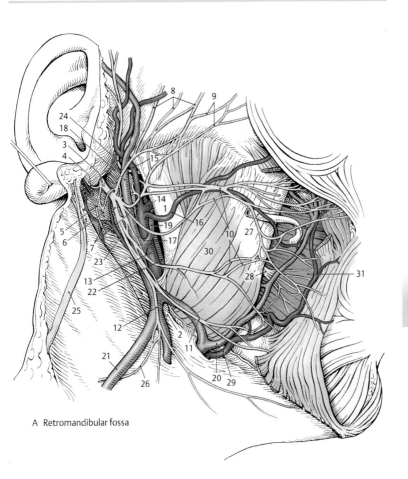

A Retromandibular fossa

Medial Cervical Region (A, B)

The division into layers produced by the cervical fasciae is particularly evident in the median cervical region.

Interfascial Space (A)

The *platysma* (**1**) is of variable size and lies directly beneath the skin. After this integumentary muscle is removed, the *superficial layer of the cervical fascia* (**2**) becomes visible, and division of that layer exposes the *pretracheal layer of the cervical fascia* (**3**) covering the infrahyoid muscles. The region is bounded inferiorly by the *sternocleidomastoids* (**4**). Just above the jugular notch, in the suprasternal space, the *jugular venous arch* (**5**) joins the *right anterior* (**6**) to the *left anterior jugular vein*. These veins may also receive deep tributaries through the pretracheal layer of the cervical fascia (**3**).

Deep Layer (B)

The pretracheal layer of the cervical fascia is removed to expose the infrahyoid muscles and the *thyroid gland* (**7**). Several muscles can be divided to afford a better view of the thyroid gland and the entire region. Most medially and superficially lies the *sternohyoid* (**8**) and lateral to it lie the *omohyoid muscle* (**9**). Deep to them lie the *thyrohyoid* (**10**) and *sternothyroid* (**11**). All the infrahyoid muscles are innervated on their respective sides by the *"deep" ansa cervicalis* (**12**) and by fibers arising from the superior root (thyrohyoid branch).

The thyroid gland (**7**) is located anterior to the cricoid cartilage and the *trachea* (**13**). Its lateral lobes (see p. 356) reach the *thyroid cartilage* (**14**). Between the thyroid and cricoid cartilages extends the *median cricothyroid ligament* (**15**), which is covered laterally by the *cricothyroid muscles* (**16**). On each side these muscles are innervated by the *external branch* (**17**) of the *superior laryngeal nerve* (**18**). The *internal branch* (**19**) of the superior laryngeal nerve perforates the *thyrohyoid membrane* (**21**). It is accompanied by the superior laryngeal artery, which arises from the *superior thyroid artery* (**20**).

The drainage of blood from the thyroid gland (see p. 356) takes place via different veins, of which the *superior thyroid vein* (**22**) and the *unpaired thyroid plexus* (**23**) are visible in this region. This plexus passes in front of the trachea as the "inferior thyroid vein," which as a general rule drains into the left brachiocephalic vein. The *brachiocephalic trunk* (**24**), which is situated directly in front of the trachea, runs obliquely upward. Lateral to the trachea and in front of the esophagus, the *recurrent laryngeal nerve* (**25**) courses toward the larynx.

Variants: The jugular venous arch can occur at any level between the hyoid bone and the jugular notch. When located just below the hyoid bone, it is designated as the subhyoid venous arch. In rare cases a vein is found that ascends from the thyroid gland, penetrates the pretracheal layer of the cervical fascia, and opens into the anterior jugular vein. In some cases a thyroidea ima artery is present, which arises from the brachiocephalic trunk or the aorta.

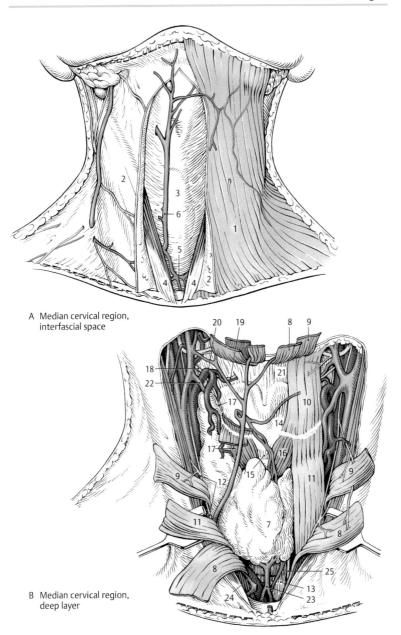

A Median cervical region,
   interfascial space

Peripheral Pathways

B Median cervical region,
   deep layer

Thyroid Region (A–G)

The **thyroid gland** consists of an *isthmus* (**1**), a *right lobe* (**2**), and a *left lobe* (**3**). Each lobe has a *superior* (**4**) and an *inferior pole* (**5**). The superior poles of both lobes reach the *thyroid cartilage* (**6**), while the isthmus is anterior to the cricoid cartilage and the trachea. Thus, the *median cricothyroid ligament* (**7**), which connects the cricoid with the thyroid cartilage, remains free, provided there is no *pyramidal lobe*. Such a lobe may sometimes ascend from the isthmus (remnant of the thyroglossal duct).

The thyroid gland receives its blood supply from the *superior* (**8**) and *inferior* (**9**) *thyroid arteries* on both sides. The superior thyroid artery originates from the *external carotid artery* (**10**) and enters the thyroid gland at its superior pole, where it divides into *anterior*, *posterior*, and *lateral glandular branches*. The anterior glandular branch gives off a variable *cricothyroid branch*, which reaches the median cricothyroid ligament. The inferior thyroid artery (**9**) is a branch of the *thyrocervical trunk* (**11**), which arises from the *subclavian artery* (**12**); it reaches the thyroid gland at its posterior surface. Of special significance is the relationship of this artery to the *recurrent laryngeal nerve* (**13**, **B–D**).

The blood returns through the *superior thyroid veins* (**14**), which open into the *internal jugular veins* (**16**) via the *common facial veins* (**15**). A *middle thyroid vein* (**17**) runs from the lateral margin of the thyroid gland directly to the internal jugular vein. At the lower end of the thyroid gland is the *unpaired thyroid venous plexus* (**18**) which, as the inferior thyroid vein, carries blood to the *left brachiocephalic vein* (**19**). Sometimes another vein may extend from the superior margin of the isthmus to the anterior jugular vein (see p. 355, Fig. B).

> **Clinical tip:** If the airways are obstructed, **a cricothyrotomy** is performed as an emergency measure. The (elastic) median cricothyroid ligament (**7**), the free portion of the conus elasticus, is divided **transversely**. This causes gapping of the incision. **Tracheotomy** is performed as an emergency procedure. Thee trachea is incised **longitudinally**. Three types are distinguished: a *superior tracheotomy* above the isthmus of the thyroid gland, a *middle tra-*

cheotomy through the isthmus, and an *inferior tracheotomy* below the isthmus. The latter type is performed in children because they exhibit a sufficiently large distance between isthmus and sternum. The two other routes are used in adults. Great care should be taken with the jugular venous arch, and the unpaired thyroid plexus (**18**) should be preserved after the pretracheal lamina of the cervical fascia has been divided. Furthermore, the **brachiocephalic trunk** (**20**) ascends from left to right and may cross the trachea at a very high level. During thyroid surgery, attention should also be given to the *thoracic duct* (**21**), because it passes the thyroid by the left inferior pole and opens into the *left venous angle* (**22**).

Variable position of the recurrent laryngeal nerve (B–D): In addition to innervating the mucosa of the subglottic space, the recurrent laryngeal nerve (**13**) innervates all the laryngeal muscles expect the cricothyroid. Except in special cases, the nerve is located either anterior to (**B**, 27%), posterior to (**C**, 36%), or between (**D**, 32%) the branches of the inferior thyroid artery (**9**), each position occurring with approximately equal frequency (according to *Lanz*). Great care must be taken when the thyroid gland is mobilized during surgery, as any stretching of the nerve may cause paralysis of the laryngeal muscles.

Variants of the inferior thyroid artery (E–G): The inferior thyroid artery is particularly variable both as to its site of origin and its course. The inferior thyroid artery (**9**) may run behind the *vertebral artery* (**23**) toward the middle (**E**). Sometimes (**F**) the artery may divide immediately after it leaves the thyrocervical trunk. One branch may then run anterior and the other posterior to the *common carotid artery* (**24**) and internal jugular vein (**16**). Finally (**G**), the inferior thyroid artery (**9**) may arise directly from the subclavian artery as its first branch (in 8% of the population). In rare cases, the inferior thyroid zartery may arise either from the vertebral artery or from the internal thoracic artery. It may be absent in approximately 3% of the population, in which case its territory is supplied by the superior thyroid artery and/or the thyroid ima artery. The latter may arise directly from the aortic arch or from the brachiocephalic trunk.

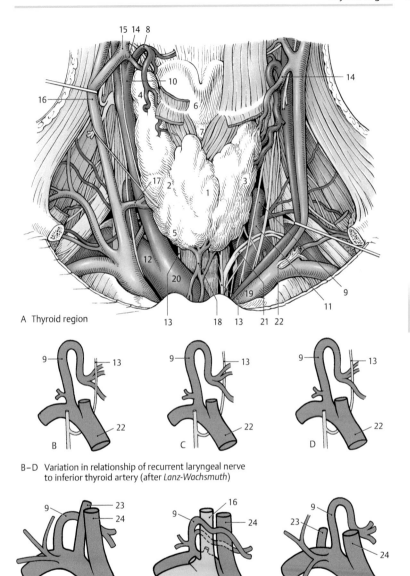

A Thyroid region

B–D Variation in relationship of recurrent laryngeal nerve to inferior thyroid artery (after *Lanz-Wachsmuth*)

E–G Variants of branches of subclavian artery (personal observations)

Anterolateral Cervical Regions (A, B)

The anterolateral cervical regions may be divided into a superficial subcutaneous region with the punctum nervosum, the lateral cervical region (posterior cervical triangle), the carotid triangle, and the sternocleidomastoid region.

The Anterolateral Subcutaneous Cervical Region (A)

Its boundaries are superiorly the *mandible*, anteriorly the median sagittal plane, posteriorly the palpable margin of the *trapezius*, and inferiorly the *clavicle* (**1**). The subcutaneous layer contains a cutaneous muscle, the platysma, large veins, and the cutaneous branches of the cervical plexus. The area in which these cutaneous branches penetrate the superficial layer of the cervical fascia is also called the **punctum nervosum**. It lies roughly where the posterior border of the platysma crosses the sternocleidomastoid. After the platysma has been removed all the superficial vessels and nerves become visible.

The *lesser occipital nerve* (**2**), which runs subcutaneously parallel to the posterior border of the sternocleidomastoid muscle, is the most cranial. This nerve, which takes part in the sensory innervation of the skin on the back of the head, may divide into two branches immediately after it has perforated the superficial layer of the cervical fascia. The largest caliber nerve is the *great auricular nerve* (**3**), which gives off an *anterior* (**4**) and a *posterior* (**5**) *branch* that ascend obliquely across the sternocleidomastoid muscle and contribute to the sensory innervation of the external ear. In approximately the same location as this nerve, the *transverse cervical nerve* (**6**) perforates the superficial layer of the cervical fascia, runs deep to the *external jugular vein* (**7**) and, together with the *cervical branch of the facial nerve* (**8**), forms the "superficial ansa cervicalis" (**9**). The platysma and overlying skin are innervated by this ansa. Caudally,

at different levels, the *medial* (**10**), *intermediate* (**11**), and *lateral* (**12**) *supraclavicular nerves* perforate the superficial layer of the cervical fascia to supply the skin of the shoulder region.

> **Clinical tip:** The Eiselsberg phenomenon occurs on the right side of the shoulder as a "false projection," meaning that pain may radiate to the right shoulder due to disease of the liver or gallbladder. Pain sensations spread out in the dermatomes (C3–C5; see Vol. 3). Diseases of the pancreas may produce referred pain in the left shoulder region.

Lateral Cervical Region, First Layer (B)

After removal of the superficial layer of the cervical fascia, the posterior border of the *sternocleidomastoid* (**13**) and the anterior border of the *trapezius* (**14**) can be seen. The *pretracheal layer of the cervical fascia* (**15**), which merges with the prevertebral layer of the cervical fascia in the lateral region of the neck, separates the first layer from the others. In addition to the structures already described above the *external branch of the accessory nerve* (**16**) and the *trapezius branch* (**17**) of the cervical plexus, both of which supply the trapezius, run in this layer. Here we also find the *superficial cervical vein* (**18**), which joins the external jugular vein, and the *superficial cervical artery* (**19**). If the superficial cervical and dorsal scapular arteries arise together from the thyrocervical trunk, their common trunk is called the transverse cervical artery. Several *superficial cervical lymph nodes* (**20**) are distributed along the veins.

Cervical plexus

– Roots: Ventral rami C1–C4
– Branches: Lesser occipital nerve
           Great auricular nerve
           Transverse cervical nerve
           Supraclavicular nerves
           Phrenic nerve

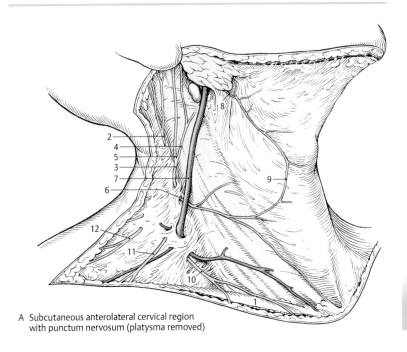

A Subcutaneous anterolateral cervical region
with punctum nervosum (platysma removed)

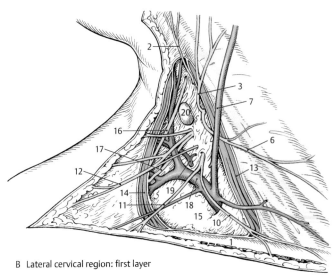

B Lateral cervical region: first layer

Anterolateral Cervical Regions, continued (A, B)

Lateral Cervical Region, Second Layer (A)

Removal of the *pretracheal layer of the cervical fascia* (**1**) exposes the *omohyoid muscle* (**2**), which is embedded within it. Above and behind the omohyoid, the pretracheal layer of the cervical fascia blends with the *prevertebral layer of the cervical fascia* (**3**). It has a firm texture only in the **omoclavicular triangle**, which is formed by the *inferior belly* (**2**) *of the omohyoid*, the *sternocleidomastoid* (**4**), and the *clavicle* (**5**).

In the omoclavicular triangle the *external jugular vein* (**6**) and the *superficial cervical vein* (**7**) combine with the *subclavian* (**8**) and *internal jugular* (**9**) *veins* at the right venous angle to form the *brachiocephalic vein*. The *suprascapular vein* (**10**) also opens into the venous angle. The order in which the veins terminate in this area shows marked variability. The *suprascapular artery* (**11**) runs with the homonymous vein just above the clavicle. The trunk of the *superficial cervical artery* (**12**) becomes visible cranial to the inferior belly of the omohyoid.

Lateral Cervical Region, Third Layer (B)

After the prevertebral layer of the cervical fascia (**3**) has been removed, the deep cervical muscles, the *scalenus anterior* (**13**), *scalenus medius* (**14**), *scalenus posterior* (**15**), *levator scapulae* (**16**), and *splenius cervicis* (**17**, one of the spinotransversales), can be seen. Within the **"scalene interval"** formed between the anterior and middle scalenes and the first rib are the *brachial plexus* (**18**) and the *subclavian artery* (**19**). In the area of the scalene interval the subclavian artery gives off the *dorsal scapular artery* (**20**), which becomes visible behind the middle scalene. This artery may also arise from the transverse cervical artery

(see p.364). The *phrenic nerve* (**21**), a branch of the cervical plexus from segment C4, obliquely crosses the anterior scalene muscle (**13**). The brachial plexus (**18**) gives off its supraclavicular branches, of which the *suprascapular* (**22**), *long thoracic* (**23**), and *dorsal scapular* (**24**) *nerves* can be seen.

The *cervical lymph nodes* (**25**) together form a lymphatic chain, the *jugular trunk*, that extends to the venous angle. The right venous angle receives lymph vessels from the right side of the head and neck, the right arm (*right subclavian trunk*), and the right half of the thorax (*right bronchomediastinal trunk*). Lymph vessels from the other body regions run to the left venous angle (see Vol. 2).

Brachial plexus

Roots: Ventral rami C5–T1
  – Superior trunk (C5, C6)
  – Middle trunk (C7)
  – Inferior trunk (C8, T1)

Branches: Supraclavicular part:
  – Dorsal scapular nerve
  – Long thoracic nerve
  – Subclavian nerve
  – Suprascapular nerve
  – Subscapular nerves
  – Thoracodorsal nerve
  – Medial pectoral nerve
  – Lateral pectoral nerve
  – Muscular branches
    (Infraclavicular part: see p.372)

Clinical tip: Lesions of the brachial plexus may have various causes (obstetric trauma, cervical rib, extrinsic pressure). They are classified as upper or lower brachial plexus lesions (see p. 372).
Upper brachial plexus paralysis (*Erb–Duchenne paralysis*), caused by lesions of the C5 and C6 nerve roots, leads to weakness of the abductors and external rotators of the shoulder joint, the elbow flexors, and the supinator. It also causes mild sensory losses affecting the shoulder and radial surface of the forearm.

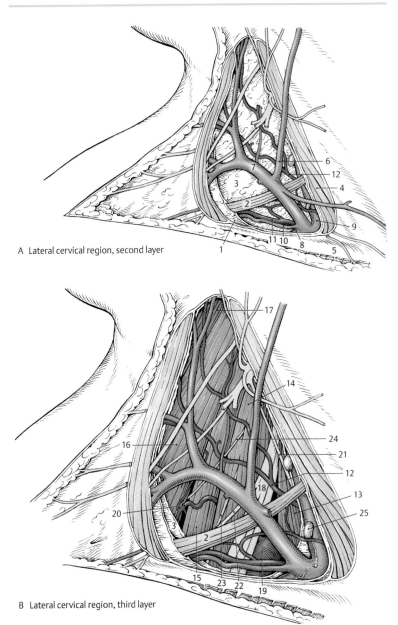

A Lateral cervical region, second layer

B Lateral cervical region, third layer

Anterolateral Cervical Regions, continued (A–F)

Carotid Triangle (A)

The boundaries of the carotid triangle are the *sternocleidomastoid* (**1**), the *omohyoid* (**2**), and the *posterior belly* (**3**) of the *digastric*. The posterior belly is fixed to the *hyoid bone* (**5**) by the stylohyoid (**4**).

The *common facial vein* (**6**) runs superficially; it receives the *vena comitans of the hypoglossal nerve* (**7**) and the *superior thyroid vein* (**8**) before joining the *internal jugular vein* (**9**). Anterior to that vein lies the *common carotid artery* (**10**) with the *carotid sinus* (**11**; see Vol. 2).

In 67% of cases, at the level of the fourth cervical vertebra, the common carotid artery divides into the *internal carotid artery* (**12**), which runs posteriorly, and the *external carotid artery* (**13**), which runs anteriorly. In approximately 20% of cases the division occurs one vertebra higher, and in 11% one vertebra lower, while in the remaining 2% there are particularly high or low divisions, perhaps located completely outside the carotid triangle.

The internal carotid artery (**12**) as a rule has no branches. The first anterior branch of the external carotid artery (**13**) is the *superior thyroid artery* (**14**), which supplies blood to the *thyroid gland* (**15**) and to the larynx through the *superior laryngeal artery* (**16**). Sometimes the superior thyroid artery also gives off a *sternocleidomastoid artery* (**17**), which more often arises directly from the external carotid artery and loops over the *hypoglossal nerve* (**18**). The *lingual artery* (**19**) is another anterior branch that extends to the tongue, medial to the *hyoglossus* (**20**). The last branch within the carotid triangle is the *facial artery* (**21**), which arises medial to the posterior belly (**3**) of the digastric and runs toward the face. The *carotid body* (**22**) lies in the angle of the carotid bifurcation. It is a paraganglion (see Vol. 2) that is reached by sympathetic and parasympathetic fibers. Parasympathetic fibers also run in the *nerve of the carotid sinus* (**23**), a branch of

the glossopharyngeal nerve, which extends to the carotid sinus (**11**) as well as to the carotid body.

The hypoglossal nerve (**18**) runs lateral to both carotid arteries and at the beginning of its arch it gives off the *superior root of the "deep" ansa cervicalis* (**24**). The fibers of this root arise from the first two cervical segments, like those of the *thyrohyoid branch* (**25**), which supplies the thyrohyoid muscle. Descending along the common carotid artery, the superior root joins the *inferior root of the "deep" ansa cervicalis* (**26**) from C2 and C3, which extends laterally or medially across the internal jugular vein, to form the *"deep" ansa cervicalis* (**27**).

This innervates the remaining infrahyoid muscles.

Medial to the external carotid artery is the *superior laryngeal nerve*, whose *internal branch* (**28**) reaches the larynx together with the superior laryngeal artery (**16**). The superior laryngeal nerve is a branch of the *vagus nerve* (**29**), which runs between the internal carotid artery and the internal jugular vein and which is only separated by the prevertebral layer of the cervical fascia from the *sympathetic trunk* (**30**) and its *superior cervical ganglion* (**31**). In the posterosuperior angle of the triangle we find the *external branch of the accessory nerve* (**32**).

■ **Variants (B–F):** Only the position of the external and internal carotid arteries and the origin of their three anterior branches are discussed here.

According to *Faller*, in 49% of cases the internal carotid artery may arise posterolateral (**B**) to the external carotid artery from the common carotid artery, and in 9% is anteromedial (**C**). All intermediate positions are possible.

A thyrolingual trunk (**D**) may be present in 4% of cases, a linguofacial trunk (**E**) in 23%, and a thyrolinguofacial trunk (**F**) in 0.6%.

Clinical tip: The common carotid artery can be pressed against the anterior tubercle of the C6 vertebra (Chassaignac tubercle) to palpate the carotid pulse or introduce a needle into the vessel.

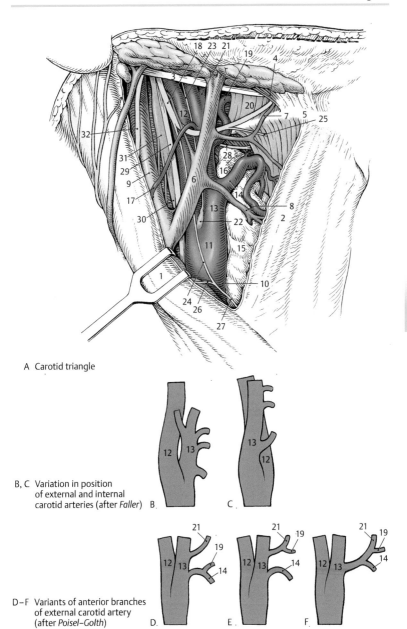

A Carotid triangle

B, C Variation in position
of external and internal
carotid arteries (after *Faller*)

D–F Variants of anterior branches
of external carotid artery
(after *Poisel–Golth*)

Peripheral Pathways

Anterolateral Cervical Regions, continued

Sternocleidomastoid Region (A)

The sternocleidomastoid region can be seen after removal of the *sternocleidomastoid* (**1**) and *omohyoid* (**2**) muscles. It joins the carotid triangle to the lateral cervical region. When the sternocleidomastoid region is exposed, the large vessels and nerves that run through the neck can be seen.

The largest artery, the *common carotid artery* (**3**), runs obliquely cephalad. It bifurcates into the *external* (**4**) and *internal* (**5**) carotid arteries. The level of this division and variations in its position are described on page 362.

The arched *inferior thyroid artery* (**6**) running to the *thyroid gland* (**7**) is covered by the common carotid artery. This artery arises from the *thyrocervical trunk* (**8**), which branches off the *subclavian artery* (**9**) just before it enters the scalene interval. The thyrocervical trunk also gives off the *suprascapular artery* (**11**), which crosses in front of the anterior scalene (**10**), the *superficial cervical artery* (**12**), which is very superficial, and the *ascending cervical artery*. The *vertebral artery* (**13**) is the first ascending branch of the subclavian artery. After the subclavian artery has entered the scalene interval, in approximately 60% of cases it gives off the *dorsal scapular artery* (**14**), which runs behind the *middle scalene* (**15**) and in front of the *posterior scalene* (**16**), and may divide into ascending and descending branches. In all other cases the dorsal scapular artery arises with the superficial cervical artery (**12**) from the thyrocervical trunk. The common origin is then called the *transverse cervical artery*.

Posterior to the common carotid artery, the large *internal jugular vein* (**17**), into which the *facial* (**18**) and *middle thyroid* (**19**) *veins* open, is seen to descend. It joins the *subclavian vein* (**20**) to form the *right brachiocephalic vein* (**21**). The *external jugular vein* (**22**), which joins the transverse *cervical vein* (**23**), and the *suprascapular vein* (**24**), also terminates at the right venous angle.

Lymphatic vessels (**25**) from the right half of the head and neck and from the right upper limb and the right half of the thorax also open into the right venous angle.

The *"deep" ansa cervicalis* (**26**), which innervates the infrahyoid muscles, lies on the common carotid artery (**3**). It is formed from a *superior root* (**27**), which, at its origin, runs together with the *hypoglossal nerve* (**28**) and the *inferior root* (**29**). Posterior to the internal jugular vein runs the *phrenic nerve* (**30**), which arises from the fourth cervical segment and uses the anterior scalene as a guide muscle. The *vagus nerve* (**31**), which gives off a *superior* (**32**) and an *inferior* (**33**) *cervical cardiac branch*, also forms part of the neurovascular bundle.

The *sympathetic trunk* (**34**) with its *superior cervical ganglion* (**35**), the sometimes absent *middle cervical ganglion* (**36**), and *inferior cervical ganglion* are separated from the vagus nerve by the prevertebral layer of the cervical fascia. The inferior cervical ganglion is usually fused with the first thoracic ganglion, forming the *stellate ganglion* (**37**), which abuts the head of the first rib medial to the vertebral artery (**13**). The sympathetic trunk (**34**) forms the *thyroid loop* (**38**) around the inferior thyroid artery (**6**) and gives off the *cervical cardiac nerves* (**39**). Deeply the *recurrent laryngeal nerve* (**40**) lies on the trachea.

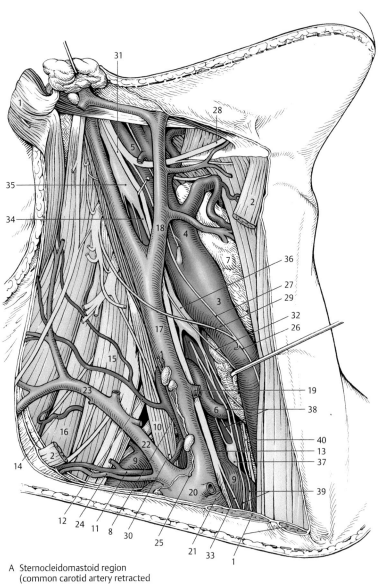

A Sternocleidomastoid region
(common carotid artery retracted
anteromedially)

Scalenovertebral Triangle (A)

The boundaries of the scalenovertebral triangle are the *longus colli* (**1**), the *anterior scalene* (**2**), and the pleural dome. The prevertebral layer of the cervical fascia covers the triangle and its contents can be seen only after the fascia is removed.

The *subclavian artery* (**3**) lies on the pleural dome, from which connective tissue bands (the costopleural ligament) run to the first rib. Its first ascending branch is the *vertebral artery* (**4**), which crosses anterior to the roots of the *brachial plexus from T1* (**5**) and *C8* (**6**), to reach the vertebral column at the transverse foramen of the sixth cervical vertebra. Behind the vertebral artery (**4**) runs the *vertebral vein* (**7**) which leaves the vertebral column at the transverse foramen of the seventh cervical vertebra. Adjacent to the vertebral artery, the *thyrocervical trunk* ascends (see p. 364), followed by the *costocervical trunk* (**8**), which gives off the *deep cervical artery* (**9**), the *highest intercostal artery* and, rarely, a *dorsal scapular artery* (**10**) of abnormal origin. The *internal thoracic artery* (**11**) turns downward, running parasternally with the *internal thoracic vein* (**12**) to reach the rectus sheath (see p. 398).

Anteriorly the subclavian artery and its branches on the left side are crossed by the *thoracic duct* (**13**), which forms a superiorly convex arch. The thoracic duct opens into the *left venous angle* (**14**), which is formed by the junction of the *internal jugular vein* (**15**) and *left subclavian veins* (**16**).

The C5 to T1 roots of the brachial plexus occupy a deep level with the *sympathetic trunk* (**17**) running superficial to them. At the level of the sixth cervical vertebra, the sympathetic trunk often contains a *middle cervical ganglion* (**18**) lying on the scalenus anterior (**2**). Caudal to the ganglion, the sympathetic trunk together with the *superior cardiac nerve* (**19**) form the *ansa thyroidea* (**20**), through which passes the inferior thyroid artery. The sympathetic trunk gives off the *ansa subclavia* (**21**), which

winds around the subclavian artery (**3**). This ansa subclavia extends to the inferior cervical ganglion, which fuses with the first thoracic ganglion to form the *stellate (cervicothoracic) ganglion* (**22**). The latter lies on the head of the first rib. The *inferior cervical cardiac nerve* (**23**) arises from it. Medial to it, the recurrent laryngeal nerve (**24**) passes upward to the larynx in the groove formed by the *trachea* (**25**) and the *esophagus* (**26**).

Clinical tip: The presence of a cervical rib may lead to **cervical rib syndrome (Naffziger syndrome)**. It is associated with complaints arising from the brachial vessels and the branches of the three fascicles, particularly in the territory of the ulnar nerve. A palpable abnormality is also found in the greater supraclavicular fossa. However, the complaints arising from the vessels and nerves may also occur in the absence of a cervical rib. This is called an anterior **scalene syndrome**, in which pain is caused by hypertrophy and hypertonicity of the anterior scalene muscle.

The supraclavicular lymph nodes, whose efferents drain directly into the junction of the left subclavian and internal jugular veins, may harbor distant lymphogenous metastases from gastric carcinoma. They are known as the **Virchow–Troisier sentinel nodes** for this type of cancer.

"Sentinel node" is the term clinicians apply to a particularly large lymph node that is the first to show metastatic spread from a malignant tumor. This principle also applies to lymph nodes in other regions.

27 Phrenic nerve
28 Left brachiocephalic vein
29 Scalenus medius
30 Scalenus posterior
31 Levator scapulae
32 Trapezius
33 Clavicular part of the pectoralis major
34 Left common carotid artery
35 Left vagus nerve

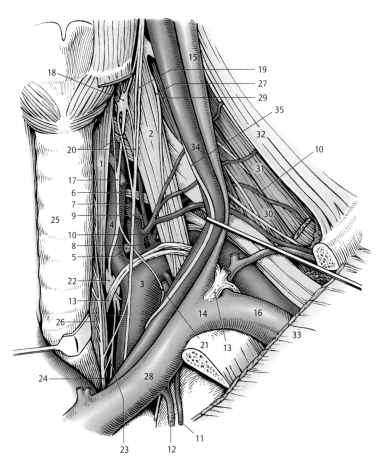

A Scalenovertebral triangle
(common carotid artery, subclavian vein,
and vagus nerve retracted laterally)

Upper Limb

Regions (A–C)

Superficially there is no clear boundary between the free upper limb or its root and the thorax, but by dissection it is possible to separate the predominantly muscular attachment of the arm and its root from the thorax. The free limb and its root must be considered together to gain a proper understanding of the topography of the peripheral neurovascular pathways. The following regional subdivisions are made for practical purposes and are not based on embryonic development.

Regions about the Shoulder

Anteriorly there is the **infraclavicular fossa** (**1**) with the *deltopectoral triangle* (**2**), through which the peripheral pathways extend to the arm, that is, the central part of the **axillary region** (**3**) with the *axillary fossa* (**4**). Lateral to the shoulder joint is the **deltoid region** (**5**), which is bounded posteriorly by the **scapular region** (**6**).

Regions of the Upper Arm

The upper arm is organized into an **anterior brachial region** (**7**), the foundation of which is the flexor muscles, and a **posterior brachial region** (**8**), occupied by the extensors. Within the anterior brachial region the *medial bicipital groove* (**9**) merits special attention because it lies in front of the medial intermuscular septum and marks the main pathway taken by the brachial vessels and nerves passing from the axilla to the cubital fossa. A *lateral bicipital groove* is described in front of the lateral intermuscular septum; in it the cephalic vein courses superficially (subcutaneously).

Regions of the Elbow

The **anterior cubital region** (**10**), the center of which is represented by the *cubital fossa*, adjoins the anterior brachial region on the flexor side. Within the cubital fossa the vascular and nerve bundles divide. The **posterior cubital region** (**11**) **at the back of the elbow** contains muscles and only small vascular networks.

Regions of the Forearm

The **anterior antebrachial region** (**12**) lies distal to the cubital fossa and contains the large vessels and nerves between the flexors. The posterior part is formed by the **posterior antebrachial region** (**13**).

Regions of the Hand

The **palm** or **palmar region** (**14**) extends from the distal intercarpal joint to the metacarpophalangeal joints. The **dorsum of the hand** (**15**) has the same boundaries. Laterally between the dorsum of the hand and the palm is the **radial fovea** (**16**), which contains the radial artery.

Regions of the Carpus

The **anterior carpal region** (**17**) lies on the palmar plane between the anterior antebrachial region and the palm of the hand. The **posterior carpal region** (**18**) lies on the dorsal plane.

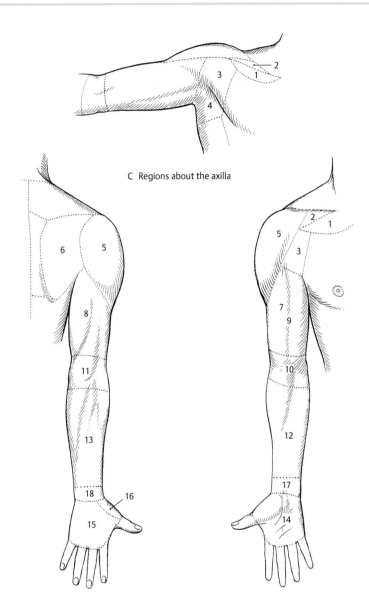

C Regions about the axilla

B Posterior view of regions
of upper limb

A Anterior view of regions
of upper limb

Deltopectoral Triangle (A, B)

The *clavicle* (**1**), the *deltoid* (**2**), and the *pectoralis major* (**3**) form the proximal, lateral, and medial boundaries of the deltopectoral triangle. It is continuous distally with the deltopectoral groove. Since the width of the base of the triangle is quite variable, it is possible to separate the *clavicular part* (**4**) *of the pectoralis major* from the clavicle and to reflect it downward.

Superficial Layer (A)

Superficially the pectoral fascia in the region of the triangle shows a slight depression. Between the clavicle (**1**), the *coracoid process* (**B 5**), and *the pectoralis minor* (**B 6**), the *clavipectoral fascia* (**7**) stretches from the deep surface of the deltoid to the deep surface of the pectoralis major. This fascia divides the triangle into two compartments.

In the superficial layer the *cephalic vein* (**8**) reaches the triangle through the deltopectoral groove. It penetrates the clavipectoral fascia to end in the *axillary vein* (**B 9**). The cephalic vein is joined by branches from the surrounding areas. Lateral to the cephalic vein, the *thoracoacromial artery*, (**B 10**), which stems from the axillary artery, pierces the clavipectoral fascia (**7**). It divides into *clavicular* (**11**), *acromial* (**12**), *deltoid* (**13**), and *pectoral* (**B 14**) *branches*. *The pectoral nerves* run together with the latter vessels and may penetrate the clavipectoral fascia as a common trunk (**15**).

Deep Layer (B)

The deep layer contains the vessels and nerve bundles that supply the upper limb. Distal to the *subclavius* (**16**) from medial to lateral are the axillary vein (**9**), *axillary artery* (**17**), and three nerve cords, which are the infraclavicular portion of the brachial plexus. They are the superficially situated *lateral cord* (**18**), which may already have divided into its branches, the *posterior cord* (**19**), and the *medial cord*

(**20**). At the upper border of the pectoralis minor (**6**) the vessels and nerves lie more deeply. The *suprascapular artery, vein, and nerve* (**21**) can be seen lying very deep in the lateral layer.

The superficial compartment sometimes contains lymph nodes (not shown in the diagram). They drain the lymph vessels that run along the cephalic vein. They are in continuity with the deep infraclavicular nodes (not shown).

▆▆ **Variants:** It is common to find a vein (**22**) looping superficially around the clavicle and interconnecting the axillary vein with the subclavian vein, producing a venous ring. The cephalic vein may sometimes be poorly developed.

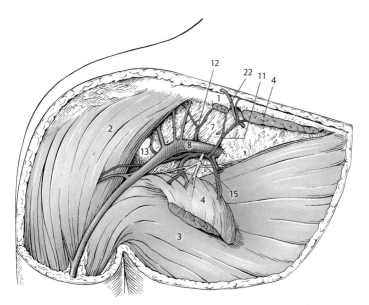

A Deltopectoral triangle, superficial layer

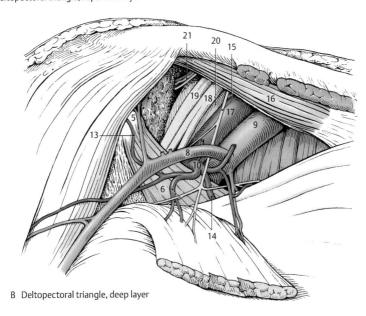

B Deltopectoral triangle, deep layer

Peripheral Pathways

Axillary Region (A)

The vessels and nerves to the upper limb run through the axilla. The boundaries of the axilla are the *pectoralis major* (**1**) and *pectoralis minor* (**2**) anteriorly and the *latissimus dorsi* (**3**) posteriorly. The chest wall with the *serratus anterior* (**4**) lies medially, and laterally there is the humerus with the *short head of the biceps brachii* (**5**) and the *coracobrachialis* (**6**).

Most medial of all is the *axillary vein* (**7**) formed from the brachial veins. It runs centrally, receiving a larger number of small veins. It is joined in the deltopectoral triangle (see p. 370) by the *cephalic vein* (**8**). The *axillary artery* (**9**), which lies lateral to the vein, gives off the *thoracoacromial artery* (**10**), with its *pectoral* (**11**), *acromial* (**12**), and *deltoid* branches. A *lateral thoracic artery* (**13**) arises from the thoracoacromial artery in approximately 10% of cases (see figure), or directly from the axillary artery. Another branch of the axillary artery, the *subscapular artery* (**14**), gives off the *thoracodorsal* (**15**) and *circumflex scapular* (**16**) arteries. The last branches of the axillary artery are the *anterior* (**17**) and *posterior* (**18**) *circumflex humeral arteries*.

At the tendinous insertion of the latissimus dorsi (**3**), the axillary artery continues as the *brachial artery* (**19**) and gives off the *profunda brachii artery* (**20**) as its first branch.

The three trunks of the brachial plexus lie in the axillary region medial, lateral, and posterior to the axillary artery, and there divide into various branches. The posterior trunk gives off the *axillary* (**21**) and *radial* (**22**) nerves. Accompanied by the posterior circumflex humeral artery and vein (**18**), the axillary nerve (**21**) passes through the quadrangular space (see p. 374) toward the *deltoid* (**23**) and *teres minor*. The radial nerve (**22**) runs in the medial bicipital sulcus accompanied by the profunda brachii artery (**20**), with which it runs into the sulcus for the radial nerve. The *medial* (**24**) and *lateral* (**25**) *trunks* form the (often du-

plicated) median bifurcation (*medial and lateral roots*), from which the *median nerve* (**26**) continues superficial to the axillary artery. The median nerve, accompanied by the brachial artery, then enters the medial bicipital groove. Other branches of the medial trunk, the *ulnar nerve* (**27**), the *medial antebrachial cutaneous nerve* (**28**), and the *medial brachial cutaneous nerve* (**29**) also reach this groove. Branches of intercostal nerves one to three join the medial cutaneous brachial nerve as *intercostobrachial nerves* (**30**).

The lateral trunks give off, apart from the lateral root of the median nerve (here duplicated), the *musculocutaneous nerve* (**31**), which pierces the coracobrachialis.

On the wall of the thorax, the *long thoracic nerve* (**32**), arising from the supraclavicular part of the brachial plexus, descends on the lateral surface of the serratus anterior and innervates it. The *subscapular nerve* (**34**) lies on the *subscapularis* (**33**) and may give off the *thoracodorsal nerve* (**35**) to innervate the latissimus dorsi (**3**).

> **Clinical tip:** Lower brachial plexus paralysis (*Déjerine-Klumpke paralysis*) is caused by lesions of the C8 and T1 nerve roots. It presents with paralysis of the short hand muscles and long digital flexors, accompanied by sensory losses on the ulnar surface of the hand and forearm.

Brachial plexus

(Roots and supraclavicular part, see p. 360)
- **Infraclavicular part**
  - Lateral trunk
    - Musculocutaneous nerve
    - Lateral root of median nerve
  - Medial trunk
    - Medial root of median nerve
    - Ulnar nerve
    - Medial brachial cutaneous nerve
    - Medial antebrachial cutaneous nerve
  - Posterior trunk
    - Axillary nerve
    - Radial nerve

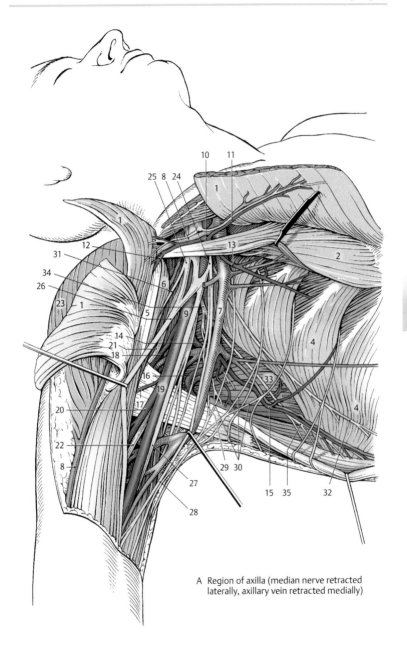

A Region of axilla (median nerve retracted
laterally, axillary vein retracted medially)

Axillary Foramina (A–D)

The slitlike opening between the *teres minor* (**1**) and *teres major* (**2**) and the *humerus* (**3**) is divided by the *long head of the triceps brachii* (**4**) into a **quadrangular space** and a **triangular space**.

Through the **quadrangular space** the *axillary (circumflex) nerve* (**5**) reaches the posterior side. This nerve supplies a branch (**6**) to the teres minor and then buries itself in the *deltoid* (**7**). It also innervates the upper lateral skin area via the *superior lateral brachial cutaneous nerve* (**8**). The axillary nerve is usually accompanied by the *posterior circumflex humeral artery* (**9**) and the commonly paired *posterior circumflex humeral veins*. The artery supplies the deltoid, the long head of the triceps brachii (**4**) and the *lateral head of the triceps brachii* (**10**).

The *circumflex scapular artery* (**11**) runs through the **triangular space** to the posterior surface of the scapula, on which it anastomoses with the suprascapular artery. The artery is accompanied by the *circumflex scapular vein*. Deeply a twig (**12**) from the subscapular nerve, which innervates the teres major (**2**), can be seen. It does not run through the triangular space.

■ Variants (B–D): The posterior circumflex humeral artery (**9**), which usually (**B**) runs through the quadrangular space, arises as one of the terminal branches of the axillary artery. It often has a common origin with the subscapular artery. Distal to the teres major tendon, the *profunda brachii artery* (**13**) arises as the first branch of the *brachial artery* (**14**). In approximately 7% of cases, according to *Lanz–Wachsmuth*, the profunda brachii artery (**13**) arises (**C**) from the posterior humeral circumflex artery (**9**). In these cases the profunda brachii artery runs distally behind the teres major tendon. In 16% of cases (**D**) the origin of the posterior circumflex humeral artery (**9**) is from a typical profunda brachii artery (**13**), and in those cases the posterior circumflex humeral artery does not traverse the quadrangular space.

15 Radial nerve

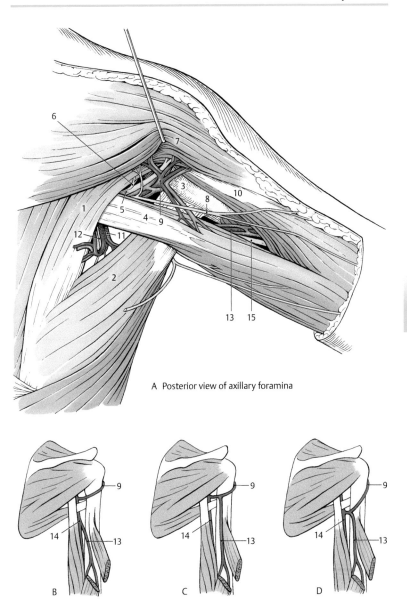

A   Posterior view of axillary foramina

B–D   Variants of arteries (after *Lanz–Wachsmuth*)

Peripheral Pathways

Anterior Brachial Region

Subcutaneous Layer (A)

The coarse, firm *brachial fascia* (**1**) surrounds the muscles of the upper arm. Medial and lateral to the humerus, the intermuscular septum blends with the fascia (see p. 180) to form two compartments, the anterior and posterior brachial compartments. The subcutaneous veins, nerves, and lymph vessels run superficially to the brachial fascia. In inflammatory conditions the lymph vessels may be visible through the skin as fine red lines.

The *cephalic vein* (**2**) runs on the lateral border of the biceps brachii. It carries blood from the radial side of the hand and the forearm via the deltopectoral groove to the deltopectoral triangle (see p. 370). The veins are accompanied by the *lateral superficial lymph vessels* (not shown), which transport lymph from the two radial digits, the radial part of the palm, and the forearm (see p. 370).

The medial bicipital groove is modeled by the brachial fascia on the medial side of the biceps brachii, and the usually well-developed *basilic vein* (**3**) runs subcutaneously in its distal half. This vein pierces the brachial fascia at the **basilic hiatus** (**4**) and runs deep to become one of the veins accompanying the brachial artery. In the subcutaneous part of its course in the upper arm it is accompanied by the *medial antebrachial cutaneous nerve* and its branches; the *anterior branch* (**5**) runs lateral to the vein and is closely apposed to it, while the *posterior branch* (**6**) lies medial and a short distance away from it.

Near the basilic hiatus, in approximately one-third of cases, *cubital* (some of them named *supratrochlear*) *lymph nodes* (**7**) are found, which act as the first filtration point for lymph from the three ulnar digits. The *medial superficial lymph vessels* run along the medial bicipital groove; they may accompany the basilic vein, or they may pass

subcutaneously to the axilla. They are usually more numerous and larger than those that accompany the cephalic vein.

Branches of the *medial brachial cutaneous nerve* (**8**) innervate the skin from the axilla downward. In addition they are joined by the *intercostobrachial nerves* (**9**) from T1 and T3, which supply a small cutaneous area on the inner surface of the upper arm.

▄▄ **Variants:** The position of the basilic hiatus is highly variable. It may lie directly at the boundary of the cubital region. The cephalic vein is sometimes absent.

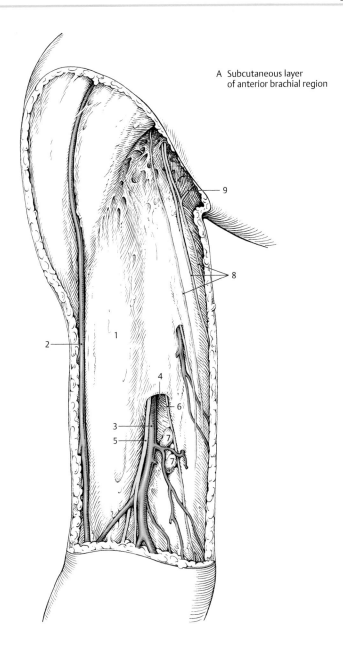

A Subcutaneous layer
of anterior brachial region

Anterior Brachial Region, continued (A–E)

Medial Bicipital Groove (A, B)

The medial bicipital groove is bounded on one side by the *biceps brachii* (**1**) and on the other by the *medial intermuscular septum* (not shown) and the *triceps brachii* (**2**). It contains the neurovascular bundle of the arm. The *medial antebrachial cutaneous nerve* (**3**) is the most superficial structure, and its anterior branch lies on the *basilic vein* (**4**). Both leave the medial bicipital groove at the basilic hiatus, which may lie at various levels. The basilic vein may drain into the *brachial veins* (**5**), or it may only join the axillary vein in the axilla (see **A**).

The most medial structure is the *ulnar nerve* (**6**), lying on the medial intermuscular septum. At the boundary between the middle and the distal thirds of the arm, the ulnar nerve penetrates the medial intermuscular septum and runs posteriorly from the septum to the posterior side of the medial epicondyle of the humerus.

The *median nerve* (**7**) runs lateral to the basilic vein and crosses the *brachial artery* (**8**) from the lateral to the medial side. The brachial artery, which is the deepest structure throughout the entire length of the medial bicipital groove, gives off a series of branches.

In addition to muscular branches (**9**), the brachial artery gives off the *profunda brachii artery* (**10**) in the proximal region of the medial bicipital groove. Here it joins the *radial nerve* (**11**) and leaves the medial bicipital sulcus with it at the level of the boundary between the proximal and middle thirds of the upper arm. Then the profunda brachii artery runs with the radial nerve in the radial groove on the posterior side of the humerus and ends as the *radial collateral artery* after giving off the *medial collateral artery*. Other branches of the brachial artery include the *superior ulnar collateral artery* (**12**), which accompanies the ulnar nerve (posterior to it) and the *inferior ulnar collateral artery* (not visible).

■ **Variants (C–E):** The relationship between the median nerve (**7**) and brachial artery (**8**) and its branches may be highly variable. Although, according to *Lanz*, the median nerve follows a typical course in 74% of cases, a *superficial brachial artery* (**13**), which arises from the brachial artery, may run superficial to the median nerve. In that case the brachial artery may be rudimentary (in 12% of cases according to *Lanz*), or it may divide into two arteries at variable levels (14%). The profunda brachii artery may arise together with the posterior circumflex humeral artery (see p. 374).

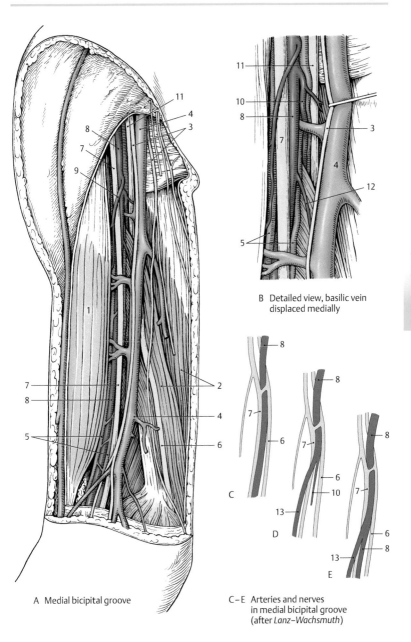

A Medial bicipital groove

B Detailed view, basilic vein displaced medially

C–E Arteries and nerves in medial bicipital groove (after *Lanz–Wachsmuth*)

Posterior Brachial Region (A, B)

Subcutaneous Layer (A)

The *deltoid fascia* (**1**) and *brachial fascia* (**2**) invest the muscles. Subcutaneously there are mainly the cutaneous nerves in addition to small arterial branches and delicate veins. Branches of the *superior lateral brachial cutaneous nerve* (**3**), which arises from the axillary nerve, pass through the fascia at the inferior border of the deltoid muscle. The branches predominantly supply the skin covering the deltoid muscle, although there is no clear dividing line marking the skin area supplied by the *inferior lateral brachial cutaneous nerve* (**4**).

The inferior lateral brachial cutaneous nerve (**4**), which branches from the *radial nerve* (**B 5**), is often accompanied by a smaller artery and vein where it passes through the fascia. It supplies the distal skin area on the lateral side up to the elbow. The *branches* (**6**) *of the posterior brachial cutaneous nerve* (**B 7**), which arises proximally from the radial nerve (**B 5**), are distributed to the posterior surface of the upper arm.

Subfascial Layer (B)

After removal of the brachial fascia, the *long head* (**8**) and *lateral head* (**9**) of the *triceps brachii* (**10**) can be divided to demonstrate the radial groove and the structures within it. The radial nerve (**5**) runs from medioproximal to laterodistal.

Its first proximal branch is the posterior brachial cutaneous nerve (**7**). In the region of the radial groove the radial nerve gives off the *muscular branches* (**11**) and, distally to these, the inferior lateral brachial cutaneous nerve (**4**).

The radial nerve is accompanied by the *profunda brachii artery* (**12**), which usually has two accompanying veins. Immediately after branching off the brachial artery (see p. 378), this artery often gives off a small branch to the deltoid muscle along with

nutrient arteries to the humerus. The *middle collateral artery* (**13**) branches off in the radial groove; it is accompanied by a muscular branch of the radial nerve (**11**). This artery, like the terminal branch of the profunda brachii, the *radial collateral artery* (**14**), reaches the articular network of the elbow. A branch of the radial collateral artery becomes visible anteriorly between the brachialis and brachioradialis muscles, together with the radial nerve, and anastomoses with the radial recurrent artery (see p. 384).

> **Clinical tip:** Fractures of the humeral shaft endanger the radial nerve. This nerve should be carefully protected during reduction of the fragments (see also p. 148).

15 Anterior and posterior branches of the medial cutaneous antebrachial nerve
16 Basilic vein
17 Middle head of triceps muscle

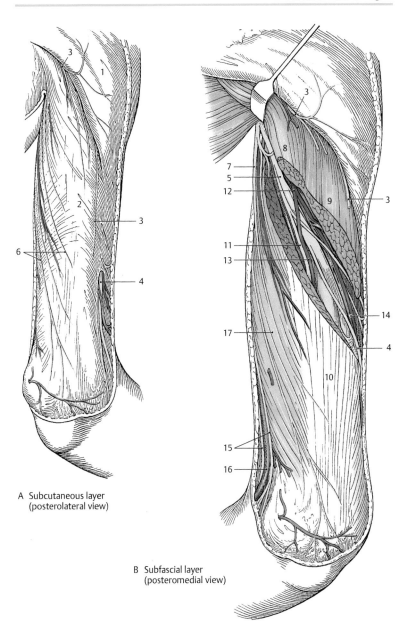

A Subcutaneous layer
(posterolateral view)

B Subfascial layer
(posteromedial view)

Cubital Fossa (A–G)

Subcutaneous Layer (A)

The anterior cubital region at the bend of the elbow is not sharply delineated from the anterior brachial region or from the forearm. Normally the term "cubital fossa" refers to an area two to three fingerwidths proximal and distal to the joint space.

Subcutaneously there is a variable amount of well-developed fatty tissue containing veins, nerves, lymphatics, and lymph nodes. The cutaneous veins of the subcutaneous layer are very important clinically as the cubital fossa is a common site for intravenous injections, drawing blood, etc.

According to the development of the venous system, the course taken by the veins, as well as their caliber, fluctuates widely.

The *basilic vein* (**1**), which is often well developed and clearly visible beneath the skin, runs medially. It is usually continuous with the *antebrachial basilic vein* (**2**), but it may also arise from the *median antebrachial vein*. Many other variants (**B–G**) are possible.

In the region of the *basilic hiatus* (**3**) the basilic vein becomes subfascial. It is accompanied by branches of the *medial antebrachial cutaneous nerve* (**4**). Often (33% of cases) there are lymph nodes near the basilic hiatus (see p. 376). The *cephalic vein* (**5**) runs along the lateral margin of the cubital fossa. It is always palpable but not always visible, and in many instances it is not as well developed as the basilic vein. The cephalic vein in the distal part of the region accompanies the *lateral antebrachial cutaneous nerve* (**6**), which is the terminal branch of the musculocutaneous nerve.

A *median cubital vein* (**7**) normally unites the basilic and cephalic veins. There is almost always a *deep median cubital vein* (**8**), which joins the superficial and deep veins.

■ **Variants (B–G):** There are numerous variants of the subcutaneous veins. Thus, the cephalic vein (**5**) and the basilic vein (**1**) may continue from one median antebrachial vein. There is also a considerable range in size of the two main cutaneous veins. The median cubital vein may sometimes be absent (**E**).

Clinical tip: Intravenous injections in the cephalic vein are less painful, as it is not closely related to any nerve. The basilic vein is closely related to the branches of the medial antebrachial cutaneous nerve.
In some individuals, particularly those with scant subcutaneous fatty tissue, the veins are easily displaced and are known clinically as "rolling veins" because they have to be secured in place for injections.
Indwelling cannulas are usually placed in a dorsal hand vein to facilitate arm movements at the elbow joint.

9   Cubital lymph nodes

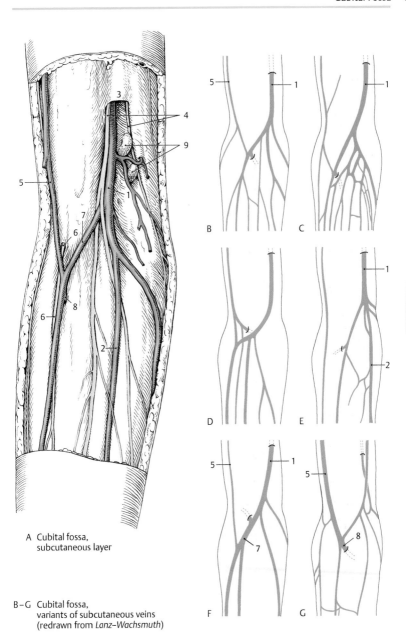

A Cubital fossa,
subcutaneous layer

B–G Cubital fossa,
variants of subcutaneous veins
(redrawn from *Lanz–Wachsmuth*)

Cubital Fossa, continued (A–E)

Deep Layer 1 (A)

After removal of the fascia, the muscles that border the cubital fossa become visible. From the proximal margin the *biceps brachii* (**1**) runs with its tendon toward the radial tuberosity, and with its *bicipital aponeurosis* (**2**) toward the antebrachial fascia. It partly covers the *brachialis* (**3**), which is inserted into the ulnar tuberosity. On the medial side, arising from the medial epicondyle, the *pronator teres* (**4**) and the superficial hand flexors run distally, and on the lateral side the fossa is bounded by the *brachioradialis* (**5**).

The neurovascular bundle, which descends from the medial bicipital groove (see p. 378), becomes divergent within the cubital fossa. The *brachial artery* (**6**), covered by the bicipital aponeurosis (**2**), gives off the radial artery. The *radial artery* (**7**) runs distally superficial to the flexors of the forearm.

In the cubital fossa the *median nerve* (**8**) leaves the brachial artery and runs distally between the two heads of the pronator teres, which it also innervates. The *ulnar nerve* (**9**) leaves the medial bicipital groove before it reaches the cubital fossa and runs posterior to the medial epicondyle. The *radial nerve* (**10**) becomes visible between the brachialis (**3**) and the brachioradialis (**5**) and divides into a smaller, sensory, *superficial branch* (**11**) and a larger, predominantly motor *deep branch* (**12**). The superficial branch supplies cutaneous fibers to the radial half of the dorsum of the hand, the thumb, and the dorsal surface of the proximal phalanges of the second and third fingers. The deep branch penetrates the *supinator* (**13**), winds laterally around the neck of the radius, innervates the radial and posterior muscles of the forearm, and terminates as the posterior interosseous nerve. This nerve provides sensory supply to the wrist joints, the interosseous membrane, and portions of the radia and ulnar periosteum.

Deep Layer 2 (B)

The bicipital aponeurosis (**2**) is divided to expose the brachial artery (**6**). Its first branch is the radial artery (**7**). The *radial recurrent artery* (**14**) branches off either from this artery or directly from the brachial artery and runs proximally along the radial nerve (**10**). It anastomoses with the anterior branch of the radial collateral artery. At the level of the proximal margin of the supinator (**13**), the brachial artery gives off the *ulnar recurrent artery* (**15**). Thereafter the brachial artery divides into the *common interosseous artery* (**16**) and the *ulnar artery* (**17**). The latter passes behind the median nerve (**8**) and the pronator teres (**4**). The individual arteries are accompanied by veins that are paired in most cases.

▪ **Variants (C–E):** The median nerve usually (approx. 95%) runs between the two heads of the pronator teres (**C**). Occasionally it pierces the *humeral head* (**18**) of the pronator teres (barely 2%; **D**). In approximately 3% of cases, the median nerve lies directly on the bone and runs deep to the two heads of the pronator teres (**E**). In such cases a fracture of the proximal part of the radius and ulna may endanger the nerve.

Variants of the brachial artery and its branches in this region have been reported, although infrequently; for example the brachial artery may run posterior to the supracondylar process when present.

Current nomenclature divides the brachial artery into a radial and an ulnar artery, the latter giving off the common interosseous artery. This nomenclature is not consistent with the embryologic development of the arteries of the arm and should be avoided, for example because of diverse variants such as a higher origin of the radial artery. For this reason the developmentally based classification has been retained (see p. 390).

> **Clinical tip:** The deep branch of the radial nerve is endangered by dislocations, lesions of the capsular ligaments, and radial neck fractures.

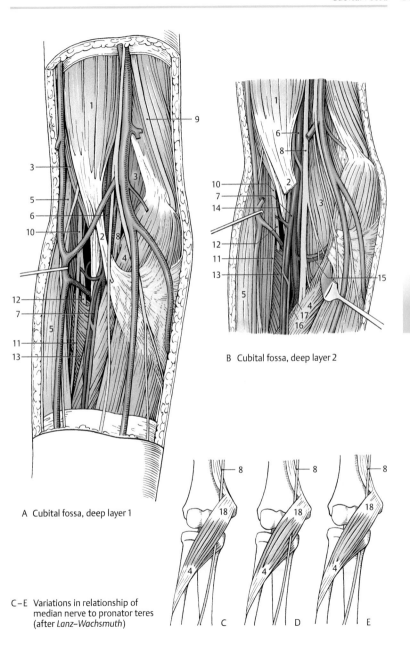

A Cubital fossa, deep layer 1

B Cubital fossa, deep layer 2

C–E Variations in relationship of
        median nerve to pronator teres
        (after *Lanz–Wachsmuth*)

Anterior Antebrachial Region (A, B)

Subcutaneous Layer (A)

In the subcutaneous fat are the well-developed cutaneous veins, which are subject to great variations in their courses. The cutaneous arteries are small and unimportant. The cutaneous nerves run independently of the veins and are very constant both in location and size.

On the radial side is the *cephalic antebrachial vein* (**1**), which mostly *anastomoses* (**2**) distally with the other veins of the forearm. Proximally it often gives off the *median cubital vein* (**3**), which sometimes may arise from the median antebrachial vein. The *lateral antebrachial cutaneous nerve* (**4**), the terminal branch of the musculocutaneous nerve, crosses beneath the cephalic vein in the cubital fossa. In the distal part of the forearm the *superficial branch of the radial nerve* (**5**) lies in close proximity to the cephalic vein.

The *antebrachial basilic vein* (**6**) runs on the medial side of the anterior region of the forearm and is accompanied medially and laterally by twigs (**7**) from the *medial antebrachial cutaneous nerve.*

In the distal third of the forearm, the *palmar branch* (**8**) of the ulnar nerve lies subcutaneously. Radial to it and just proximal to the anterior carpal region, the *palmar branch* (**9**) of the median nerve pierces the fascia.

Subfascial Layer (B)

After division of the firm antebrachial fascia which is reinforced proximally and medially by the bicipital aponeurosis, the deeply situated vessels and nerves come into view. These vessels and nerves are arranged essentially into three bundles or tracts consisting of a radial, middle, and ulnar bundle.

The **radial neurovascular bundle**, consisting of the *radial artery* (**10**) and *radial veins* (**11**), proceeds distally between the *brachioradialis* (**12**) and *flexor carpi radialis*

(**13**) and is accompanied in its proximal segment by the *superficial branch of the radial nerve* (**14**). The *deep branch of the radial nerve* (**15**), which gives off the *posterior interosseous nerve* in the forearm, penetrates the *supinator* (**16**) within the cubital fossa.

The **middle neurovascular bundle**, which is situated between the superficial and deep flexors, houses the *median nerve* (**17**), sometimes accompanied by a *median artery* (variant, see p. 390). The median nerve usually travels between the two heads of the *pronator teres* (**18**) and, at the level of the wrist, lies radial to the *flexor digitorum superficialis tendons* (**19**). The *anterior interosseous artery* and the *anterior interosseous nerve*, a branch of the median nerve, occupy a deep compartment of the middle bundle between the deep flexors and the interosseous membrane.

The **ulnar neurovascular bundle** lies in the middle and distal thirds of the forearm between the flexor digitorum superficialis (**19**) and *flexor carpi ulnaris* (**20**). It consists of the *ulnar nerve* (**21**), the *ulnar artery* (**22**), and the *ulnar veins* (divided in the illustration, **23**). After arising from the brachial artery, the ulnar artery crosses proximally under the median nerve (**17**), the pronator teres (**18**), and the common head of the superficial flexors. The flexor carpi ulnaris (**20**) serves as a landmark for locating the ulnar nerve (**21**).

Clinical tip: The **pulse** is palpated at a typical site (see also p. 164) on the distal forearm. The radial artery runs on the palmar side, medial to the brachioradialis tendon, passing just in front of the radial styloid process.
Note: If occlusive vascular disease is suspected, the pulse should be taken at additional sites such as the dorsal pedal artery (see p. 438), superficial temporal artery, facial artery (see p. 340), and common carotid artery (see p. 364).

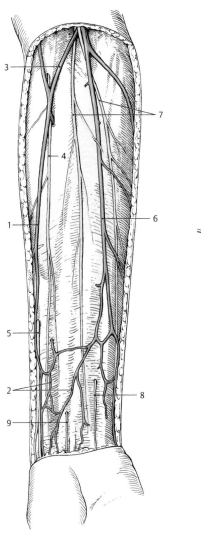

A Anterior antebrachial region,
   subcutaneous layer

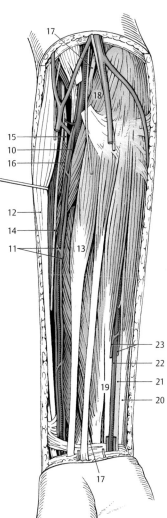

B Anterior antebrachial region,
   subfascial layer

Anterior Carpal Region (A)

The wrist is bounded distally by the flexor retinaculum. The proximal boundary is visible on the skin only as the proximal skin crease of the wrist.

Proximal to the *flexor retinaculum* there are strong fiber strands in the *antebrachial fascia* (**1**), which also form a deep layer (**2**) and are connected to the bones of the forearm. The veins and nerves run superficially as described previously on page 386, as does the tendon of the *palmaris longus* (**3**). Deeply the most radial structure is the *radial artery* (**5**) and its accompanying veins lying on the *pronator quadratus* (**4**).

On the ulnar side of the artery lies the tendon of the *flexor carpi radialis* (**6**) within its own sheath, followed next by the tendon sheath of the *flexor pollicis longus* (**7**). Between this muscle and the common tendon sheath (**8**) for the *flexor digitorum superficialis* and *flexor digitorum profundus* runs the *median nerve* (**9**). The structures run through the carpal tunnel (canalis carpi, see p. 124) to the palm of the hand.

> **Clinical tip: Carpal tunnel syndrome** often results from a transligamentous origin (23%) of the thenar branch of the median nerve. In any case a disproportion exists between the carpal tunnel and its contents, causing severe pain in the thenar region as well as hypo- and paresthesias.

The *ulnar artery* (**10**) with its accompanying veins and the *ulnar nerve* (**11**) lie radial to the flexor carpi ulnaris (**12**) and run to the palm of the hand superficial to the flexor retinaculum. They lie between the deep (**2**) and superficial layers of the antebrachial fascia. The superficial layer is usually strengthened by tendinous fiber bands from the flexor carpi ulnaris (see p. 160) so that the ulnar artery and nerve reach the palm in their own fascial ulnar tunnel (**Guyon's box**).

Palm of Hand

Superficial Layer (B)

The palm of the hand is subdivided into three regions: the thenar eminence, the central compartment (metacarpal region), and the hypothenar eminence. Fascia encloses these lateral regions, while the central compartment is covered by the coarse, firm *palmar aponeurosis* (**13**). This represents the continuation of the palmaris longus (**A 3**) and on its ulnar border it radiates into the *palmaris brevis* (**14**), which is highly variable in its development.

The palmar aponeurosis is divided into longitudinal (**15**) and transverse (**16**; see p. 178) *fascicles*. At the radial, ulnar, and distal margins of the palmar aponeurosis, the *common palmar digital arteries* (**17**) and homonymous nerves become subcutaneous. The arteries divide into the *proper palmar digital arteries* (**18**), which, accompanied by the *proper palmar digital nerves* (**18**), extend to the distal phalanges of the fingers. The *proper palmar digital veins* reach the *superficial palmar venous arch*, which lies superficially at the root of the fingers.

In the forearm (see p. 386) the ulnar nerve gives off the *palmar branch* to supply the skin of the ball of the little finger.

> **Clinical tip:** The nerves at the sides of the fingers can be anesthetized by the **Oberst nerve block**. It is important to remember that the skin on the distal phalanx of the thumb and the middle and distal phalanges of the index and middle fingers is also innervated on its dorsal surface by the *proper palmar digital branches* of the median nerve.

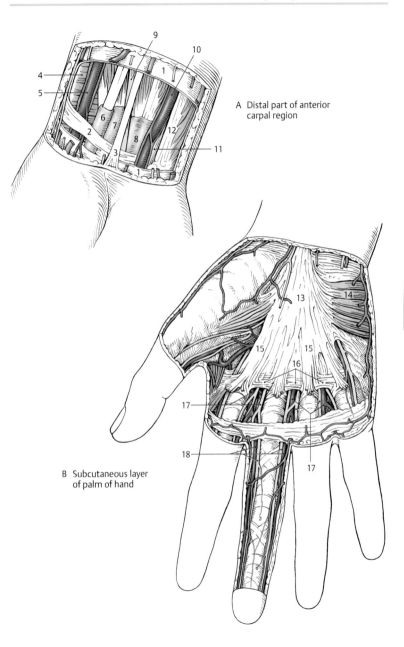

A Distal part of anterior carpal region

B Subcutaneous layer of palm of hand

Palm of Hand, continued (A–H)

Deep Layer, Superficial Palmar Arch (A)

After removal of the fascia and palmar aponeurosis, the superficial palmar arch (**1**) and the muscles of the thenar and hypothenar eminences become visible. The *superficial palmar arch* (**1**) is mainly formed by the *ulnar artery* (**2**), which runs superficial to the *flexor retinaculum* (**3**). It is connected with the *superficial palmar branch of the radial artery* (**4**). The superficial palmar arch gives off the common *palmar digital arteries* (**5**), which run at first superficial to the tendons of the long flexors (**6**) and at the roots of the fingers between these tendons.

The ulnar artery, which gives off a *deep palmar branch* (**7**), accompanies the *ulnar nerve* (**8**), which with its *superficial branch* (**9**) medial to the artery reaches the palm of the hand. The superficial branch of the ulnar nerve innervates the skin of the ulnar two and one-half digits. It is often connected with the branches of the *median nerve* (**11**) by an *anastomotic branch* (**10**). In the region of the flexor retinaculum (**3**), the *deep branch* (**12**) becomes separated from the *ulnar nerve* and penetrates deeply between the *abductor digiti minimi* (**13**) and the *flexor digiti minimi brevis* (**14**).

Already in the carpal tunnel (see p.124) the median nerve has often divided into the *common palmar digital nerves* (**15**). It gives off branches to the thenar muscles (excluding the deep head of the flexor pollicis brevis and the adductor pollicis).

Deep Palmar Arch (B)

When the tendons of the finger flexors (**6**) are removed, the *deep palmar arch* (**18**) appears lying on the *interossei* (**16**) and usually running proximal to the *transverse head* (**17**) *of the adductor pollicis*. This arch is formed by the deep palmar branch of the ulnar artery (**7**) and radial artery and gives

off the *palmar metacarpal arteries* (**19**). It is accompanied by the deep branch of the ulnar nerve (**12**).

Variants of the Superficial Palmar Arch (C–H)

The superficial palmar arch is highly variable in its development. The typical palmar arch (**C**) is present in only 27% of cases (*Lanz–Wachsmuth*). In the same proportion of subjects (27%) the arch is formed solely by the ulnar artery (**D**).

In some cases, the artery accompanying the median nerve is retained as the "original" median artery and may, either by anastomosing with the ulnar artery or without forming the arch (**E**), together with the ulnar artery, give off the artery to the digits. During embryonic development of the blood supply to the hand, the "original" median artery takes over from its predecessor, the common interosseous artery. In lower mammals this stage of development persists longer, while in primates the radial and ulnar arteries arise from the median artery. Embryologically a persistent median artery is an atavism.

Sometimes (6%) not all the digital arteries arise from a superficial palmar arch, which is formed only by the ulnar artery (**F**). A superficial palmar arch may be completely absent and then the arteries of the digits are given off by the radial artery as well as by the ulnar artery (4.5%, **G**) or (12%) the arteries of the digits arise from the deep palmar arch and the ulnar artery (**H**).

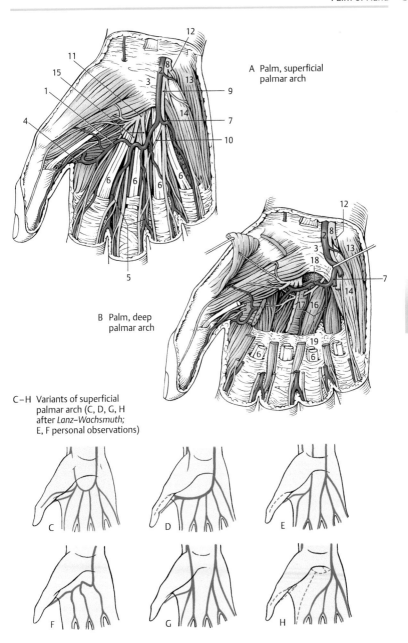

A Palm, superficial palmar arch

B Palm, deep palmar arch

C–H Variants of superficial palmar arch (C, D, G, H after *Lanz–Wachsmuth;* E, F personal observations)

Dorsum of the Hand (A, B)

Subcutaneous Layer (A)

The proximal boundary of the dorsum of the hand is the *extensor retinaculum* (**1**), a part of the fascia that is strengthened by a large number of transverse fibers.

Subcutaneously the veins coming from the digits (usually two joined by anastomoses) are continued in the *dorsal metacarpal veins* (**2**), of which three are usually particularly well developed. The largest are the dorsal metacarpal veins at the root of the fourth finger, which, after combining, run as the *accessory cephalic vein* (= vena salvatella, **3**) to the forearm. The *dorsal metacarpal vein of the fifth finger* (**4**) represents the beginning of the basilic vein, while the first dorsal metacarpal vein is called the *cephalic vein of the thumb* (**5**). A large number of anastomoses interconnect all the veins to form the *venous network of the dorsal hand* (**6**). On the ulnar side, covered by veins, runs the *dorsal branch of the ulnar nerve* (**7**), while radially the terminal parts of the *superficial branch of the radial nerve* (**8**) are found.

Subfascial Layer (B)

After removal of the fascia, the extensor tendons and the branches of the *radial artery* (**9**) become visible. In the region of the radial fovea, the radial artery gives off the *dorsal carpal branch* (**10**) and runs between the heads of the *first dorsal interosseous* (**11**) into the palm of the hand. The dorsal carpal branch gives off the *dorsal metacarpal arteries* (**12**), which again divide into the *dorsal digital arteries* (**13**).

Radial Fovea, "Anatomical Snuff Box" (C)

The triangular radial fovea, or anatomical snuff box, is bounded dorsally by the tendon of the *extensor pollicis longus* (**14**) and on the palmar side by the tendons of the *extensor pollicis brevis* (**15**) and abductor pollicis longus (**16**). The scaphoid and trapezium bones form the floor. Proximally the extensor retinaculum (**1**) completes the depression. It contains the tendons of the *extensor carpi radialis longus* (**17**) and *extensor carpi radialis brevis* (**18**), and the radial artery (**9**). In the fovea, the radial artery gives off its dorsal carpal branch (**10**). The branches of the superficial part (**8**) of the radial nerve cross the radial fovea superficially.

> **Clinical tip:** The term "anatomical snuff box" is a misnomer. The fovea is a site where blood is sampled from the *radial artery* (**9**) to determine the O_2/CO_2 ratio (Astrup method of blood gas analysis).

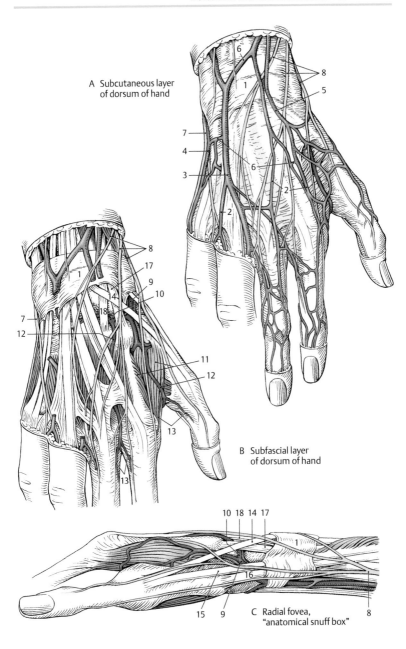

A Subcutaneous layer
   of dorsum of hand

B Subfascial layer
   of dorsum of hand

C Radial fovea,
   "anatomical snuff box"

Trunk

Regions (A, B)

Superficially there are no distinct features marking the divisions between the trunk and the upper and lower limbs. The subdivision into regions has a purely practical purpose and has no developmental basis. The lack of demarcation results in some overlap in the transitional regions between the trunk and limbs. The trunk regions are subdivided into the regions of the thorax and those of the abdomen.

Regions of the Thorax

The *deltoid region* (**1**), the *infraclavicular fossa* (**2**) with the *clavipectoral triangle* (**3**), and the *axillary region* (**4**) are described on page 368 as transitional regions of the free upper limb.

The **mammary region** (**5**) includes the area of the mammary gland. The **inframammary region** (**6**) lies inferiorly, and the **lateral pectoral region** (**7**) lies laterally. These three regions are collectively known as **pectoral region**. The lateral pectoral region connects with the axillary region. The **presternal region** (**8**) connects the left and right mammary and inframammary regions.

The **vertebral region** (**9**) follows the midline of the back, and lateral to it are the **suprascapular region** (**10**), the **interscapular region** (**11**), the **scapular region** (**12**), and the **infrascapular region** (**13**).

Regions of the Abdomen

The transitional region between the thorax and abdomen, the **hypochondrium** (**14**) lies laterally. Between the two hypochondriac regions, in the area of the infrasternal angle, is the **epigastric region** (**15**). These three regions are bounded below by the *transpyloric plane*, which is the transverse plane through the midpoint between the jugular notch of the sternum and the upper edge of the symphysis. The **umbilical region** (**16**) covers the area between the two mid-clavicular lines, the transpyloric plane, and the plane running through the anterior superior iliac spines. The latter plane is called the *interspinous plane* and contains the *interspinous distance* (see p. 190).

The umbilical region is flanked by the **lateral abdominal regions** (**17**). Just below that the **inguinal regions** (**18**) adjoin laterally to the inguinal sulcus, and the **pubic region** (**19**) adjoins medially to the upper edge of the symphysis and the pubic crests.

In the posterior midline, below the vertebral region, lies the **sacral region** (**20**), which includes the area above the sacrum. On both sides of these regions lie the **lumbar regions** (**21**), which merge into the gluteal regions at the iliac crests.

Adjacent to the pubic region is the urogenital region (not illustrated), which adjoins the anal region (not illustrated). These two regions are known collectively as the **perineal region**; they connect the abdominal regions with those of the back.

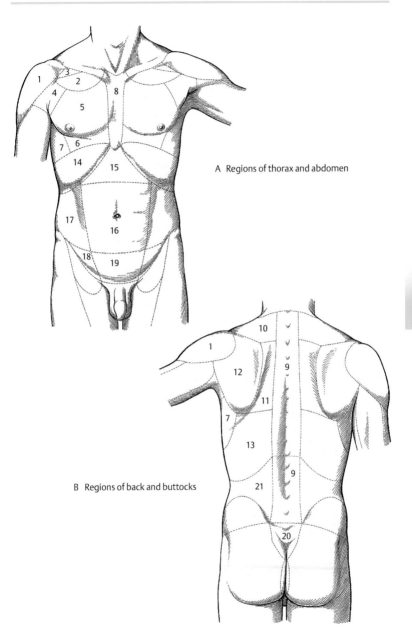

A Regions of thorax and abdomen

B Regions of back and buttocks

Regions of the Thorax (A, B)

Anterior Thoracic Regions (A)

Of special importance in the female are the tissues in the subcutaneous layer of the mammary region. The breast rests upon the *pectoral fascia* (**1**). It consists of the mammary gland, fibrous connective tissue, and adipose tissue, which are collectively known as the *body of the breast* (**2**). A process of varying size, the *axillary process* (**3**), extends into the axilla. The fibrous tissue forms the suspensory ligaments of the breast, which connect the pectoral fascia with the skin and are located between the lobes of the gland.

The *areola of the mammary gland* (**4**) is surrounded by a delicate venous plexus, the *areolar venous plexus* (**5**). From this plexus the blood drains via the *anterior cutaneous branches* (**6**) to the anterior intercostal veins and, laterally, to the *thoracoepigastric vein* (**7**) and *lateral thoracic vein* (**8**). Blood is supplied both laterally and medially. Branches of the lateral thoracic artery, the *lateral mammary branches* (**9**) penetrating the *axillary fascia* (**10**), extend laterally to the body of the breast. The internal thoracic artery gives off perforating branches that reach the subcutaneous layer through the first to sixth intercostal spaces near the sternum. Larger perforating branches supply the breast medially as *medial mammary branches* (**11**).

Lateral to the breast are the paramammary lymph nodes, and in the axilla there are the *axillary lymph nodes* (**12**).

Crossing the clavicle from above, the *medial* (**13**) and *intermediate* (**14**) *supraclavicular nerves* from the cervical plexus reach the clavipectoral triangle and the infraclavicular fossa, respectively. The mammary region is innervated by the *medial mammary branches* (**15**) from the *anterior cutaneous branches* (**16**) of the second to fourth intercostal nerves and by the *lateral mammary branches* (**17**) from the *lateral cutaneous branches* (**18**) of the second to fourth intercostal nerves. One or two *intercostobrachial nerves* (**19**), usually from the second (and third) intercostal nerve, ex-

tend to the upper arm through the axillary region.

The pectoralis major muscle with its three parts is visible in the subfascial layer. The cephalic vein runs laterally through the clavipectoral triangle (see p. 370).

Posterior Thoracic Regions (B)

In the subcutaneous layer on the thoracic fascia there are cutaneous branches of arteries, veins, and nerves. It is important to note that the scapular line represents the boundary between the territories of the posterior and anterior branches of the spinal nerves.

The following muscles can be demonstrated in the subfascial layer: *trapezius* (**20**), *latissimus dorsi* (**21**), and *rhomboideus major* (**22**). The *infraspinatus muscle* (**23**) lies on the scapula, the *teres minor* (**24**) originates from the lateral margin of the scapula, while the *teres major* (**25**) originates inferior to the teres minor. Between the two teres muscles and the *long head of the triceps* (**26**) is the medial axillary foramen (see p. 374) with the *circumflex vein and circumflex artery of the scapula* (**27**). From the *scapular spine* (**28**), the *spinal part of the deltoid muscle* (**29**) extends to the upper arm.

Clinical tip: The lymphatic drainage of the breast is of special importance because of the high incidence of **breast cancer**. The lymph drains via several vessels, usually four, into the venous angle. One lymphatic vessel reaches the axillary lymph nodes either directly or via the paramammary lymph nodes. From there it drains into the venous angle via infraclavicular and supraclavicular nodes.

The second vessel extends from the parammary lymph nodes directly to the infraclavicular lymph nodes and finally into the venous angle via the supraclavicular lymph nodes.

The third vessel reaches the infraclavicular and supraclavicular nodes, frequently involving also the interpectoral lymph nodes.

The fourth vessel comes from the medial portions of the gland and runs through the parasternal lymph nodes alongside the internal thoracic arteries and veins to the venous angle. The first lymph node affected by metastasis is called the **sentinel node** (see p. 366).

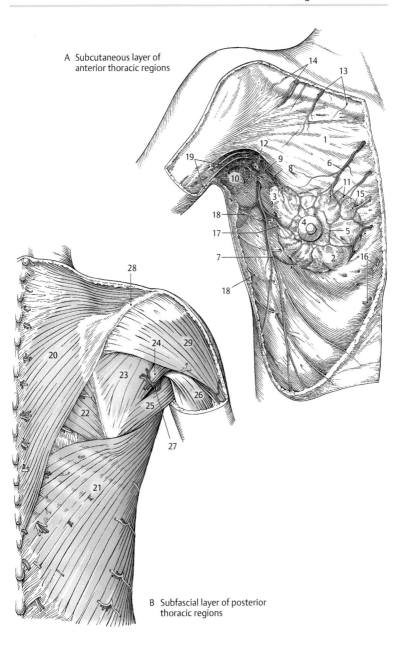

A Subcutaneous layer of
anterior thoracic regions

Peripheral Pathways

B Subfascial layer of posterior
thoracic regions

Regions of the Abdomen (A)

Upon removal of the subcutaneous tissue of the abdomen (see p. 92), the subcutaneous vessels and nerves become visible on the delicate *(superficial) abdominal fascia*. Especially noteworthy are the *paraumbilical veins* circling the navel; they anastomose with the *superficial epigastric veins* (**1**) and with the *thoracoepigastric veins*.

The superficial epigastric vein, which is accompanied by a delicate artery of the same name, crosses the inguinal ligament and joins the femoral vein in the saphenous hiatus (see p. 416). The thoracoepigastric vein ascends laterally from the navel and opens into the axillary vein. The *superficial circumflex iliac artery and vein* (**2**) ascend in the lateral area of the inguinal ligament.

In the paramedian region, the *anterior cutaneous branches* (**3**) of the *eighth to 12th intercostal nerves* (**4**) penetrate the rectus sheath and the fascia. The *lateral cutaneous branches* (**5**) of the *ninth to 12th intercostal nerves* are visible lateral to them.

Just superior to the superficial inguinal ring, the *anterior branch of the iliohypogastric nerve* (see p. 400) becomes subcutaneous. The *lateral branch of the iliohypogastric nerve* (**6**) penetrates the fascia in the area of the anterior superior iliac spine.

Upon removal of the fascia and subsequent incision of the anterior layer of the rectus sheaths (see p. 88), the *rectus abdominis* (**7**) becomes visible on both sides. Posterior to the rectus abdominis, but inside the rectus sheath, run the *inferior epigastric artery and vein* (**8**), which anastomose above the navel with the *superior epigastric artery and vein* (**9**).

The rectus sheath contains the rectus abdominis, which is attached to the anterior layer at the *tendinous intersections* (**10**). The inferior and superior epigastric arteries and veins also run inside the rectus sheath, and so do the eighth to 12th intercostal nerves, which enter through the *posterior layer* (**11**) *of the rectus sheath*.

Clinical tip: The paraumbilical veins extend along the *round ligament of the liver* (see Vol. 2) to the left branch of the portal vein; they connect with the superficial epigastric veins and the thoracoepigastric veins. This creates a subcutaneous portosystemic anastomosis. If there is a backflow of blood due to liver disease, these veins become dilated and visible underneath the skin. This condition is referred to as **"caput medusae"** (Medusa head).

Other portosystemic anastomoses of clinical importance are the submucosal plexus in the distal third of the esophagus and the submucosal plexus in the rectum. Retroperitoneal anastomoses are also present. Note, however, that the subcutaneous anastomosis described here is the *only* anastomosis that communicates directly with the left branch of the portal vein; all the others drain into the portal vein trunk. **The presence of a "Medusa head" generally indicates congestion of the left hepatic lobe.**

12 Linea alba
13 Pyramidalis
14 External oblique
15 Transversalis fascia
16 Arcuate line
17 Anterior layer of the rectus sheath

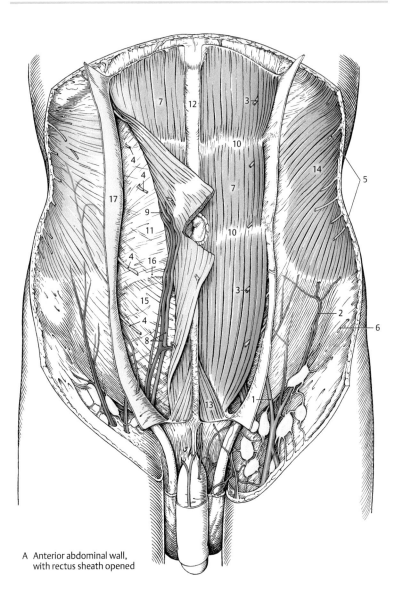

A Anterior abdominal wall,
   with rectus sheath opened

Inguinal Region

Inguinal Canal (A–C)

First Layer (A)

The inguinal region and the pubic region are superficially covered by the subcutaneous fascia of the abdomen (see p. 92). Only upon removal of the connective tissue membrane is it possible to view the subcutaneous vessels and nerves. Running over the *(superficial) abdominal fascia* (**1**) and crossing the inguinal canal are the *superficial epigastric artery and vein* (**2**), while the *superficial circumflex iliac artery and vein* (**3**) run laterally.

Both vascular bundles extend to the saphenous hiatus in the subinguinal region (see p. 416). The *external pudendal artery and vein* (**4**), which are frequently duplicated, also connect to the saphenous hiatus. After crossing the *spermatic cord* (**5**), they reach the pudendal region.

Superior to the *superficial inguinal ring* (**6**) the *anterior cutaneous branch of the iliohypogastric nerve* (**7**) can be viewed, while the *ilioinguinal nerve* (**8**) runs together with the spermatic cord (or round ligament of the uterus, respectively) and gives off sensory branches to supply the proximal inner surface of the thigh, the mons pubis, the scrotal skin in the male, and the labium majus in the female.

> **Clinical tip:** Of special importance are the *superficial inguinal lymph nodes* (**9**). In the female, they are reached through the inguinal canal by lymph vessels **from the fundus and body of the uterus**. They play a major role in the lymphogenous spread of **endometrial carcinoma**. (See also "sentinel node," p. 366.)
> Other lymphatics drain to the interiliac lymph nodes and directly to the aortic lymph nodes. The **cervix** never (!) drains to the inguinal nodes. It drains to the iliac, interiliac, gluteal, sacral, and rectal lymph nodes and directly to the aortic nodes.

Second Layer (B, C)

After sharp dissection of the superficial inguinal ring (**6**) in the male, the outer sheath of the spermatic cord (**5**), the *external spermatic fascia* (**10**), is opened. This exposes the external inguinal ring with the *lateral crus* (**11**), the *medial crus* (**12**), the *intercrural fibers* (**13**), and the *reflected ligament* (**14**).

Upon dividing the aponeurosis of the *external oblique* (**15**), the *internal oblique* (**16**) can be viewed. Its inferior fibers extend as the *cremaster muscle* (**17**) on the spermatic cord and form its middle sheath, the *cremasteric fascia and muscle* (**18**). It is accompanied by the *genital branch* (**19**) of the genitofemoral nerve, which supplies the cremaster muscle and also participates in the sensory supply of the ilioinguinal nerve. The delicate cremasteric artery and vein are embedded in the muscle and are, therefore, hardly visible.

Resting on the internal oblique, the *iliohypogastric nerve* (**20**) extends medially and penetrates with its anterior cutaneous branch (**7**) the external oblique aponeurosis and the fascia above the external inguinal ring. Sometimes the nerve splits into two branches before it penetrates. It provides sensory supply to the skin in the inguinal region.

After sharp dissection of the superficial inguinal ring (**6**) in the female, the round ligament of uterus can be seen. It radiates into the connective tissue of the labium majus. Closely adjoining this band are the delicate artery and vein of the uterine round ligament and the genital branch of the genitofemoral nerve. The uterine round ligament is accompanied by the ilioinguinal nerve (**8**).

When the inguinal canal is opened, some fibers of the internal oblique are revealed; they merge with the round ligament of the uterus. They are referred to as the round ligament part of the internal oblique and correspond to the cremaster muscle in the male.

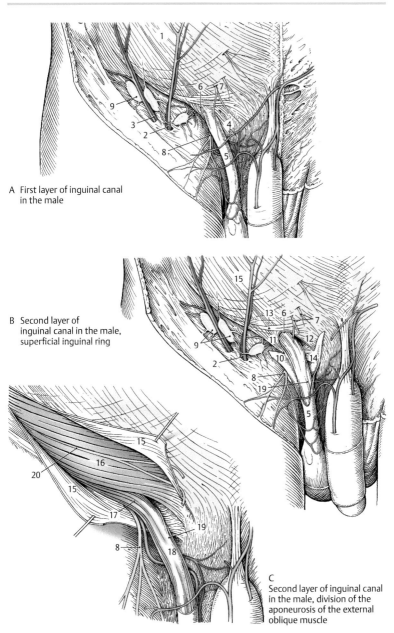

A First layer of inguinal canal in the male

B Second layer of inguinal canal in the male, superficial inguinal ring

C Second layer of inguinal canal in the male, division of the aponeurosis of the external oblique muscle

Inguinal Region, continued

Inguinal Canal, continued (A–C)

Third Layer (A, B)

After division of the *cremasteric fascia and muscle* (**1**) the last and very thin sheath of the spermatic cord, the *internal spermatic fascia* (**2**), becomes visible. Further incision of the *internal oblique* (**3**) exposes the roof of the inguinal canal, the *transverse abdominis* (**4**), and also the posterior wall, the *transversalis fascia* (**5**). The internal spermatic fascia (**2**) evaginates as a continuation of the transversalis fascia, thus making it possible to determine the position of the *deep inguinal ring* (**6**). The variably developed *interfoveolar ligament* (**7**) lies medial to the deep inguinal ring (see also p. 98).

Fourth Layer (C)

Opening the internal spermatic fascia (**2**) exposes the contents of the spermatic cord and also opens the deep inguinal ring. The spermatic cord contains the white, round *ductus deferens* (**8**), the *testicular artery* (**9**), and the *pampiniform plexus* (**10**).

The ductus deferens, called also the *vas deferens* (**8**), is the continuation of the *duct of the epididymis* and extends through the inguinal canal into the lesser pelvis. Here it unites, together with its ampulla of the ductus deferens, with the *excretory duct* of the *seminal gland (seminal vesicle)* to form the *ejaculatory duct*. The testicular artery (**9**) originates directly from the abdominal aorta. The pampiniform plexus (**10**) continues as the *testicular vein*. On the left side, the testicular vein extends across the *left renal vein* to the inferior vena cava. The right testicular vein drains directly into the inferior vena cava.

If parts of the transversalis fascia are removed when the internal spermatic fascia is opened, the preperitoneal structures are exposed: the *inferior epigastric artery and vein* (**11**) and the *cord of the umbilical artery* (**12**). The weak sites in this region of the abdominal wall also become visible. These are the peritoneal fossae: lateral to the inferior epigastric artery and vein is the *lateral inguinal fossa* (**13**); the deep inguinal ring projects into it. The *medial inguinal fossa* (**14**) lies between the chorda of the umbilical artery and the inferior epigastric artery and vein, while the *supravesical fossa* (**15**) lies medial to the chorda of the umbilical artery. The superficial inguinal ring projects into the latter two fossae.

> **Clinical tip:** The three fossae constitute areas of weakness in the abdominal wall, creating sites of predilection for inguinal hernias (see p. 100). Three types of inguinal hernia are distinguished based on the location of the internal opening:
> a) **Indirect (lateral) inguinal hernia:** pulsation of the inferior epigastric artery medial to the hernia.
> b) **Direct (medial) inguinal hernia:** pulsation lateral to the hernia.
> c) **Supravesical hernia:** no pulsations because the hernial opening is medial to the lateral umbilical ligament.
> The hernial opening can be quickly identified by endoscopic examination, which aids in classifying the hernia.

16 External oblique aponeurosis
17 Inguinal ligament
18 Iliohypogastric nerve
19 Reflected ligament
20 Ilioinguinal nerve

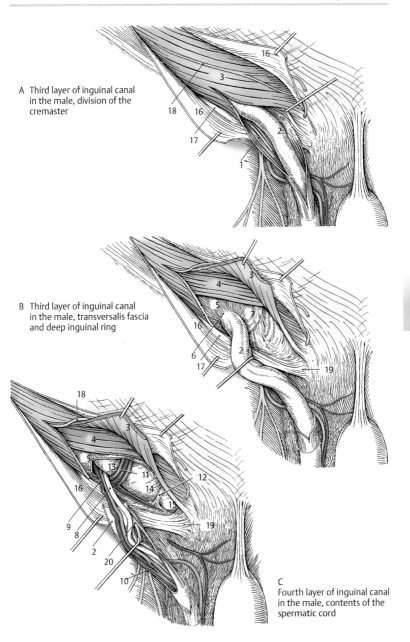

A Third layer of inguinal canal in the male, division of the cremaster

B Third layer of inguinal canal in the male, transversalis fascia and deep inguinal ring

C Fourth layer of inguinal canal in the male, contents of the spermatic cord

Lumbar Region (A, B)

First Layer (A)

Upon removal of the abdominal viscera, the parietal abdominal fascia can be mobilized chiefly to expose branches of the lumbar plexus.

At the inferior margin of the 12th rib (**1**) runs the *subcostal nerve* (**2**) as the last of the anterior branches of the thoracic nerves. It is partially covered by the portion of the *lumbar part of the diaphragm* (**4**) originating from the *lateral arcuate ligament* (**3**). The *quadratus lumborum* (**5**) is visible underneath the lateral arcuate ligament, while the portion of the *psoas major* (**7**) originating from the 12th thoracic vertebra is visible underneath the *medial arcuate ligament* (**6**).

The first branch of the lumbar plexus, the *iliohypogastric nerve* (**8**), is visible at the lateral margin of the psoas major. It crosses the quadratus lumborum and penetrates the abdominal muscles above the iliac crest. Almost parallel to it and penetrating the psoas major runs the *ilioinguinal nerve* (**9**), which extends to the deep inguinal ring. Next, the *genitofemoral nerve* (**10**) penetrates the psoas major and divides at varying levels into the *genital branch* (**11**) and the *femoral branch* (**12**). The former extends to the inguinal canal, while the latter passes through the vascular space to reach the subinguinal region.

At the lateral margin of the psoas major and near the iliac fossa is another branch of the lumbar plexus, the *lateral femoral cutaneous nerve* (**13**). It extends laterally, near the anterior superior iliac spine, to the muscular space. The most prominent branch, the *femoral nerve* (**14**), runs in the groove between the *iliacus muscle* (**15**) and psoas major (**7**) and passes through the muscular lacuna to reach the thigh. The last branch, the *obturator nerve* (**16**), is the only one running medial to the psoas major; after crossing the *external iliac artery and vein* (**17**) it reaches the *obturator canal*.

Second Layer (B)

Removal of the superficial part of the psoas major exposes the *anterior branches* (**18**) of the first four lumbar nerves. These lie on the *deep part* (**19**) of the psoas major and form the lumbar plexus. The branch of the fourth lumbar nerve divides into a *superior* and an *inferior branch* (**20**). The latter unites with the anterior branch of the fifth lumbar nerve to form the *lumbosacral trunk*, which participates in the formation of the sacral plexus.

Medial to the emerging anterior branches runs the *sympathetic trunk* (**21**) and, on the right side, also the *inferior vena cava* (**22**). The segmental *lumbar arteries and veins* (**23**) adjoin the vertebral column. They pass underneath the anterior branches and the deep portion of the psoas major.

24 Internal iliac artery
25 Inferior epigastric artery
26 Deep circumflex iliac artery and vein

(The sacral plexus and the lumbar plexus can be considered a unit that is called the **lumbosacral plexus.**)

Lumbar plexus

– Roots: Ventral rami (L1 – L4)
– Branches: – Iliohypogastric nerve
 – Ilioinguinal nerve
 – Genitofemoral nerve
 – Lateral femoral cutaneous nerve
 – Obturator nerve
 – Femoral nerve

Sacral plexus

– Roots: Ventral rami (L4 – S3)
– Branches: – Gluteal nerves
 – Muscular branches
 – Inferior clunial nerves
 – Posterior femoral cutaneous nerve
 – Pudendal nerve
 – Coccygeal nerve
 – Sciatic nerve

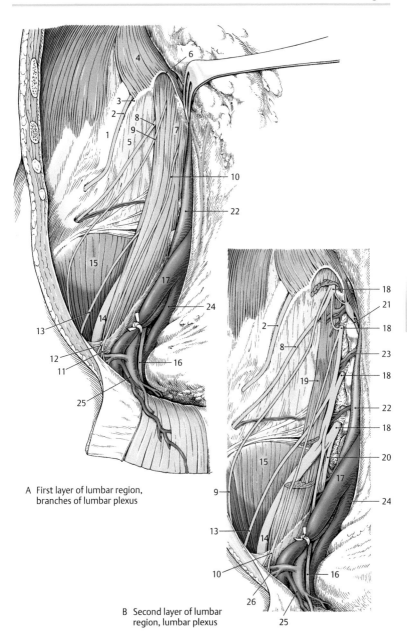

A First layer of lumbar region, branches of lumbar plexus

B Second layer of lumbar region, lumbar plexus

Perineal Region in the Female (A, B)

The perineal region is divided into the urogenital region anteriorly and the anal region posteriorly. Fasciae and muscles allow for definition of several structural layers.

Superficial and Middle Layers (A)

Urogenital region: In the lateral area along the inferior pubic ramus and the ramus of the ischium, the *superficial perineal fascia* (**1**) is divided into two layers, a fatty *outer layer* and a membranous *inner layer* (right side of the dissection). The two layers unite near the *vestibule of the vagina* (**2**). Removal of the superficial perineal fascia exposes the *superficial perineal space* (left side of the dissection). *Posterior labial branches* (**3**), which originate from the *perineal artery* (**4**) and are accompanied by veins of the same name, extend to the vestibule of vagina and the *perineal body* (**5**). The perineal artery often penetrates the inner layer of the superficial perineal fascia. The *perineal nerves* (**6**) cross the posterior margin of the urogenital diaphragm (see p. 106) and extend together with the arterial branches to the vestibule of the vagina and the perineal body.

The superficial perineal space contains the following muscles: the *bulbospongiosus* (**7**) medially, the *ischiocavernosus* (**8**) laterally, and the *transversus perinei superficialis* (**9**) posteriorly.

Anal region: The *obturator fascia* (**10**) borders laterally on the *ischioanal (ischiorectal) fossa*. This fossa extends to the front and lies then between the urogenital diaphragm and the pelvic diaphragm with the *inferior fascia of the pelvic diaphragm* (**11**). It contains copious abdominal fat, the *fat pad of the ischioanal fossa*. In a fold of the obturator fascia (**10**) lies the *pudendal canal* (**12**). The *inferior rectal artery* (**13**) and the *inferior rectal nerve* (**14**) supply the *external anal sphincter* (**15**) and the anal skin. There may be additional perineal branches (not shown) for the labial skin and a perforating cutaneous

nerve for the anal skin. Both originate from the posterior cutaneous nerve of the thigh. Numerous *inferior rectal veins* (**16**), which anastomose with the medial rectal veins, extend to the internal pudendal vein.

Removal of the inferior fascia of the pelvic diaphragm (**11**) exposes the external anal sphincter (**15**) and the *levator ani* (**17**) (left side of the preparation). Posterior to the *anus* (**18**) in the median plane is the *anococcygeal ligament* (**19**); parts of the levator ani muscles radiate into it. The *internal pudendal arteries and veins* (**20**) and the *pudendal nerve* (**21**) pass through the lesser sciatic foramen and then run inside the pudendal canal (*Alcock's canal*).

Deep Layer (B)

Urogenital region: Removal of the bulbospongiosus and ischiocavernosus muscles (**8**) with the inferior fascia of the urogenital diaphragm (perineal membrane) opens the deep perineal space. In addition to muscles, it contains the *crura of the clitoris* (**22**); they unite to form the *body of the clitoris* (**23**), which terminates in the *glans of the clitoris* (**24**).

On each side lateral to the vestibule of the vagina (**2**) is an erectile body, the *bulb of the vestibule* (**25**); the two bulbs are connected by the *commissure of the bulbs* (**26**) between the crura of the clitoris. On both sides lies the *great vestibular gland* (**27**) covered by the bulb of the vestibule inside the urogenital diaphragm. It opens via a secretory duct between the labium minus and the vaginal orifice into the vestibule of the vagina (**28**).

> **Clinical tip:** The inner layer of the superficial perineal fascia is often called the deep perineal fascia, although the term "*perineal membrane*" is generally used today.
> Unfortunately the term "urogenital diaphragm" has been dropped from modern anatomical (but not clinical!) usage, although it is technically correct.

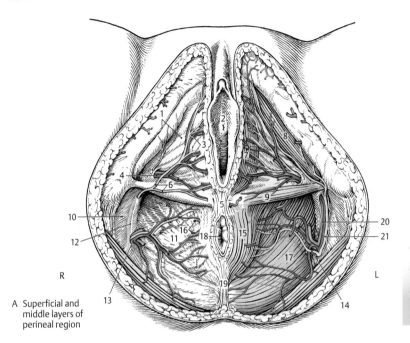

A Superficial and middle layers of perineal region

R   L

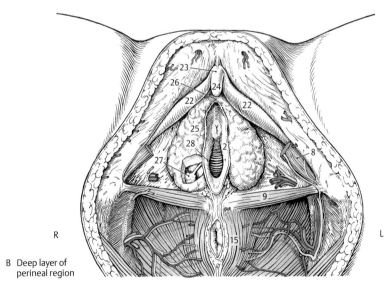

B Deep layer of perineal region

R   L

Perineal Region in the Male (A)

Superficial Layer (Right Side of Specimen)

Urogenital region: The *superficial perineal fascia* (**1**) with its outer layer and inner layer (deep perineal fascia or Colles fascia) continues on the thigh as the *fascia lata* (**2**) and on the *penis* (**3**) as the *superficial fascia of the penis* (**4**). Together with the superficial abdominal fascia it also forms the tunica dartos.

The *perineal artery* (**5**), which originates from the internal pudendal artery, often penetrates the urogenital diaphragm near its posterior margin and gives off *posterior scrotal branches* (**6**). These are accompanied by *posterior scrotal veins* (**7**). *Scrotal* and *muscular branches* (**8**) from the pudendal nerve extend to the scrotum and to the skin and muscles of the urogenital region. *Perineal branches* (**9**) from the posterior cutaneous nerve of the thigh also extend to the scrotum, while the *inferior clunial nerves* (**10**) are distributed to the skin in the lower part of the gluteal region.

Anal region: The *obturator fascia* (**11**) bounds the region laterally, the *gluteus maximus* (**12**) with the *gluteal fascia* posteriorly, and the *perineal body* (**13**), the *anus* (**14**), and the *anococcygeal ligament* (**15**) medially. The deep ischioanal fossa is filled with fatty tissue, the *fat pad of the ischioanal fossa*. Its roof is formed by the *inferior fascia of the pelvic diaphragm* (**16**).

Middle Layer (Left Side of Specimen)

Urogenital region: Removal of the superficial perineal fascia opens the superficial perineal space. Medially the *bulbospongiosus* (**17**) lies on the corpus spongiosum of the penis (urethra) and the corpus cavernosum of the penis. The *ischiocavernosus* (**18**), which originates from the ramus of the ischium, lies laterally. The space is bounded posteriorly by the *superficial transverse perineal* (**19**) while the *perineal membrane* (*inferior fascia of the urogenital diaphragm*) (**20**) forms the roof.

The *internal pudendal artery and vein* (**21**) penetrate the urogenital diaphragm and give off the above-mentioned branches. The *pudendal nerve* (**22**) extends to the superficial perineal space (*Colles space*) at the posterior margin of the urogenital diaphragm.

Anal region: Removal of the inferior fascia of the pelvic diaphragm (**16**) exposes the *levator ani* (**23**) and the *coccygeus* (**24**) *muscles*.

Inside the *pudendal canal* (*Alcock's canal*) (**25**), the internal pudendal artery gives off the *inferior rectal artery* (**26**), which often divides into two branches. It is accompanied by the *inferior rectal veins* (**27**), which extend to the pudendal vein. *Inferior rectal nerves* (**28**) supply the *external anal sphincter* (**29**) and the anal skin.

Clinical tip: The posterior urethra is usually approached surgically through the perineum, especially in the treatment of strictures. The perineal approach is also used for radical prostatectomy. In all cases the central tendon of the perineum, the **perineal body**, must be divided.

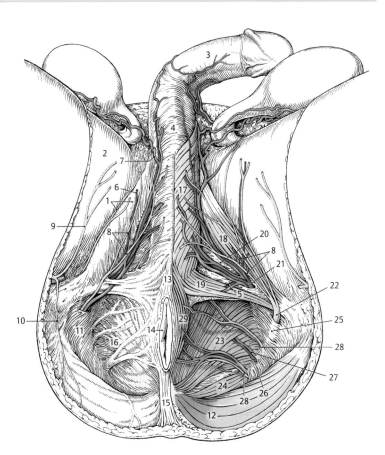

A Superficial and middle layers
of perineal region

Perineal Region in the Male, continued (A, B)

Deep Layer (A, B)

Urogenital region: Removal of the perineal membrane (inferior fascia of the urogenital diaphragm; right side of specimen) opens the deep perineal space. The *transversus perinei profundus* (**1**) extends to the urogenital hiatus and, with its most posterior fibers, to the *perineal body* (**2**). The *ischiocavernosus* (**3**), which originates from the ramus of the ischium, radiates into the tunica albuginea of the *crus of the penis* (**4**).

The *internal pudendal artery* (**5**) gives off the *perineal artery* (**6**) at the posterior margin of the urogenital diaphragm. Covered by the crus penis, it runs anteriorly and gives off the urethral artery where the crura of the penis unite. The artery is accompanied by the *internal pudendal vein* (**7**), which receives the *posterior scrotal veins* (**8**).

Removal of the *bulbospongiosus* (**9**) exposes the *corpus spongiosum of the penis* (**10**; left side of specimen). Posterior to the *bulb of the penis* (**11**), the posterior end of the corpus spongiosum, there is a pea-sized *bulbourethral gland* (**12**) on both sides.

Anal region: Removal of the *obturator fascia* (**13**) opens the pudendal canal and exposes the internal pudendal artery and vein and also the *pudendal nerve* (**14**). Alongside the *internal obturator* (**15**) the *tendinous arch of the levator ani* (**16**) extends to the *ischial tuberosity* (**17**). The *sacrospinal ligament* (**18**) reaches from there to the sacrum and forms together with the lesser sciatic notch the lesser sciatic foramen.

The *levator ani* (**19**) extends together with the *puborectalis* (**20**), *pubococcygeus* (**21**), and *iliococcygeus* (**22**) to the external anal sphincter and *anococcygeal ligament* (**23**). The most anterior fibers of the puborectal muscle, the *prerectal fibers = puboperineal muscle* (**24**), bound the *urogenital hiatus* (**25**) on both sides and radiate into the perineal body (**2**). The *prostate* (**26**) is visible inside the urogenital hiatus. The *external anal sphincter* (**27**) surrounds the *anus* (**28**) with three parts. The *coccygeus* (**29**) forms the pelvic diaphragm together with the levator ani muscle.

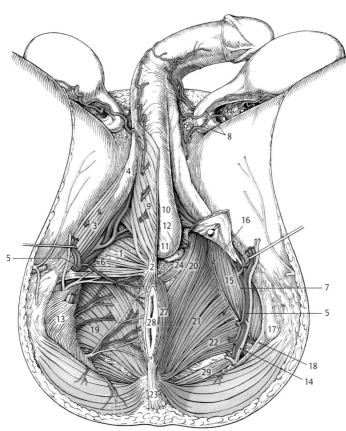

A **Deep layer**
of perineal region

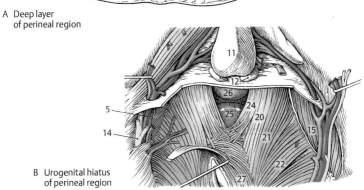

B **Urogenital hiatus**
of perineal region

Lower Limb

Regions (A, B)

As in the upper limb, the boundaries between the regions of the lower limb are somewhat arbitrary and have been drawn from a practical viewpoint.

Regions about the Hip

Anteriorly the regions around the hip joint also represent subdivisions of the thigh. The *subinguinal region* (**1**) is bounded by the inguinal ligament and the sartorius and pectineus muscles as part of the large femoral triangle. The **femoral triangle** (**2**) extends farther distally and is bounded by the inguinal ligament, the sartorius, and the adductor longus. Posteriorly is the **gluteal region** (**3**), which corresponds roughly to the region of the gluteus maximus and extends to the gluteal sulcus.

Regions about the Thigh

The femoral trigone represents one part of the **anterior femoral region** (**4**), which extends distally to the knee and laterally to the tensor fasciae latae. Posteriorly the **posterior femoral region** (**5**) lies next to the gluteal region and ends above the popliteal fossa.

Regions about the Knee

In front, the **anterior region of the knee** (**6**) extends from the lower margin of the anterior femoral region to the tibial tuberosity. The **posterior region of the knee** (**7**) lies posteriorly. The middle part of this region is also called the **popliteal fossa**.

Regions of the Lower Leg

The **anterior crural region** (**8**) extends from the tibial tuberosity to the malleoli. Medially this region, at the part of the tibia palpable through the skin, continues into the **posterior crural region** (**9**), which has its proximal and distal borders at the same level as those of the anterior region. Behind the medial malleolus is the **medial retromalleolar region**, and behind the lateral malleolus lies the **lateral retromalleolar region** (**10**).

Regions of the Foot

The **calcaneal region** (**11**) lies posterior to the retromalleolar regions. Anteriorly and superiorly is the **dorsum (dorsal region) of the foot** (**12**), and inferiorly the **sole (plantar region) of the foot** (**13**).

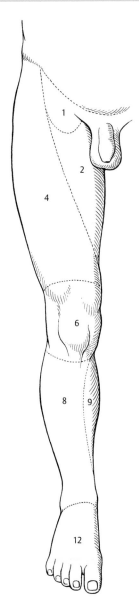

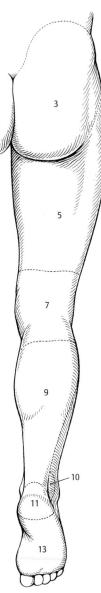

A Anterior view
   of lower limb regions

B Posterior view
   of lower limb regions

Subinguinal Region

Subcutaneous Layer (A, B)

The abundant subcutaneous fat is divided by dense *connective tissue lamellae = membranous layer* (**1**) into two layers. The connective tissue lamellae, which were formerly known as the superficial femoral fascia or *Scarpa's fascia*, partly cover the subcutaneous vessels and nerves and extend below the saphenous hiatus. Only after removal of all the subcutaneous fat and connective tissue layers can the *fascia lata* (**2**) be seen. Most of the fascia lata is generally of an aponeurotic character, except in the region of the saphenous hiatus, where there is a looser, reticular structure called the *cribriform fascia* (**3**; see p. 254).

The subcutaneous veins, which reach this region in a stellate pattern, pierce the cribriform fascia. The largest and the most regularly occurring vessel is the *great (long) saphenous vein* (**4**). It runs from the thigh to the cribriform fascia (**3**). Often a *lateral accessory saphenous vein* (**5**) accompanies it. The *external pudendal veins* (**6**) run from the pubic region and the *superficial epigastric vein* (**7**) runs from the umbilical region to the cribriform fascia. The *superficial circumflex iliac vein* (**8**) runs parallel to the inguinal ligament. The junction of all these veins is very variable and will be discussed on page 416. Smaller arteries are the *external pudendal artery* (**9**), the *superficial epigastric artery* (**10**), and the *superficial circumflex iliac artery* (**11**), accompanied by the homonymous veins.

The *superficial inguinal lymph nodes* (see p. 400), which can be divided into two groups, lie on the cribriform fascia. One group, the *horizontal tract*, lies parallel to the inguinal ligament, whereas the other group, the *vertical tract*, is parallel to the great saphenous vein. The horizontal tract is organized into the *superomedial* (**12**) and the *superolateral* (**13**) *superficial inguinal lymph nodes*. The lymph nodes of the vertical tract are called the *inferior superficial inguinal lymph nodes* (**14**).

The cutaneous nerves in this region arise from the *femoral branch* (**15**) of the *genitofemoral nerve*. In the male the *spermatic cord* (**16**), accompanied by the *ilioinguinal nerve* (**17**), courses in the inguinal region above the inguinal ligament and reaches the scrotum. The skin lateral to the cribriform fascia is innervated by anterior cutaneous rami of the femoral nerve.

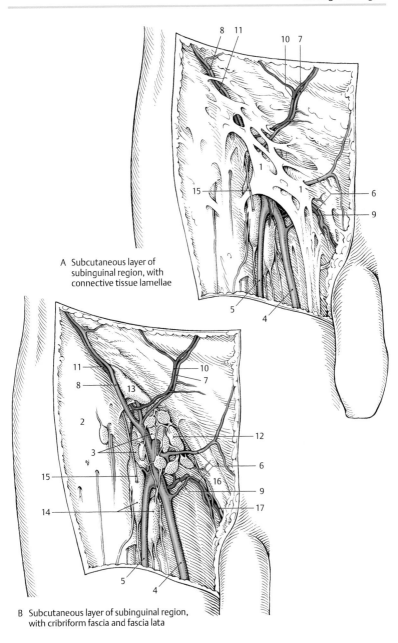

A Subcutaneous layer of
subinguinal region, with
connective tissue lamellae

B Subcutaneous layer of subinguinal region,
with cribriform fascia and fascia lata

Saphenous Hiatus (A–R)

The **saphenous hiatus,** bounded by the *falciform margin* (**1**) with its *superior* (**2**) and *inferior* (**3**) *horns*, becomes visible after removal of the cribriform fascia. Within the opening lie medially the *deep inguinal lymph nodes* (**4**), next to them the *femoral vein* (**5**), and most laterally the *femoral artery* (**6**). In or lateral to the saphenous hiatus the *femoral branch* (**7**) of the genitofemoral nerve becomes subcutaneous. Still farther laterally the *anterior cutaneous branches* (**8**) *of the femoral nerve* perforate the fascia lata.

According to *Lanz–Wachsmuth*, in the region of the saphenous hiatus in 37% of cases the following veins open into the femoral vein (**A**): the *great saphenous vein* (**9**), the *lateral accessory saphenous vein* (**10**), the *superficial circumflex iliac vein* (**11**), the *superficial epigastric vein* (**12**), and one or more *external pudendal veins* (**13**). This venous confluence shows a great many variations, which are shown in the various detailed diagrams.

Variants (B–R)

Lateral Accessory Saphenous Vein (B–E). In 1% of cases this vein may join the femoral vein proximal to the hiatus (**B**). In 9% of cases there is a common junction with a trunk consisting of the superficial circumflex iliac vein and the superficial epigastric vein (**C**). In the same proportion there is a common termination of the lateral accessory saphenous vein and superficial circumflex iliac vein (**D**). Rarely the lateral accessory saphenous vein and the superficial epigastric vein join at their termination (**E**).

The **great saphenous vein (F–G)** may receive a *medial accessory saphenous vein* (**14**). Either it perforates the fascia (**F**) distal to the saphenous hiatus (in 1%) or it reaches the femoral vein (**G**) in the saphenous hiatus.

In 1% of cases the **external pudendal veins** (**H–I**) join a medial accessory saphenous vein (**H**), while in 2% of cases they combine with the superficial epigastric vein (**I**).

The position of the **superficial epigastric vein** (**J–N**) is particularly variable. It may open with the superficial external pudendal vein into the great saphenous vein (**J**). Sometimes (1%) it opens proximal to the saphenous hiatus into the femoral vein (**K**). In 9% of cases it may form a common trunk with the superficial circumflex iliac vein and this opens into the lateral accessory saphenous vein (**L**), which reaches the great saphenous vein in the saphenous hiatus. Sometimes the superficial epigastric and the superficial circumflex iliac veins join the superficial external pudendal vein and the lateral accessory saphenous vein to form a common trunk, which opens into the great saphenous vein within the saphenous hiatus (**M**). In 6% of cases, the superficial epigastric vein runs into the superficial circumflex iliac vein and this trunk opens directly into the femoral vein (**N**).

As has already been described, in 9% of cases the **superficial circumflex iliac vein** (**O–R**) may open with the superficial epigastric vein and the lateral accessory saphenous vein into the great saphenous vein (**O**), and in another 9% the lateral accessory saphenous vein also opens into it (**P**). Sometimes the superficial circumflex iliac vein opens into the great saphenous vein together with the superficial epigastric vein (**R**).

The variants described above represent a summary of the author's many observations, as well as those of *Lanz–Wachsmuth*.

Clinical tip: Intra-arterial injections into the femoral artery are performed at a site approximately 1 cm below the inguinal ligament. Locate the midpoint of a straight line between the anterior superior iliac spine and pubic tubercle. Measure approximately 0.5 cm lateral from that point, and insert the needle vertically. Vertical movement of the needle will be noted when the needle tip comes in contact with the pulsating artery. When the needle pierces the artery wall, a pulsatile surge of blood will enter the syringe.

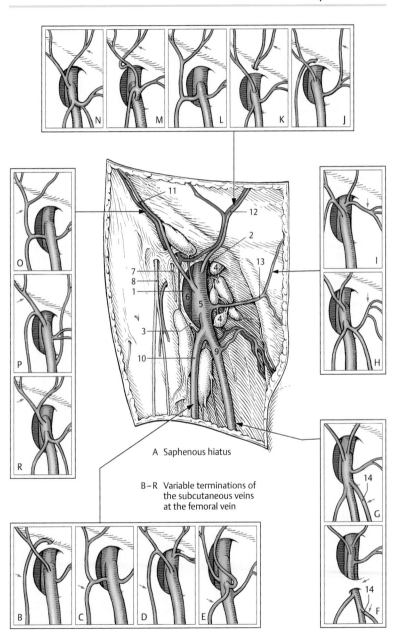

A Saphenous hiatus

B – R Variable terminations of
the subcutaneous veins
at the femoral vein

Peripheral Pathways

Gluteal Region (A, B)

Subcutaneous Layer (A)

The *gluteal fascia* (**1**) becomes evident after removing the skin and the fatty subcutaneous tissue. At the upper border of the gluteus maximus, this fascia becomes continuous with the firm *gluteal aponeurosis* (**2**).

The skin is innervated by the clunial nerves and by the *lateral cutaneous branch* (**3**) *of the iliohypogastric nerve.* The upper portion is supplied by the *superior clunial nerves* (**4**), which are the dorsal rami of spinal nerves L1 to L3. The middle area of the skin of the gluteal region is innervated by the *middle clunial nerves* (**5**), which are the dorsal rami of spinal nerves S1 to S3. *Inferior clunial branches* (**6**) that arise directly or indirectly from the sacral plexus loop around the inferior border of the gluteus maximus. (Indirect branches would be twigs from the posterior femoral cutaneous nerve.)

The blood supply of the skin is derived mainly from branches of the superior and inferior gluteal arteries. In the medial region it involves a twig from the lumbar arteries, whereas laterally, in the region of the greater trochanter, the arterial branches arise from the first perforating artery (from the profunda femoris artery).

Subfascial Layer (B)

The *gluteus maximus* (**7**) and the ischiocrural muscle group at its inferior border become visible after removal of the gluteal fascia. The latter muscle group comprises muscles originating from the ischial tuberosity: the *adductor magnus* (**8**), *semimembranosus* (**9**), *semitendinosus* (**10**), and the *long head of the biceps* (**11**). The *posterior femoral cutaneous nerve* (**12**) runs lateral to the biceps and crosses over it superficially.

The *sciatic nerve* (**13**) runs distally to deeper levels and can be tracked relatively easily if one draws a line from the ischial tuberosity to the greater trochanter and divides it into thirds. The sciatic nerve can then be found at the lower margin of the gluteus maximus between the prolonged border of the medial and middle thirds of this line. Lateral to the sciatic nerve, the *first perforating artery* (**14**) and its accompanying veins descend obliquely while crossing over the *adductor minimus* (**15**).

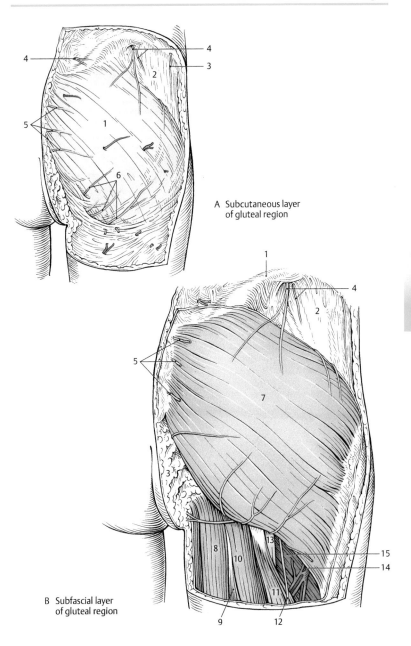

A Subcutaneous layer
of gluteal region

B Subfascial layer
of gluteal region

Gluteal Region, continued (A–C)

Deep Layer (A)

After the *gluteus maximus* (**1**) has been divided, the vessels and nerves that traverse the suprapiriform and infrapiriform foramina come into view.

The two foramina are formed by the *piriformis* (**2**), which subdivides the **greater sciatic foramen**.

The *superior gluteal artery and vein* (**3**) and the *superior gluteal nerve* (**4**) pass through the **suprapiriform foramen** laterally. The artery sends a branch (**5**), accompanied by a vein, to the gluteus maximus (**1**), and then, together with a vein and the nerve, it runs between the *gluteus medius* (**6**) and *gluteus minimus* (**7**). The superior gluteal nerve innervates the gluteus medius and minimus and the tensor fasciae latae.

The *inferior gluteal artery and vein* (**8**) and the *inferior gluteal nerve* (**9**) run through the **infrapiriform foramen** to the gluteus maximus (**1**). The *internal pudendal artery and vein* (**10**) and the *pudendal nerve* (**11**) wind around the ischial spine and reach the ischiorectal (ischioanal) fossa through the lesser sciatic foramen. They run posterior to the *superior gemellus* (**12**) and then adjoin the *obturator internus* (**13**). The *posterior femoral cutaneous nerve* (**14**) and the *sciatic nerve* (**15**) leave the lesser pelvis through the infrapiriform foramen and reach the thigh by passing posterior to the *superior gemellus* (**12**), the *obturator internus* (**13**), the *inferior gemellus* (**16**), and the *quadratus femoris* (**17**).

The posterior cutaneous femoral nerve (**14**) gives off the *inferior clunial nerves* (**18**) and then *a perineal branch* (**19**) soon after it emerges from the infrapiriform foramen. It then passes superficial to the *long head of the biceps muscle* (**20**), while the sciatic nerve (**15**) runs between that muscle and the *adductor magnus* (**21**).

Variants: In approximately 85 % of cases the sciatic nerve runs through the infrapiriform foramen (**A**) as a trunk. In approximately 15 % of cases, the sciatic nerve already divides within the pelvis into its two branches, the tibial nerve and the common peroneal nerve. In approximately 12 % the common peroneal nerve perforates the piriform muscle, while in 3 % it even leaves the pelvis through the suprapiriform foramen.

Clinical tip (B, C): The gluteal region is an ideal site for **intramuscular (intragluteal) injections**. Intragluteal injections are usually given into the superolateral quadrant (cross-hatched in blue) of the gluteal region (**B**) into the gluteus maximus (**1**) or the gluteus medius (**6**). There is, however, danger of injecting too superficially, that is, subcutaneously, or too deep between the gluteus maximus and gluteus medius into the intermuscular fat, posing a risk to the superior gluteal nerve (**4**). Injury to this nerve causes paralysis of the gluteus medius, gluteus minimus, and tensor fasciae latae. A. v. Hochstetter has recommended injecting from the side (**C**) in a triangular field (hatched in red), behind the anterior superior iliac spine, into the gluteus medius and gluteus minimus. The muscles should be in a relaxed state (aided by slight anteversion of the hip and slight flexion of the knee), as this allows for a painless injection.

Besides sternal puncture (see p. 66), it is now (2012) common practice to use the hip bone as a site for bone marrow sampling. A bone marrow biopsy needle (Jamshidi) is introduced close to the iliac crest and advanced toward the anterior superior iliac spine.

22 Sacrotuberous ligament
23 Trochanteric bursa of gluteus maximus

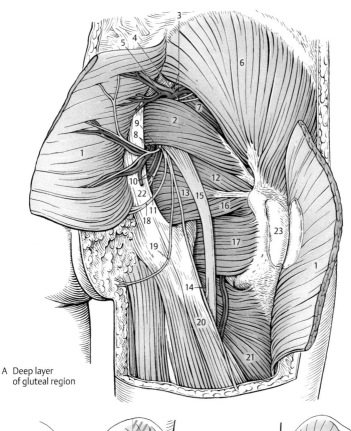

A Deep layer
of gluteal region

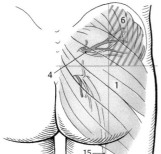

B Diagram of vessels and nerves potentially
endangered by intragluteal injections

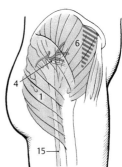

C Intragluteal injection site as
recommended by *A. v. Hochstetter*

Anterior Femoral Region

Subcutaneous Layer (A)

The various areas of the subcutaneous tissue of the anterior thigh region differ in their structure. The proximal part, in the subinguinal region, has strong connective tissue lamellae = membranous layer (see p. 414), which divide the subcutaneous fatty tissue into two layers. In addition, the **saphenous hiatus** (**1**) is covered by a loose connective tissue layer, the cribriform fascia.

When the cribriform fascia is removed, the sharp edge of the saphenous hiatus, the falciform margin, becomes visible. The falciform margin merges into the fascia lata medially in the superior and inferior horns (see p. 254). The **fascia lata** (**2**), which is continuous but for the saphenous hiatus, is also variable in structure. In the lateral thigh it is taut and kept stretched by the tensor fasciae latae, which radiates into it. This part of the fascia is also called the *iliotibial tract* (**3**). The fascia is looser in the medial part of the thigh.

The *great saphenous vein* (**4**) runs subcutaneously and is often joined by the *lateral accessory saphenous vein* (**5**) and less often by the *medial accessory saphenous vein* (**6**). Other veins entering the saphenous hiatus were described previously on page 416.

Laterally, near the junction of the proximal and middle thirds, the *lateral femoral cutaneous nerve* (**7**) becomes epifascial while the *anterior cutaneous branches of the femoral nerve* (**8**) perforate the fascia at various levels. The *femoral branch* (**9**) *of the genitofemoral nerve* either runs through the saphenous hiatus or lateral to it through the fascia lata. A small area of skin on the upper medial side of the thigh is innervated by the *ilioinguinal nerve* (**10**).

11 Superolateral and inferior superficial inguinal lymph nodes
12 Deep inguinal lymph nodes
13 Femoral vein
14 Femoral artery
15 Superficial epigastric artery and vein
16 Superficial circumflex iliac artery and vein
17 External pudendal artery and vein

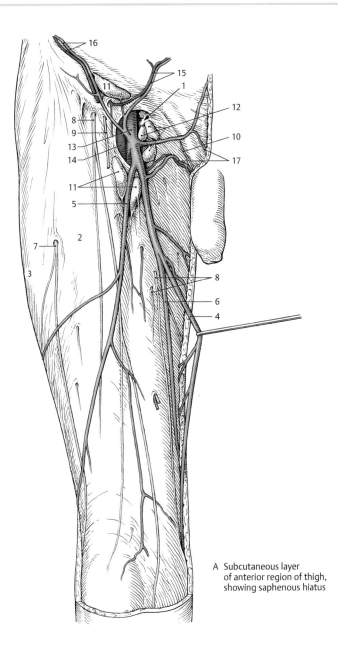

A Subcutaneous layer
of anterior region of thigh,
showing saphenous hiatus

Anterior Femoral Region, continued (A–H)

Deep Layer (A)

The large vessels and nerves are seen after removal of the fascia lata. Within the **femoral triangle**, which is limited by the *inguinal ligament*, the *sartorius* (**1**), the *adductor longus* (**2**), the lymphatics, the *femoral vein* (**3**), and *the femoral artery* (**4**) reach the thigh through the vascular lacuna, and the *femoral nerve* (**5**) and the *iliopsoas* (**6**) through the muscular lacuna.

After having given off its superficial branches (see p. 414), the femoral artery (**4**) gives rise to muscular branches, and a particularly large one, the *profunda femoris artery* (**7**), is buried deeply in the muscles. In 58% of cases the profunda femoris artery gives off the *medial circumflex femoral artery* (**8**) to the adductors and the femoral head, and the *lateral circumflex femoral artery* (**9**), which sends an *ascending branch* (**10**) to the femoral head and a *descending branch* (**11**) to the *quadriceps femoris* (**12**). The profunda femoris artery usually terminates in three *perforating arteries* (**13**) that are distributed to the adductor muscles and the muscles of the thigh. Medial to the femoral artery, the femoral vein (**3**) enters the vascular lacuna. It collects the veins that accompany the arteries in addition to the subcutaneous veins (see p. 416).

The femoral nerve (**5**) passes through the muscular lacuna into the thigh and, after giving off the anterior femoral cutaneous branches, innervates the sartorius (**1**), the quadriceps femoris (**12**), and the *pectineus* (**14**). Its longest, purely sensory branch is the *saphenous nerve* (**15**), which runs lateral to and together with the femoral artery (**4**) and femoral vein to reach the **adductor canal**. These structures lie on the adductor longus (**2**), which takes part in forming the anteromedial intermuscular septum (= vastoadductor membrane), and the posterior wall of the adductor canal. Apart from the adductor longus, the *vastus*

medialis (**16**), the *adductor magnus* (**17**), and the *anteromedial intermuscular septum* (= *vastoadductor membrane*, **18**) are involved in formation of the adductor canal. The saphenous nerve usually (62%) perforates this membrane together with the *descending genicular artery* (**19**) to extend onto and innervate the medial surface of the lower leg. It gives off an *infrapatellar branch* (**20**).

Variants (B–H)

There is great variability in the origin of the saphenous nerve (**15**) from the femoral nerve and its course in the thigh (*Sirang*). Very often it arises from the femoral nerve (**B 5**) proximal to the lateral circumflex femoral artery (**9**). It may embrace the lateral circumflex femoral artery (**C**) with two roots. Somewhat less commonly it only arises from the femoral nerve after crossing the lateral circumflex femoral artery (**D, E**). It reaches the adductor canal, perforates the anteromedial intermuscular septum (= vastoadductor membrane, **18**) and may give off its infrapatellar branch either medial (**B, C**) or lateral (**D**) to or through the sartorius (**E**). In rare cases (**E**), the infrapatellar branch also receives fibers from the cutaneous branch of the *anterior branch of the obturator nerve* (**21**).

The branches from the femoral artery (**4**) are also variable. Most commonly (58% according to *Lippert*) the medial (**8**) and lateral (**9**) circumflex femoral arteries arise from the profunda femoris artery (**F 7**). In 18% of cases (according to *Lippert*, **G**) the lateral circumflex femoral artery (**9**) arises from the profunda femoris artery (**7**), while, according to the same author, the medial circumflex femoral artery (**8**) arises from the profunda femoris artery (**7**) in only 15% of cases (**H**). The remaining 8% are distributed among much rarer variants.

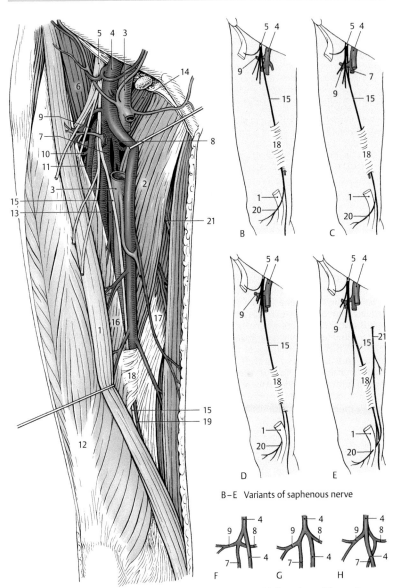

A Subfascial layer of anterior femoral region, with femoral artery retracted medially

B–E Variants of saphenous nerve

F–H Variable branching of femoral artery in subinguinal region (after *Lanz-Wachsmuth*)

Posterior Femoral Region (A, B)

After removal of the fascia, leaving the *iliotibial tract* (**1**) intact, at the lower margin of the *gluteus maximus* (**2**) the subfascial part of the *posterior femoral cutaneous nerve* (**3**) can be identified as it runs superficial to the *long head of the biceps femoris* (**4**).

Between the *long head* (**4**) and the *short head* (**5**) *of the biceps femoris*, the *sciatic nerve* (**6**) runs distally. At variable levels it divides into the *tibial* (**7**) and *common fibular (peroneal) nerves* (**8**). Proximal to this division the sciatic nerve gives off another branch (**9**) to the biceps femoris. The tibial nerve runs between the heads of the *gastrocnemius* (**10**), giving off various branches (see p. 430). The common fibular nerve follows the posterior border of the *biceps femoris* (**11**).

The *primary perforating artery* (**12**), a branch of the deep femoral artery, reaches the posterior side of the thigh. It passes between the pectineus and adductor brevis muscles and then pierces the adductor minimus and magnus muscles. With its accompanying veins, it crosses the sciatic nerve anteriorly (but posterior to the adductor minimus and magnus) and gives off branches to the long head of the biceps femoris (**4**) and the *semitendinosus* (**13**). On the posterior surface of the adductor magnus, the primary perforating artery anastomoses with branches of the *secondary perforating artery* (**14**) and the latter anastomoses with branches of the *tertiary perforating artery*. The tertiary perforating artery is the terminal branch of the profunda femoris artery and penetrates the adductor magnus close to the hiatus of the adductor tendon. It supplies the semimembranosus and the short head of the biceps.

The *semimembranosus* (**15**) is displaced, bringing the *adductor hiatus* (**16**) into view. The adductor hiatus (**B**) is bounded by the two parts of the *adductor magnus* (**17**). One part is inserted into the medial lip of the linea aspera and the other into the adduc-

tor tubercle of the medial epicondyle. The femoral artery, which runs through the adductor canal, passes through the adductor hiatus to reach the popliteal fossa and becomes the *popliteal artery* (**18**) on the back of the thigh. In addition to muscular branches, it also gives off the medial and lateral superior genicular arteries. The popliteal artery is generally accompanied by the usually paired *popliteal veins* (**19**).

■ **Variant:** Very occasionally there is one *sciatic artery*, which developmentally is the primary vascular supply to the leg. Remnants remain as the *comitans artery* of the *sciatic nerve*.

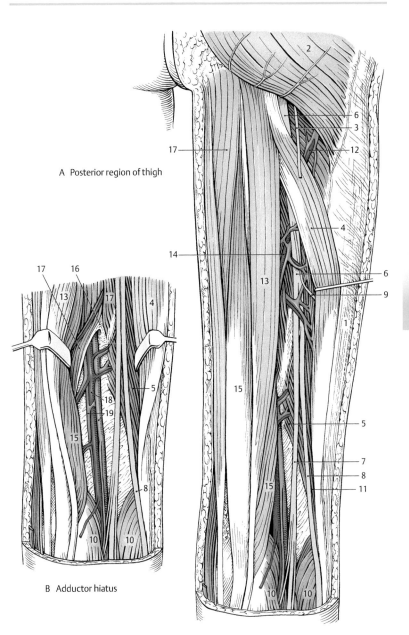

A Posterior region of thigh

B Adductor hiatus

Posterior Region of the Knee (A–K)

Subcutaneous Layer (A)

The *great saphenous vein* (**1**) lies in the subcutaneous layer at the medial border of the posterior knee region. In the leg it is accompanied by the *saphenous nerve* (**2**), which becomes subcutaneous at the inferior border of the popliteal fossa. The *small saphenous vein* (**3**) sometimes (see below) perforates the fascia at the inferior border of the popliteal fossa. It is accompanied by the *medial sural cutaneous nerve* (**4**), which is continued as the sural nerve (see p. 434). In addition, the *posterior femoral cutaneous nerve* with its branches (**5**) terminates in the popliteal fossa.

Variations in the Course of the Small Saphenous Vein (B–E)

The small saphenous vein, which is very important in phlebology, runs a variable course in relation to the crural fascia. According to *Moosmann–Hartwell* the small saphenous vein (**3**) perforates the crural fascia in the distal third of the leg in 7% of cases (**B**), runs subfascially to the popliteal fossa and then turns deep to join the *popliteal vein* (**6**). Most commonly (51.5%) the small saphenous vein (**3**) perforates the fascia in the middle third of the lower leg (**C**).

The second most common site (32.5%) for the small saphenous vein (**3**) to perforate the fascia is in the proximal third (**D**). It perforates the fascia within the posterior knee region (**E**) in only 9% of cases.

Variations in the Termination of the Small Saphenous Vein (F–K)

Mercier et al also reported great variability in the termination of the small saphenous vein (**3**). In addition to its typical opening (**F**) into the popliteal vein (**6**), the small saphenous vein (**3**) may also give off a branch to the great saphenous vein (**G**, **1**). In the presence of this branch, the small saphenous vein (**3**) may also open directly into the *femoral vein* (**H**, **7**). Other variants include an opening solely into either the great saphenous vein (**I**) or the femoral vein (**J**), which in the latter termination may also be delta-shaped (**K**).

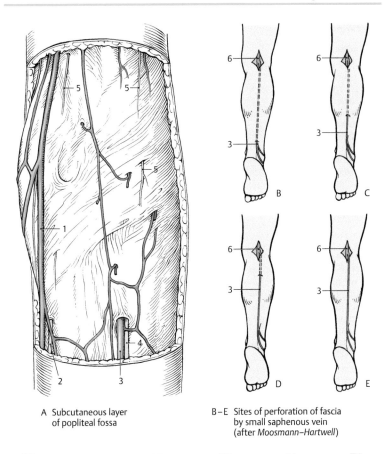

A Subcutaneous layer of popliteal fossa

B–E Sites of perforation of fascia by small saphenous vein (after *Moosmann–Hartwell*)

F–K Variable terminations of the small saphenous vein (after *Mercier* et al.)

Popliteal Fossa (A–G)

Deep Layer (A)

After removal of the fascia, the rhomboidal popliteal fossa bounded by muscles is seen. The popliteal fossa is bounded medially and proximally by the *semimembranosus* (**1**), laterally and proximally by the *biceps femoris* (**2**), and distally by the *lateral head* (**3**) and *medial head* (**4**) *of the gastrocnemius*. The sciatic nerve and its branches can be seen proximally between the semimembranosus and the biceps femoris.

The *common fibular (peroneal) nerve* (**5**) descends superficially along the posterior border of the biceps femoris, while the second branch, the *tibial nerve* (**6**), extends distally between the two heads of the gastrocnemius. The tibial nerve gives off *muscular branches* (**7**) and a *medial sural cutaneous nerve* (**8**), which, together with the communicating peroneal branch, forms the sural nerve (see p. 434).

Deep in the popliteal fossa we find the *popliteal artery* (**10**) accompanied by the *popliteal veins* (**9**). At a variable level (see below) this artery gives off the *anterior tibial artery* (**11**). The small saphenous vein usually reaches the popliteal vein but, as in the specimen shown, it may not open into a larger vein until it is proximal to the popliteal fossa.

Variants of the Arterial Branches (B–G)

In 90% of cases (**B**) the popliteal artery (**10**) gives off as its first branch the anterior tibial artery (**11**) posterior to the *popliteus muscle* (**12**), dividing farther distally into the *posterior tibial* (**13**) and *fibular (peroneal)* (**14**) *arteries*. In approximately 4% of cases (**C**) the arteries arise together. It is unusual (1%) for the anterior tibial artery and the peroneal artery (*anterior peroneotibial trunk*, **15**) to originate together at the distal border of the popliteus (**D**).

In 3% of cases the popliteal artery (**10**) gives off the anterior tibial artery just proximal to the popliteus (**E**, see also **A**).

In 1% of individuals the anterior tibial artery (**11**) arises at the same high level with the presence of an anterior peroneotibial trunk (**F 15**) or, in another variant, the course of the anterior tibial artery (**11**) runs anterior to the popliteus (**G 12**).

Clinical tip: Atypical or additional origins of gastrocnemius muscle fibers from the popliteal fascia, from the medial lip of the linea aspera, and from the connective tissue covering the popliteal vessels may lead to a **"popliteal compression syndrome."** This syndrome may also appear in those rare instances in which the anterior tibial artery runs anterior to the popliteus muscle (**G**).

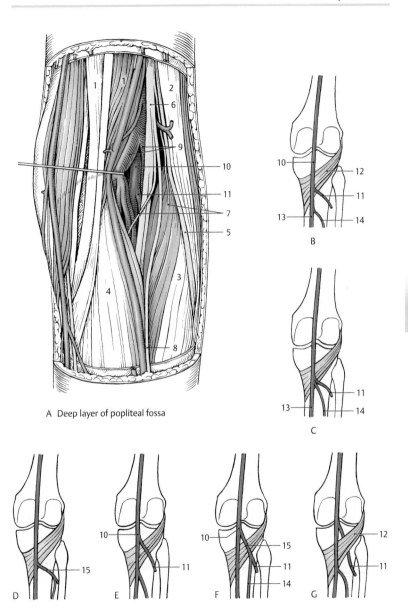

A   Deep layer of popliteal fossa

B

C

D   E   F   G

B–G   Variable branches of the popliteal artery (after *Lanz–Wachsmuth*)

Anterior Crural Region (A, B)

The subcutaneous neurovascular bundles run predominantly on the medial side of the leg.

The *great saphenous vein* (**1**) collects blood from the medial side and the dorsum of the foot and ascends to the triceps surae with the *saphenous nerve* (**2**). This nerve supplies the skin on the medial surface of the leg as far as the medial border of the foot, and with its *infrapatellar branch* (**3**) it innervates the skin of the *infrapatellar region*. Later it gives off the *medial crural cutaneous branches* (**4**).

After removal of the crural fascia, the *tibialis anterior* (**5**) is seen proximal to the tibia (**6**) in the lateral part of the field. The *extensor digitorum longus* (**7**) lies lateral to the tibialis anterior, and deeply in between them is the *extensor hallucis longus* (**8**). Laterally the *fibularis (peroneus) longus* (**9**) and the *fibularis (peroneus) brevis* (**10**) can also be seen.

The *superficial fibular (peroneal) nerve* (**11**) runs distally between the extensor digitorum longus (**7**) and the fibular muscles and branches on the dorsum of the foot. It perforates the fascia of the distal half of the lower leg. Deep between the tendon of the tibialis anterior (**5**) and the extensor hallucis longus muscle (**8**) runs the *anterior tibial artery* (**12**) with its accompanying veins, the *anterior tibial veins* (**13**), and the *deep fibular (peroneal) nerve* (**14**), which together with its motor fibers also carries sensory fibers from the skin area between the first and second toes.

Clinical tip: The stress of prolonged marching may cause an **"anterior tibial syndrome,"** characterized by sharp pain lateral to the tibia due to damage to the anterior tibial artery and the tibialis anterior. There is usually associated damage to the deep fibular (peroneal) nerve, which may be misdiagnosed as peroneal paralysis.

15 Fibularis tertius

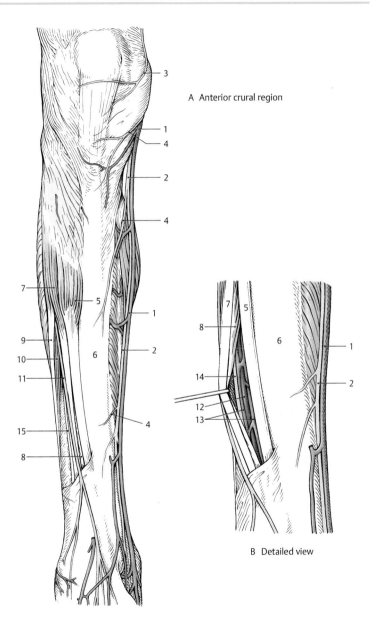

A Anterior crural region

B Detailed view

Posterior Crural Region (A–E)

Of the larger structures only veins and nerves are visible subcutaneously. The region is supplied with blood deeply through branches of the posterior tibial artery. The appearance is not greatly altered by removal of the crural fascia, although the *triceps surae* (**1**) does become visible with the two heads of the *gastrocnemius* (**2**) and the *soleus* (**3**). The triceps surae is attached to the calcaneus by the *calcaneal tendon* (**4**).

The *saphenous nerve* (**5**) and *great saphenous vein* (**6**) are visible medially. The largest structure is the *small saphenous vein* (**7**), which begins at the lateral border of the foot and ascends toward the popliteal fossa. Its relationship to the fascia is described on p.428. The long and short saphenous veins are interconnected by numerous anastomoses. There are also the *perforating veins* (**8**), which join the subcutaneous veins to the deep veins (anterior and posterior tibial and fibular = peroneal veins). Valves direct the flow of blood from the superficial to the deep veins.

The *medial cutaneous sural nerve* (**9**) is accompanied by the small saphenous vein and usually perforates the fascia in the middle of the leg. It joins the *peroneal communicating branch* (**10**) to form the *sural nerve* (**11**), which innervates the skin of the posterior crural region. With its continuation, the *lateral dorsal cutaneous nerve* (**12**), it innervates the lateral side of the dorsum of the foot, and with the *lateral calcaneal branches* (**13**) it supplies the lateral calcaneal area. *Medial calcaneal branches* (**14**) arise directly from the tibial nerve and innervate the skin in the medial region of the calcaneal area. The *common fibular (peroneal) nerve* (**15**) descends just posterior to the fibular head. It is always susceptible to injury because of its superficial position.

After removal of the *medial head of the gastrocnemius* (**16**), the *popliteus* (**17**) becomes visible; it is covered by fascia.

In this way the *popliteal artery* (**18**), *popliteal veins* (**19**), and *tibial nerve* (**20**) can be viewed until they enter the *tendinous arch of the soleus muscle* (**21**). The entrance may be hidden by the *plantaris muscle* (**22**). Deep in the posterior crural region, covered by the soleus (**3**), are the *posterior tibial artery* (**23**) and the *fibular (peroneal) artery* (**24**). The posterior tibial artery is the continuation of the *popliteal artery* (**18**) after it has given off the *anterior tibial artery* (**25**).

█ Variants (C–E): As at other sites, the arteries show a number of variants, knowledge of which is important for clinical purposes (arteriography, ligations, etc.). As a rule (**C**) the posterior tibial artery (**23**) descends on the posterior surface of the tibia, reaches the medial retromalleolar region (see p.436), and divides into the plantar arteries. The fibular (peroneal) artery (**24**) descends near the fibula, giving off a *perforating branch* (**26**) that pierces the interosseous membrane and ends at the level of the lateral malleolus. Sometimes (**D**) the phylogenetically older fibular (peroneal) artery (**24**) may replace a poorly developed posterior tibial artery (**23**). In rare cases (**E**) the posterior tibial artery is completely absent and the fibular (peroneal) artery (**24**) takes over the blood supply for the entire region usually supplied by this artery.

> **Clinical tip:** A practical distinction is drawn between **communicating veins** and **perforator veins** in the leg. Communicating veins establish a direct connection between the superficial (epifascial) and deep (subfascial) venous systems, while perforator veins establish an indirect connection via muscular veins. All the veins have valves that normally direct blood flow from the superficial veins to the deep veins. When the valves are incompetent, the flow direction is reversed and **varicose veins** develop. **Thrombosis** occurs exclusively in the deep veins (!) and may lead to varicosity, edema, and crural ulcer.

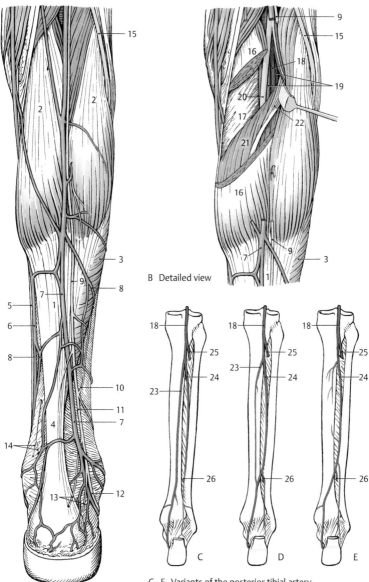

B   Detailed view

C–E   Variants of the posterior tibial artery
and fibular (peroneal) artery
(redrawn from *Lanz–Wachsmuth*)

A   Posterior crural region

Medial Retromalleolar Region (A, B)

The medial retromalleolar region includes the area between the medial malleolus and the Achilles tendon. It is bounded distally by the **flexor retinaculum** (laciniate ligament), which consists of a *superficial* and a *deep layer* (see below).

The *superficial layer* (**1**) is a thickening of the *crural fascia* (**2**). It extends from the medial malleolus to the posterior surface of the Achilles tendon and calcaneal tuberosity. Its proximal and distal boundaries are indistinct.

Subcutaneous Layer (A)

This layer contains veins, cutaneous nerves, and small cutaneous arteries (not shown). The *great saphenous vein* (**3**) runs near the malleolus and is clearly visible through the thin skin. It receives blood from the cutaneous venous network and from deep veins (**4**). The *saphenous nerve* (**5**) branches in this region to supply sensory innervation to the skin.

Subfascial Layer (B)

After removal of the crural fascia, the neurovascular bundle and the long plantar muscles of the foot can be seen proximal to the flexor retinaculum. Also visible is the *deep layer* (**6**) of the flexor retinaculum, which extends from the medial malleolus to the calcaneus and complements the bony grooves in creating fibro-osseous canals for the long muscles of the foot.

Immediately behind the medial malleolus runs the tendon of the *tibialis posterior* (**7**) and adjacent to it the tendon of the *flexor digitorum longus* (**8**). The tendon of the *flexor hallucis longus* (**9**) lies deeper and is displaced slightly backward by the medial tubercle of the posterior process of the talus. All three muscles have their own tendon sheaths (see p. 279), which are not pictured here.

Between the superficial (**1**) and deep (**6**) layers runs the neurovascular bundle for the sole of the foot. Adjacent to the tendon of the flexor digitorum longus (**8**) runs the *posterior tibial artery* (**10**) with its accompanying *posterior tibial veins* (**11**). Posterior to these veins lies the *tibial nerve* (**12**), which usually divides between the two layers into its terminal branches, the *medial* and *lateral plantar nerves.*

Sometimes this division may occur proximal to the flexor retinaculum and then the medial plantar nerve lies just posterior to the flexor digitorum longus.

> **Clinical tip:** The loose, highly mobile skin in this region permits tissue fluid to accumulate, and **edema** may occur. Finger pressure will then produce lasting indentations (pitting), which indicate fluid retention in the body. The pulse of the posterior tibial artery is also palpable in this region.

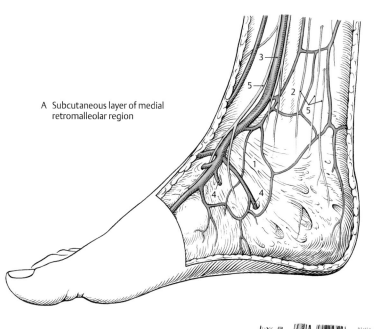

A Subcutaneous layer of medial retromalleolar region

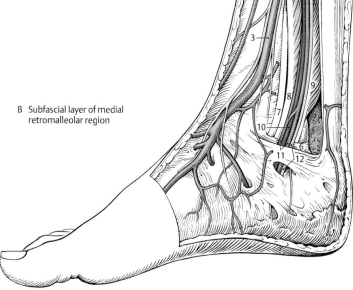

B Subfascial layer of medial retromalleolar region

Dorsum of the Foot (A–G)

Subcutaneous Layer (A)

A dense network of veins, the *dorsal venous network of the foot* (**1**), forms a *dorsal venous arch* (**2**) over the metatarsal bones. These superficial veins drain not only the *superficial dorsal metatarsal veins* (**3**) but also deep veins, the *perforating veins* (**4**) and the *intercapitular veins* (**5**). The blood is drained mainly through the *great saphenous vein* (**6**) and only a smaller portion travels via the *lateral malleolar network* (**7**) to the small saphenous vein.

Only small branches from the deep arteries reach the subcutaneous tissue, and the *first dorsal metatarsal artery* (**8**), which has a variable origin (see below), is the only one that is visible.

The *medial dorsal cutaneous nerve* (**9**) supplies the medial skin on the dorsum of the foot, in many cases supplemented by the *saphenous nerve* (**10**), which innervates the medial border of the foot. Sometimes the saphenous nerve (**10**) terminates in the region of the medial malleolus. Only the adjacent skin areas of the first and second toes are innervated by the *deep fibular (peroneal) nerve* (**11**), which may anastomose with branches of the *medial dorsal cutaneous nerve* (**12**). The *intermediate dorsal cutaneous nerve* (**13**) supplies the lateral half of the skin of the dorsum of the foot, supplemented at its lateral margin by the terminal branch of the sural nerve, the *lateral dorsal cutaneous nerve* (**14**).

Subfascial Layer (B)

Removing the fascia while preserving the inferior extensor retinaculum reveals the *dorsal pedal artery* (**15**). It runs onto the dorsum of the foot, accompanied by the deep fibular (peroneal) nerve (**11**). With the tendon of the *tibialis anterior* (**16**) passing beneath the medial ends of the inferior extensor retinaculum, the dorsal pedal artery and accompanying veins and nerve lie between the tendons of the *extensor hallucis longus* (**17**) and *extensor digitorum*

longus (**18**). The dorsal pedal artery gives off the lateral tarsal artery in the region of the retinaculum and forms an *arcuate artery* (**19**) from which the *dorsal metatarsal arteries* (**20**) arise. These give origin not only to the *dorsal digital arteries* (**21**) but also to the perforating branches to the sole of the foot, of which the *deep plantar branch* (**22**) to the first interosseous space is particularly important. The dorsal pedal artery is accompanied by veins that communicate with the superficial veins.

> **Clinical tip:** The dorsal pedal artery pulse is palpable just lateral to the extensor hallucis longus tendon. The loose subcutaneous tissue on the dorsum of the foot becomes filled with fluid if circulation is impaired, resulting in edema.

■ **Variants of the Arteries (C–G):** The dorsal metatarsal arteries, and thus the arcuate artery, are highly variable. Only in 20% of cases (**C**) do the dorsal metatarsal arteries arise from the dorsal pedal artery, while in 6% (**D**) the fourth metatarsal artery is supplied by a perforating branch from the sole of the foot. In 40% (**E**) only the first metatarsal artery originates from the dorsal pedal artery, and the remainder of the dorsal metatarsal arteries arise from plantar arteries. In 10% (**F**) all the dorsal metatarsal arteries come from the sole of the foot, and in 5% of cases (**G**) the first dorsal metatarsal artery alone arises from a plantar artery.

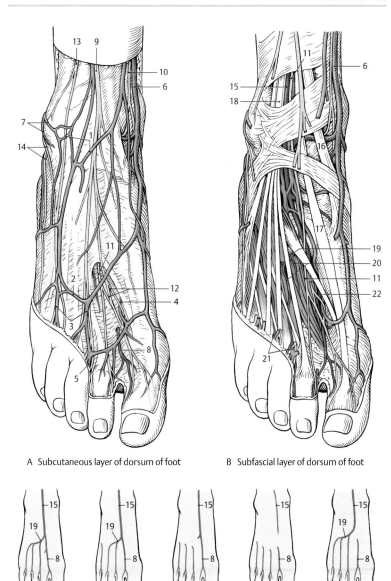

A Subcutaneous layer of dorsum of foot

B Subfascial layer of dorsum of foot

C–G Variants of arteries of dorsum of foot (after *Lippert*)

Peripheral Pathways

Sole of the Foot (A–G)

Superficial Layer (A)

Except for the sides of the foot, the *plantar aponeurosis* (**1**) covers the deep structures of the sole including the main trunks of the peripheral pathways. As the plantar skin has a particularly rich blood supply, there are a large number of *plantar cutaneous arteries* (**2**) and *plantar cutaneous veins* (**3**). In the calcaneal region the arteries form a network, the *rete calcaneum (calcaneal anastomosis)*, which is supplied by branches from the *posterior tibial* and *fibular arteries*. Additional branches stem from the *medial plantar* and the *lateral plantar arteries*. The *medial plantar artery* gives off a *superficial branch* (**4**), which becomes visible at the medial edge of the plantar aponeurosis, accompanied by the *first proper plantar digital nerve* (**5**). Lateral to the aponeurosis there is often a subcutaneous *branch* (**6**) *of the lateral plantar artery* accompanied by the *proper plantar digital nerve* (**7**) for innervating the lateral border of the small toe.

Between the longitudinal bundles of the plantar aponeurosis (**1**), the *common plantar digital arteries* (**8**) and the *common plantar digital nerves* (**9**) become subcutaneous. The common plantar digital arteries, which divide into *proper plantar digital arteries* (**10**), usually represent a continuation of the plantar metatarsal arteries (see p. 442), but may (very rarely) arise from a **"superficial" plantar arch**.

Often the superficial branch (**4**) of the medial plantar artery can take over the blood supply to the medial side of the big toe as the *first proper plantar digital artery* (**11**). The common plantar digital nerves (**9**) divide subcutaneously into the *proper digital nerves* (**12**).

Variants of the Deep Plantar Arch (B–G)

In 27% of cases (**B**) the four plantar metatarsal arteries are supplied by the *deep plantar branch* (**13**) of the dorsal pedal artery, while in 26% (**C**) the *deep plantar arch* (**14**) is formed entirely by the deep plantar branch. In 19% (**D**), the fourth plantar metatarsal artery arises from the *deep branch* (**15**) of the lateral plantar artery, and in 13% (**E**) the third plantar metatarsal artery does so as well, while the others stem from the deep plantar branch (**13**). In only 7% of cases (**F**) do all the plantar metatarsal arteries arise from a deep plantar arch (**14**) that is formed entirely from the deep branch (**15**) of the lateral plantar artery. In 6% (**G**) the second to fourth plantar metatarsal arteries arise from a deep plantar arch (**14**), while the first plantar metatarsal artery arises from the deep plantar branch (**13**).

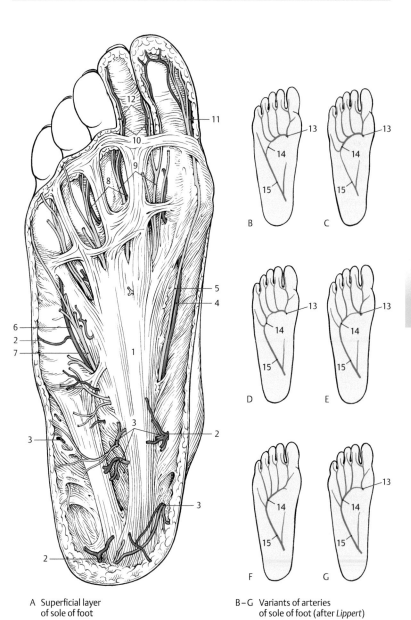

A Superficial layer
of sole of foot

B–G Variants of arteries
of sole of foot (after *Lippert*)

Sole of the Foot, continued (A, B)

Deep Layer (A)

After removal of the *plantar aponeurosis* and the *flexor digitorum brevis* (**1**), the medial and lateral neurovascular bundles of the sole of the foot are revealed. Medially, lying next to the *abductor hallucis* (**2**), the *medial plantar artery* (**3**), its accompanying veins, and the *medial plantar nerve* (**4**) supply the sole of the foot. The medial plantar artery (**3**), which may run laterally (more frequently) or medially (less frequently) to the nerve, divides into a *superficial branch* (**5**), which runs superficially to the *flexor hallucis brevis* (**6**), and a *deep branch. The superficial branch may (rarely) continue as the first proper plantar digital artery* (**7**), accompanied by the *first proper plantar digital nerve* (**8**), which may have divided proximally from the medial plantar nerve (**4**). The medial plantar nerve divides in sequence into the *first, second, and third common plantar digital nerves* (**9**), which give off branches (**10**) to the lumbricals. The first to third common plantar digital nerves continue as the *proper plantar digital nerves* (**11**). Sometimes, the *proper plantar digital nerve* (**12**) to the lateral side of the fourth toe may arise from the medial plantar nerve. Usually this region is innervated by branches of the *lateral plantar nerve* (**13**).

The lateral neurovascular bundle, which extends toward the toes medial to the *abductor digiti minimi* (**14**), consists (from medial to lateral) of the *lateral plantar nerve* (**13**) and the *lateral plantar artery* (**15**) and its *accompanying veins* (**16**). The lateral plantar artery divides into a *superficial branch* (**17**) and a *deep branch* (**18**). The superficial branch supplies the lateral border of the foot and the small toe, while the deep branch takes part in the formation of the *deep plantar arch* (**19**).

The lateral plantar nerve (**13**) gives off muscular branches to the muscles that arise from the calcaneus, and also cutaneous branches to the lateral border of the foot. It divides into a *superficial* (**20**) and a *deep* (**21**) *branch.* The superficial part distributes muscular branches to the *flexor digiti minimi brevis* (**22**) and the *fourth lumbricalis* (**23**) and the skin areas above them. The skin of the small toe and usually the lateral surface of the fourth toe are innervated by the *common plantar digital nerves* (**24**), which divide into *proper plantar digital nerves* (**25**). The deep branch (**21**) accompanies the deep plantar arch and innervates the *adductor hallucis longus* and the *opponens digiti minimi* as well as the interossei.

Deep Plantar Arch (B)

The deep plantar arch (**19**) can be seen after removal of the *quadratus plantae* (**26**) and the tendons of the *flexor digitorum longus* (**27**) and the *oblique head* (**28**) of the adductor hallucis.

Running at a deep level, the arch is closely apposed to the interossei muscles and anastomoses with the *deep plantar branch* (**29**) of the dorsal pedal artery. Three or four *metatarsal plantar arteries* (**30**) arise from the plantar arch. They usually give off the *common plantar digital arteries* (**31**), which divide into the *proper plantar digital arteries* (**32**).

For variants of the deep plantar arch, see p. 440.

33 Interossei plantar muscles
34 Transverse head of adductor hallucis

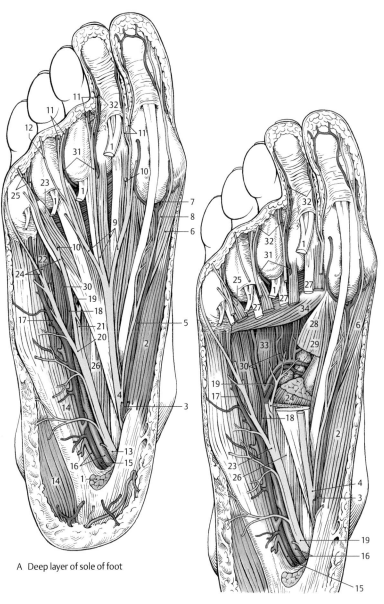

A Deep layer of sole of foot

B Deep plantar arch

Anatomical Terms and their Latin Equivalents

Peripheral Pathways

Topography of Peripheral Nerves and Vessels	
Anterior (posterior) crural region	Regio cruris anterior (posterior)
Anterior (posterior) femoral region	Regio femoris anterior (posterior)
Arm	Brachium
Artery (vein) of uterine round ligament	A.(V.) ligamenti teretis uteri
Buccal fat pad	Corpus adiposum buccae
Cord (lateral, etc.)	Fasciculus (lateralis, etc.)
Deltopectoral (clavipectoral) triangle	Trigonum clavipectorale
Falciform margin	Margo falciformis (arcuatus)
Fat pad of ischioanal fossa	Corpus adiposum fossae ischioanalis
Forearm	Antebrachium
Calcaneal (heel) region	Regio calcanea
Hip joint	Articulatio coxae
Medial (lateral) bicipital groove	Sulcus bicipitalis medialis (lateralis)
Muscular lacuna	Lacuna musculorum
Parotid plexus	Plexus intraparotideus
Parotid region	Regio parotideomasseterica
Quadrangular space	Foramen axillare laterale
Region of knee	Regio genus
Region of wrist	Regio carpalis
Saphenous hiatus	Hiatus saphenus
Seminal gland	Glandula vesiculosa
Sole of foot	Planta pedis
Suboccipital triangle	Trigonum a. vertebralis
Triangular space	Foramen axillare mediale
Upper limb	Membrum superius (Extremitas superior)
Vascular space	Lacuna vasorum

For Those Who Want to Learn More

There is a centuries-old tradition in medicine of naming a structure, disease, diagnostic method, or condition after the person who first described it. Unfortunately an ignorance of history has resulted in a tendency to use the name of the second, third, or even fourth describer rather than the discoverer. For example, some 15 scientists described the inguinal ligament over the centuries, and their pupils honored them by assigning various regional names to the ligament. This resulted in terms such as the Poupart ligament and Cooper's ligament, even though Vesalius and his pupil Fallopio actually discovered the ligament in the 16th century.

Failure to apply the first-describer rule has led to a chaotic situation in which some terms have even been associated with the wrong names. By the mid-20th century, this led European anatomists to *dispense with eponyms* and use only Latin terms for descriptions and findings. Unfortunately this practice was not adhered to, nor was it adopted in the United States, so the author has appended a list of the proper names that pertain to the locomotor system.

Index of Proper Names

HESSELBACH, Franz (1759–1816). Anatomist and surgeon, Würzburg (p. 98, Hesselbach ligament, called also Blumberg ligament or Heymann ligament = internal inguinal ligament = interfoveolar ligament; p. 98, Hesselbach's triangle = inguinal triangle).

LISFRANC, Jacques (1790–1847). Surgeon, Paris (p. 222, Lisfranc joint line = tarsometatarsal joint line).

LUSCHKA, von Hubert (1820–1875). Anatomist, Tübingen (p. 58, Luschka joints = uncovertebral joints).

NAFFZIGER, Howard (1884–1956). Surgeon, San Francisco (p. 36, Naffziger syndrome = cervical rib triad).

NELATON, Auguste (1807–1873). Surgeon, Paris (p. 196, Roser–Nélaton line = imaginary line between anterior superior iliac spine, tip of greater trochanter, and ischial tuberosity).

OBERST, Maximilian (1849–1925). Surgeon, Halle, Saale (p. 388, Oberst nerve block = anesthesia of the fingers).

ROSENMÜLLER, Johann (1771–1820). Surgeon and anatomist, Leipzig (see under CLOQUET).

ROSER, Wilhelm (1817–1888). Surgeon, Marburg, Lahn (see under NELATON).

SCARPA, Antonio (1752–1832). Anatomist and surgeon, Modena (p. 92, Scarpa's fascia = membranous layer of subcutaneous abdominal fascia; p. 254, Scarpa ligament = superior horn of falciform margin).

SCHMORL, Christian Georg (1861–1932). Pathologist, Dresden (p. 54, Schmorl's nodes = protrusions of intervertebral disk material into adjacent vertebrae).

TRENDELENBURG, Friedrich (1844–1924). Surgeon, Rostock, Bonn, and Leipzig (p. 246, Trendelenburg sign = pelvic tilt in one-legged stance, waddling gait).

TROISIER, Charles Emile (1844–1919). Physician, Paris (p. 366, Virchow–Troisier sentinel nodes = enlarged left supraclavicular lymph nodes due to metastasis from gastric malignancies).

VESALIUS, Andreas (1514–1564). Anatomist, Padua, Basel, and Madrid (p. 84 and 188, Vesalius ligament = inguinal ligament; first described by Vesalius and his pupil Gabriele FALLOPIO (1523–1562), anatomist, Ferrara, Pisa, and Padua).

VIRCHOW, Rudolf (1821–1902). Pathologist, Würzburg and Berlin (p. 366, see under TROISIER).

VOLKMANN, Alfred (1800–1879). Physiologist, Halle, Saale (p. 14, Volkmann's canals = oblique vascular canals in bone).

References

Textbooks, Handbooks

Bardeleben, K.: Handbuch der Anatomie des Menschen, Bd. II. Fischer, Jena 1908–1912

Benninghoff: Makroskopische Anatomie, Embryologie und Histologie des Menschen, 15. Aufl., Bd. I, hrsg. von D. Drenckhahn u. W. Zenker. Urban & Schwarzenberg, München–Wien–Baltimore 1994

Braus, H.: Anatomie des Menschen, 3. Aufl., Bd. I, hrsg. von C. Elze. Springer, Berlin 1954

Bucher, O., H. Wartenberg: Cytologie, Histologie und mikroskopische Anatomie des Menschen, 11. Aufl. Huber, Bern 1989

Feneis, H.: Anatomisches Bildwörterbuch, 7. Aufl. Thieme, Stuttgart 1993

Figge, F. H. J., W. J. Hild: Atlas of Human Anatomy. Urban & Schwarzenberg, München 1974

Frick, H., H. Leonhardt, D. Starck: Taschenlehrbuch der gesamten Anatomie, Bd. 1 u. 2, 3. Aufl. Thieme, Stuttgart 1987

Gardner, E., J. D. Gray, R. O'Rahilly: Anatomy, 4. Aufl. Saunders, Philadelphia 1975

Grosser, O.: Grundriß der Entwicklungsgeschichte des Menschen, 7. Aufl., hrsg. von R. Ortmann. Springer, Berlin 1970

Hafferl, A.: Lehrbuch der topographischen Anatomie, 3. Aufl., hrsg. von W. Thiel. Springer, Berlin 1969

Hollinshead, W. H.: Functional Anatomy of the Limbs and Back, 4. Aufl. Saunders, Philadelphia 1976

Kremer, K., W. Lierse, W. Platzer, H. W. Schreiber, S. Weller (Hrsg.): Chirurgische Operationslehre. Vol. 1–10. Thieme, Stuttgart 1987–1999

Lang, J., W. Wachsmuth: Praktische Anatomie, Bein und Statik, Bd. I/4, 2. Aufl. Springer, Berlin 1972

Langmann, J.: Medizinische Embryologie, 8. Aufl. Thieme, Stuttgart 1989

von Lanz, T., W. Wachsmuth: Praktische Anatomie, Bd. I/2: Hals. Springer, Berlin 1955

von Lanz, T., W. Wachsmuth: Praktische Anatomie, Bd. I/3: Arm, 2. Aufl. Springer, Berlin 1959

Leonhardt, H.: Histologie, Zytologie und Mikroanatomie des Menschen, 8. Aufl. Thieme, Stuttgart 1990

Mc Gregor, A. L., J. du Plessis: A Synopsis of Surgical Anatomy, 3. Aufl. Wright, Bristol 1969

Montgomery, R. L., M. C. Singleton: Human Anatomy Review. Pitman Medical, London 1975

Nishi, S.: Topographical Atlas of Human Anatomy, Bd. I–IV. Kanehara Shuppan, Tokyo 1974–1975

Pernkopf: Anatomie, Bd. 1 u. 2, hrsg. von W. Platzer, 3. Aufl. Urban u. Schwarzenberg, München–Wien–Baltimore 1987–1989

Rauber/Kopsch: Lehrbuch und Atlas der Anatomie des Menschen, Bd. I: Bewegungsapparat, hrsg. von B. Tillmann, G. Töndury. Thieme, Stuttgart 1987

Reiffenstuhl, G., W. Platzer, P.-G. Knapstein: Die vaginalen Operationen, 2. Aufl. Urban & Schwarzenberg, München–Wien–Baltimore 1994

Saegesser, M.: Spezielle chirurgische Therapie, 10. Aufl. Huber, Bern 1976

Sobotta: Atlas der Anatomie, Bd. 1 u. 2, 20. Aufl., hrsg. von R. Putz, u. R. Pabst, Urban und Schwarzenberg. München–Wien–Baltimore 1994

Starck, D.: Embryologie, 3. Aufl. Thieme, Stuttgart 1975

Tischendorf, F.: Makroskopisch-anatomischer Kurs, 3. Aufl. Fischer, Stuttgart 1979

Tittel, K.: Beschreibende und funktionelle Anatomie des Menschen, 8. Aufl. Fischer, Stuttgart 1978

Töndury, G.: Angewandte und topographische Anatomie, 5. Aufl. Thieme, Stuttgart 1981

Williams, P. L., R. Warwick, M. Dyson, L. H. Bannister: Gray's Anatomy, 37. Aufl. Churchill, Livingstone, Edinburgh 1989

General Anatomy

Barnett, C. H.: The structure and functions of synovial joints. In: Clinical Surgery, hrsg. von Rob, C., R. Smith. Butterworth, London 1966 (S. 328–344)

Barnett, C. H., D. V. Davies, M. A. MacConaill: Synovial Joints, Their Structure and Mechanics. Longmans, London 1961

Basmajian, J. V.: Muscles Alive, 3. Aufl. Williams & Wilkins, Baltimore 1974

Bernstein, N.: The Coordination and Regulation of Movements. Pergamon Press, Oxford 1967

Bourne, G. H.: Biochemistry and Physiology of Bone, 2. Aufl., Bd. I: Structure. Academic Press, New York 1972

Bourne, G. H.: The Structure and Function of Muscle, 2. Aufl., Bd. I: Structure. Academic Press, New York 1972

Brookes, M.: The Blood Supply of Bone. Butterworth, London 1971

Dowson, D., V. Wright, M. D. Longfield: Human joint lubrication. Bio-med. Engng 4 (1969) 8–14, 160–165, 517–522

Freeman, M. A. R.: Adult Articular Cartilage. Pitman, London 1973

Haines, R. W., A. Mohiudin: The sites of early epiphyseal union in the limb girdles and major long bones of man. J. Anat. (Lond.) 101 (1967) 823–831

Hancox, N. M.: Biology of Bone. Cambridge University Press, London 1972

Jonsson, B., S. Reichmann: Reproducibility in kinesiologic EMG-investigation with intramuscular electrodes. Acta Morphol. Neerl.-Scand. 7 (1968) 73–90

Joseph, J.: Man's Posture: Electromyographic Studies. Thomas, Springfield/Ill. 1960

Kapandji, I. A.: The Physiology of Joints, 2. Aufl., Bd. I–III. Longman, London 1970/71/74

MacConaill, M. A., J. V. Basmajian: Muscles and Movements. Williams & Wilkins, Baltimore 1969

Mysorecar, V. R.: Diaphyseal nutrient foramina in human long bones. J. Anat. (Lond.) 101 (1967) 813–822

Rasch, P. J., R. K. Burke: Kinesiology and Applied Anatomy, 5. Aufl. Lea & Febiger, Philadelphia 1974

Russe, O. A., J. J. Gerhardt, O. J. Russe: Taschenbuch der Gelenkmessung mit Darstellung der Neutral-Null-Methode und SFTR-Notierung. 2. Aufl. Huber, Bern 1982

Serratrice, G., J. Eisinger: Innervation et circulation osseuses diaphysaires. Rev. Rhum. 34 (1967) 505–519

Smith, D. S.: Muscle. Academic Press, New York 1972

Trunk

Beck, A., J. Killus: Mathematisch statistische Methoden zur Untersuchung der Wirbelsäulenhaltung mittels Computer. Biomed. Techn. 19 (1974) 72–74

Bowden, R., H. El-Ramli: The anatomy of the oesophageal hiatus. Brit. J. Surg. 54 (1967) 983–989

Cavallotti, C.: Morfologia del trigoni lombocostali del diaframma umano. Acta Med. Rom 6 (1968) 21–29

Condor, R. E.: Surgical anatomy of the transversus abdominis and transversalis fascia. Ann. Surg. 173 (1971) 1–5

Danburg, R.: Functional anatomy and kinesiology of the cervical spine. Manu. Med. 9 (1971) 97–101

Diaconescu, N., C. Veleanu: Die Wirbelsäule als formbildender Faktor. Acta anat. (Basel) 73 (1969) 210–241

Donisch, E. W., W. Trapp: The cartilage endplates of the human vertebral column (some considerations of postnatal development). Anat. Rec. 169 (1971) 705–716

Doyle, J. F.: The superficial inguinal arch. A reassessment of what has been called the inguinal ligament. J. Anat. (Lond.) 108 (1971) 297–304

Drexler, L.: Röntgenanatomische Untersuchungen über Form und Krümmung der Halswirbelsäule in den verschiedenen Lebensaltern. Hippokrates, Stuttgart 1962

Epstein, B. S.: The Vertebral Column. Year Book Medical Publishers, Chicago 1974

François, R. J.: Ligament insertions into the human lumbar body. Acta anat. (Basel) 91 (1975) 467–480

Groeneveld, H. B.: Metrische Erfassung und Definition von Rückenform und Haltung des Menschen. Hippokrates, Stuttgart 1976

Helmy, I. D.: Congenital diaphragmatic hernia (A study of the weakest points of the diaphragm by dissection and a report of a case of hernia through the right foramen of Morgagni. Alexandria Med. J. 13 (1967) 121–132

Hesselbach, A. K.: Die Erkenntnis und Behandlung der Eingeweidebrüche, Bauer u. Raspe, Nürnberg 1840

Johnson, R. M., E. S. Crelin, A. A. White et al.: Some new observations on the functional anatomy of the lower cervical spine. Clin. Orthop. 111 (1975) 192–200

Kapandji, I. A.: L'Anatomie fonctionelle du rachis lombo sacre. Acta Orthop. Belg. 35 (1969) 543–566

Krämer, J.: Biomechanische Veränderungen im lumbalen Bewegungssegment. Hippokrates, Stuttgart 1973

Krmpotic-Nemanic, J., P. Keros: Funktionale Bedeutung der Adaption des Dens axis beim Menschen. Verh. Anat. Ges. (Jena) 67 (1973) 393–397

Langenberg, W.: Morphologie, physiologischer Querschnitt und Kraft des M. erector spinae im Lumbalbereich des Menschen. Z. Anat. Entwickl.-Gesch. 132 (1970) 158–190

Liard, A. R., M. Latarjet, F. Crestanello: Precisions anatomiques concernant la partie superieure du muscle grand droit de l'abdomen et de sa gaine. C. R. Ass. Anat. 148 (1970) 532–542

Ludwig, K. S.: Die Frühentwicklung des Dens epistrophei und seiner Bänder beim Menschen. Morphol. Jb. 93 (1953) 98–112

Ludwig, K. S.: Die Frühentwicklung des Atlas und der Occipitalwirbel beim Menschen. Acta Anat. (Basel) 30 (1957) 444–461

Lytle, W. J.: The inguinal and lacunar ligaments. J. Anat. (Lond.) 118 (1974) 241–251

MacVay, C. B.: The normal and pathologic anatomy of the transversus abdominis muscle in inguinal and femoral hernia. Surg. Clin. N. Amer. 51 (1971) 1251–1261

Mambrini, A., M. Argeme, J. P. Houze, H. Isman: A propos de l'orifice aortique du diaphragme. C. R. Ass. Anat. 148 (1970) 433–441

Nathan, H., B. Arensburgh: An unusual variation in the fifth lumbar and sacral vertebrae: a possible cause of vertebral canal narrowing. Anat. Anz. 132 (1972) 137–148

Niethard, F. U.: Die Form-Funktionsproblematik des lumbosakralen Überganges. Hippokrates, Stuttgart 1981

Okada, M., K. Kogi, M. Ishii: Endurance capacity of the erectores spinae muscles in static work. J. Anthrop. Soc. Nippon 78 (1970) 99–110

Pierpont, R. Z., A. W. Grigoleit, M. K. Finegan: The transversalis fascia. A practical analysis of an enigma. Amer. Surg. 35 (1969) 737–740

Platzer, W.: Funktionelle Anatomie der Wirbelsäule. In: Erkrankungen der Wirbelsäule, hrsg. von R. Bauer. Thieme, Stuttgart 1975 (S. 1–6)

Platzer, W.: Die zervikokraniale Übergangsregion in Kopfschmerzen, hrsg. von H. Tilscher et al. Springer, Berlin 1988

Prestar, F. L., R. Putz: Das Lig. longitudinale posterius – Morphologie und Funktion. Morphol. Med. 2 (1982) 181–189

Putz, R.: Zur Manifestation der hypochordalen Spangen im cranio-vertebralen Grenzgebiet beim Menschen. Anat. Anz. 137 (1975) 65–74

Putz, R.: Charakteristische Fortsätze – Processus uncinati – als besondere Merkmale des 1. Brustwirbels. Anat. Anz. 139 (1976) 442–454

Putz, R.: Zur Morphologie und Rotationsmechanik der kleinen Gelenke der Lendenwirbel. Z. Orthop. 114 (1976) 902–912

Putz, R.: Funktionelle Anatomie der Wirbelgelenke. Thieme, Stuttgart 1981

Putz, R., A. Pomaroli: Form und Funktion der Articulatio atlanto-axialis lateralis. Acta Anat. (Basel) 83 (1972) 333–345

Radojevic, S., E. Stolic, S. Unkovic: Le muscle cremaster de l'homme (Variations morphologiques et importance partique). C. R. Ass. Anat. 143 (1969) 1383–1386

Reichmann, S., E. Berglund, K. Lundgren: Das Bewegungszentrum in der Lendenwirbelsäule bei Flexion und Extension. Z. Anat. Entwickl.-Gesch. 138 (1972) 283–287

Schlüter, K.: Form und Struktur des normalen und des pathologisch veränderten Wirbels. Hippokrates, Stuttgart 1965

Shimaguchi, S.: Tenth rib is floating in Japanese. Anat. Anz. 135 (1974) 72–82

de Sousa, O. M., J. Furlani: Electromyographic study of the m. rectus abdominis. Acta Anat. (Basel) 88 (1974) 281–298

Steubl, R.: Innervation und Morphologie der Mm. levatores costarum. Z. Anat. Entwickl.-Gesch. 128 (1969) 211–221

Takebe, K., M. Vitti, J. v. Basmajian: The functions of semispinalis capitis and splenius capitis muscles. An electromyographic study. Anat. Rec. 179 (1974) 477–480

Taylor, A.: The contribution of the intercostal muscles to the effort of respiration in man. J. Physiol. (Lond.) 151 (1960) 390–402

Taylor, J. R.: Growth of human intervertebral discs and vertebral bodies. J. Anat. (Lond.) 120 (1975) 49–68

Töndury, G.: Entwicklungsgeschichte und Fehlbildungen der Wirbelsäule. Hippokrates, Stuttgart 1958

v. Torklus, D., W. Gehle: Die obere Halswirbelsäule, 3. Aufl. Thieme, Stuttgart 1987

Veleanu, C., U. Grun, M. Diaconescu, E. Cocota: Structural peculiarities of the thoracic spine. Their functional significance. Acta Anat. (Basel) 82 (1972) 97–107

Witschel, H., R. Mangelsdorf: Geschlechtsunterschiede am menschlichen Brustbein. Z. Rechtsmed. 69 (1971) 161–167

Zaki, W.: Aspect morphologique et fonctionnel de l'annulus fibrosus du disque intervertebral de la colonne cervicale. Bull. Ass. Anat. 57 (1973) 649–654

Zukschwerdt, L., F. Emminger, E. Biedermann, H. Zettel: Wirbelgelenk und Bandscheibe. Hippokrates, Stuttgart 1960

Upper Limb

Basmajian, J. V., W. R. Griffin jr.: Function of anconeus muscle. An electromyographic study. J. Bone Jt. Surg. 54-A (1972) 1712–1714

Basmajian, J. V., A. Travill: Electromyography of the pronator muscles in the forearm. Anat. Rec. 139 (1961) 45–49

Bearn, J. G.: An electromyographical study of the trapezius, deltoid, pectoralis major, biceps and triceps, during static loading of the upper limb. Anat. Rec. 140 (1961) 103–108

Bojsen-Møller, F., L. Schmidt: The palmar aponeurosis and the central spaces of the hand. J. Anat. (Lond.) 117 (1974) 55–68

Christensen, J. B., J. P. Adams, K. O. Cho, L. Miller: A study of the interosseous distance between the radius and ulnar during rotation of the forearm. Anat. Rec. 160 (1968) 261–271

Čihák, R.: Ontogenesis of the Skeleton and the Intrinsic Muscles of the Hand and Foot. Springer, Berlin 1972

Clarke, G. R., L. A. Willis, W. W. Fish, P. J. R. Nichols: Assessment of movement at the glenohumeral joint. Orthopaedics (Oxford) 7 (1974) 55–71

Dempster, W. T.: Mechanisms of shoulder movement. Arch. Phys. Med. 46 (1965) 49–70

Doody, S. G., L. Freedman, J. C. Waterland: Shoulder movements during abduction in the scapular plane. Arch. Phys. Med. 51 (1970) 595–604

Dylevsky, I.: Ontogenesis of the M. palmaris longus in man. Folia Morphol. (Prague) 17 (1969) 23–28

Franzi, A. T., E. Spinelli, G. Ficcarelli: Variazione del muscolo palmare lungo: Contributo alla casistica. Quad. Anat. Prat. 25 (1969) 71–76

Garn, S. M., C. G. Rohman: Variability in the order of ossification of the bony centers of the hand and wrist. Amer. J. Phys. Anthropol. (N.S.) 18 (1960) 219–230

Glasgow, E. F.: Bilateral extensor digitorum brevis manus. Med. J. Aust. 54 (1967) 25

Hohmann, G.: Hand und Arm, ihre Erkrankungen und deren Behandlung. Bergmann, München 1949

Jonsson, B., B. M. Olofsson, L. C. Steffner: Function of the teres major, latissimus dorsi and pectoralis major muscles. A preliminary study. Acta Morphol. Neerl.-Scand. 9 (1972) 275–280

Kaneff, A.: Über die wechselseitigen Beziehungen der progressiven Merkmale des M. extensor pollicis brevis beim Menschen. Anat. Anz. 122 (1968) 31–36

Kapandji, A.: La rotation du pouce sur son axe longitudinal lors de l'opposition. Rev. Chir. Orthop. 58 (1972) 273–289

Kauer, J. M. G.: The interdependence of carpal articulation chains. Acta Anat. (Basel) 88 (1974) 481–501

Kauer, J. M. G.: The articular disc of the hand. Acta Anat. (Basel) 93 (1975) 590–605

Kiyosumi, M.: New ligaments at articulations manus. Kumamoto Med. J. 18 (1965) 214–227

Krmpotic-Nemanic, J.: Über einen bisher unbeachteten Mechanismus der Fingergrundgelenke. Gegenseitige Längsverschiebung der Finger bei der Flexion. Z. Anat. Entwickl.-Gesch. 126 (1967) 127–131

Kuczynski, K.: Carpometacarpal joint of the human thumb. J. Anat. (Lond.) 118 (1974) 119–126

Landsmeer, J. M. F.: Atlas of the Hand. Churchill, Livingstone, Edinburgh 1976

Lewis, O. J., R. J. Hamshere, T. M. Bucknill: The anatomy of the wrist joint. J. Anat. (Lond.) 106 (1970) 539–552

Long, C.: Intrinsic-extrinsic muscle control of the fingers. Electromyographic studies. J. Bone Jt. Surg. 50-A (1968) 973–984

McClure, J. G., R. Beverly: Anomalies of the scapula. Clin. Orthop. 110 (1975) 22–31

Metha, H. J., W. U. Gardner: A study of lumbrical muscles in the human hand. Amer. J. Anat. 109 (1961) 227–238

Mrvaljevic, D.: Sur les insertions et la perforation du muscle coracobrachial. C. R. Ass. Anat. 139 (1968) 923–933

Murata, K., K. Abe, G. Kawahara et al.: The M. serratus anterior of the Japanese. The area of its origin and its interdigitation with the M. obliquus externus abdominis. Acta Anat. Nippon. 43 (1968) 395–401

Neiss, A.: Sekundäre Ossifikationszentren. Anat. Anz. 137 (1975) 342–344

Pauly, J. E., J. L. Rushing, L. E. Scheving: An electromyographic study of some muscles crossing the elbow joint. Anat. Rec. 159 (1967) 47–54

Poisel, S.: Die Anatomie der Palmaraponeurose. Therapiewoche 23 (1973) 3337

Ravelli, A.: Die sogenannte Rotatorenmanschette. Öst. Ärzteztg. 13/14 (1974)

Renard, M., B. Brichet, A. Fonder, P. Poisson: Rôle respectif des muscles sous-èpineux et petit rond dans l cinématique de l'humerus. C. R. Ass. Anat. 139 (1968) 1266–1272

Renard, M., A. Fonder, C. Mentre, B. Brichet, J. Cayotte: Contribution à l'étude de la fonction du muscle sousépineux. Communication accompagnée d'un film. C. R. Ass. Anat. 136 (1967) 878–883

Roche, A. F.: The sites of elongation of the human metacarpals and metatarsals. Acat. Anat. (Basel) 61 (1965) 193–202

Schmidt, H.-M.: Die Guyon'sche Loge. Ein Beitrag zur klinischen Anatomie der menschlichen Hand. Acta Anat. 131 (1988) 113–121

Schmidt, H.-M., U. Lanz: Chirurgische Anatomie der Hand. Hippokrates Verlag, Stuttgart 1992

Shrewsbury, M. M., R. K. Johnson: The fascia of the distal phalanx. J. Bone Jt. Surg. 57 A (1975) 784–788

Shrewsbury, M. M., M. K. Kuczynski: Flexor digitorum superficialis tendon in the fingers of the human hand. Hand 6 (1974) 121–133

Shrewsbury, M. M., R. K. Johnson, D. K. Ousterhout: The palmaris brevis. A reconstruction of its anatomy and possible function. J. Bone Jt. Surg. 54-A (1972) 344–348

Soutoul, J. H., J. Castaing, J. Thureau, E. De Giovanni, P. Glories, M. Jan, J. Barbat: Les rapports tête humérale-glène scapulaire dans d'abduction du membre supérieur. C. R. Ass. Anat. 136 (1967) 961–971

Stack, H. G.: The Palmar Fascia. Churchill, Livingstone, London 1973

Strasser, H.: Lehrbuch der Muskel- und Gelenkmechanik, Bd. IV: Die obere Extremität. Springer, Berlin 1917

Weston, W. J.: The digital sheaths of the hand. Aust. Radiol. 13 (1969) 360–364

Lower Limb

Ahmad, I.: Articular muscle of the knee: articularis genus. Bull. Hospit. Dis. (N. Y.) 36 (1975) 58–60

Altieri, E.: Aplasia bilaterale congenita della rotula. Boll. Soc. Tosco-Umbra Chir. 28 (1967) 279–286

Asang, E.: Experimentelle und praktische Biomechanik des menschlichen Beins. Med. Sport (Berl.) 13 (1973) 245–255

Aumüller, G.: Über Bau und Funktion des Musculus adductor minimus. Anat. Anz. 126 (1970) 337–342

Basmajian, J. V., T. P. Harden, E. M. Regenos: Integrated actions of the four heads of quadriceps femoris: An electromyographic study. Anat. Rec. 172 (1972) 15–20

Bojsen Møller, F., V. E. Flagstadt: Plantar aponeurosis and internal architecture of the ball of the foot. J. Anat. (Lond.) 121 (1976) 599–611

Bowden, R. E. M.: The functional anatomy of the foot. Physiotherapy 53 (1967) 120–126

Bubic, I.: Sexual signs of the human pelvis. Folia Med. (Sarajevo) 8 (1973) 113–115

Candiollo, L., G. Gautero: Morphologie et fonction des ligaments méniscofémoraux de l'articulation du genou chez l'homme. Acta Anat. (Basel) 38 (1959) 304–323

Ching Jen Wang, P. S. Walker: Rotatory laxity of the human knee joint. J. Bone Jt. Surg. 56-A (1974) 161–170

Čihák, R.: Ontogenesis of the Skeleton and Intrinsic Muscles of the Human Hand and Foot. Springer, Berlin 1972

Dahhan, P., G. Delephine, D. Larde: The femoropatellar joint. Anat. Clin. 3 (1981) 23–39

Detenbeck, L. C.: Function of the cruciate ligaments in knee stability. J. Sports Med. 2 (1974) 217–221

Didio, L. J. A., A. Zappalá, W. P. Carney: Anatomico-functional aspects of the musculus articularis genu in man. Acta Anat. (Basel) 67 (1967) 1–23

Emery, K. H., G. Meachim: Surface morphology and topography of patello-femoral cartilage fibrillation in Liverpool necropsies. J. Anat. (Lond.) 116 (1973) 103–120

Emmett, J.: Measurements of the acetabulum. Clin. Orthop. 53 (1967) 171–174

Gluhbegovic, N., H. Hadziselimovic: Beitrag zu den vergleichenden anatomischen Untersuchungen der Bänder des lateralen Meniskus. Anat. Anz. 126 Suppl. (1970) 565–575

Goswami, N., P. R. Deb: Patella and patellar facets. Calcutta Med. J. 67 (1970) 123–128

Heller, L., J. Langman: The menisco-femoral ligaments of the human knee. J. Bone Jt. Surg. 46-B (1964) 307–313

Hoerr, N. L., S. J. Pyle, C. C. Franciss: Radiographic Atlas of Skeletal Development of Foot and Ankle. Thomas, Springfield/Ill. 1962

Hohmann, G.: Fuß und Bein, ihre Erkrankungen und deren Behandlung. 5. Aufl. Bergmann, München 1951

Hooper, A. C. B.: The role of the iliopsoas muscle in femoral rotation. Irish J. Med. Sci. 146 (1977) 108–112

Jacobsen, K.: Area intercondylaris tibiae: osseous surface structure and its relation to soft tissue structures and applications to radiography. J. Anat. (Lond.) 117 (1974) 605–618

Janda, V., V. Stará: The role of thigh adductors in movements patterns of the hip and knee joints. Courrier, Centre Internat. de l'Enfance 15 (1965) 1–3

Jansen, J. C.: Einige nieuwe functioneelanatomische aspecten von de voet. Ned. T. Geneesk. 112 (1968) 147–155

Johnson, C. E., J. V. Basmajian, W. Dasher: Electromyography of sartorius muscle. Anat. Rec. 173 (1972) 127–130

Joseph, J.: Movements at the hip joint. Ann. R. Call. Surg. Engl. 56 (1975) 192–201

Kaplan, E. B.: The iliotibial tract, clinical and morphological significance. J. Bone Jt. Surg. 40-A (1958) 817–831

Kaufer, H.: Mechanical function of the patella. J. Bone Jt. Surg. 53-A (1971) 1551–1560

Kennedy, J. C., H. W. Weinberg, A. S. Wilson: The anatomy and function of the anterior cruciate ligament. As determined by clinical morphological studies. J. Bone Jt. Surg. 56-A (1974) 223–235

Knief, J.: Materialverteilung und Beanspruchungsverteilung im coxalen Femurende. Densitometrische und spannungsoptische Untersuchungen. Z. Anat. Entwickl.-Gesch. 126 (1967) 81–116

Kummer, B.: Die Biomechanik der aufrechten Haltung. Mitt. Naturforsch. Ges. Bern 22 (1965) 239–259

Kummer, B.: Funktionelle Anatomie des Vorfußes. Verh. dtsch. arthrop. Ges. 53 (1966) 483–493

Kummer, B.: Die Beanspruchung der Gelenke, dargestellt am Beispiel des menschlichen Hüftgelenks. Verh. dtsch. Ges. orthop. Traumatol. 55 (1968) 302–311

Lesage, Y., R. Le Bars: Etude electromyographique simultanée des differents chefs du quadriceps. Ann. Méd. Phys. 13 (1970) 292–297

Loetzke, H. H., K. Trzenschik: Beitrag zur Frage der Varianten des M. soleus beim Menschen. Anat. Anz. 124 (1969) 28–36

Marshall, J. L., E. G. Girgis, R. R. Zelko: The biceps femoris tendon and its functional significance. J. Bone Jt. Surg. 54-A (1972) 1444–1450

Martin, B. F.: The origins of the hamstring muscles. J. Anat. (Lond.) 102 (1968) 345–352

Menschik, A.: Mechanik des Kniegelenkes. I. Z. Orthop. 112 (1974) 481–495

Menschik, A.: Mechanik des Kniegelenkes. II. Z. Orthop. 113 (1975) 388–400

Mörike, K. D.: Werden die Menisken im Kniegelenk geschoben oder gezogen? Anat. Anz. 133 (1973) 265–275

Morrison, J. B.: The mechanics of the knee joint in relation to normal walking. J. Biochem. 3 (1970) 51–61

Novozamsky, V.: Die Form der Fußwölbung unter Belastung in verschiedenen Fußstellungen. Z. Orthop. 112 (1974) 1137–1142

Novozamsky, V., J. Buchberger: Die Fußwölbung nach Belastung durch einen 100-km-Marsch. Z. Anat. Entwickl.-Gesch. 131 (1970) 243–248

Oberländer, W.: Die Beanspruchung des menschlichen Hüftgelenks. Z. Anat. Entwickl.-Gesch. 140 (1973) 367–384

Ogden, S. A.: The anatomy and function of the proximal tibiofibular joint. Clin. Orthop. 101 (1974) 186–191

Olbrich, E.: Patella emarginata – Patella partita. Forschungen und Forscher der Tiroler Ärzteschule 2 (1948–1950) 69–105

Pauwels, F.: Gesammelte Abhandlungen zur funktionellen Anatomie des Bewegungsapparates. Springer, Berlin 1965

Pheline, Y., S. Chitour, H. Issad, G. Djilali, J. Ferrand: La région soustrochantérienne. C. R. Ass. Anat. 136 (1967) 782–806

Platzer, W.: Zur Anatomie des Femoropatellargelenks. In: Fortschritte in der Arthroskopie, hrsg. v. H. Hofer. Enke, Stuttgart 1985

Platzer, W.: Zur funktionellen und topographischen Anatomie des Vorfußes. In: Hallux valgus, hrsg. von N. Blauth. Springer, Berlin 1986

Raux, P., P. R. Townsend, R. Miegel et al.: Trabecular architecture of the human patella. S. Biomech. 8 (1975) 1–7

Ravelli, A.: Zum anatomischen und röntgenologischen Bild der Hüftpfanne. Z. Orthop. 113 (1975) 306–315

Renard, M., B. Brichet, J. L. Cayotte: Analyse fonctionelle du triceps sural. C. R. Ass. Anat. 143 (1969) 1387–1394

Rideau, Y., P. Lacert, C. Hamonet: Contribution à l'etude de l'action des muscles de la loge postérieure de la cuisse. C. R. Ass. Anat. 143 (1969) 1406–1415

Rideau, Y., C. Hamonet, G. Outrequin, P. Kamina: Etude électromyographique de l'activité fonctionelle des muscles de la loge postérieure de la cuisse. C. R. Ass. Anat. 146 (1971) 597–603

Rother, P., E. Luschnitz, S. Beau, P. Lohmann: Der Ursprung der ischiokruralen Muskelgruppe des Menschen. Anat. Anz. 135 (1974) 64–71

Sick, H., P. Ring, C. Ribot, J. G. Koritke: Structure fonctionelle des menisques de articulation du genou. C. R. Ass. Anat. 143 (1969) 1565–1571

Sirang, H.: Ein Canalis alae ossis illii und seine Bedeutung. Anat. Anz. 133 (1973) 225–238

Stern jr., J. T.: Anatomical and functional specializations of the human gluteus maximus. Amer. J. Phys. Anthropol. 36 (1972) 315–339

Strasser, H.: Lehrbuch der Muskel- und Gelenkmechanik, Bd. III: Die untere Extremität. Springer, Berlin 1917

Strauss, F.: Gedanken zur Fuß-Statik. Acta Anat. (Basel) 78 (1971) 412–424

Suzuki, N.: An electromyographic study of the role of muscles in arch support of the normal and flat foot. Nagoya Med. J. 17 (1972) 57–79

Takebe, K., M. Viti, J. V. Basmajian: Electromyography of pectineus muscle. Anat. Rec. 180 (1974) 281–284

Tittel, K.: Funktionelle Anatomie und Biomechanik des Kniegelenks. Med. Sport (Berl.) 17 (1977) 65–74

von Volkmann, R.: Wer trägt den Taluskopf wirklich, und inwiefern ist der plantare Sehnenast des M. tibialis post. als Bandsystem aufzufassen? Anat. Anz. 131 (1972) 425–432

von Volkmann, R.: Zur Anatomie und Mechanik des Lig. calcaneonaviculare plantare sensu strictiori. Anat. Anz. 134 (1973) 460–470

Zivanovic, S.: Menisco-meniscal ligaments of the human knee joint. Anat. Anz. 135 (1974) 35–42

Head and Neck

Bochu, M., G. Crastes: La selle turcique normale etude radiographique. Lyon Méd. 231 (1974) 797–805

Buntine, J. A.: The omohyoid muscle and fascia; morphology and anomalies. Aust. N. Z. J. Surg. 40 (1970) 86–88

Burch, J. G.: Activity of the accessory ligaments of the mandibular joint. J. Prosth. Dent. 24 (1970) 621–628

Campell, E. J. M.: The role of the scalene and sternomastoid muscles in breathing in normal subjects. An electromyographical study. J. Anat. (Lond.) 89 (1955) 378–386

Carella, A.: Apparato stilo ioideo e malformazioni della cerniera atlo occipitale. Acta neurol. (Napoli) 26 (1971) 466–472

Couly, G., C. Brocheriou, J. M. Vaillant: Les menisques temporomandibulaires. Rev. Stomat. (Paris) 76 (1975) 303–310

Fischer, C., G. Ransmayr: Ansatz und Funktion der infrahyalen Muskulatur. Anat. Anz. 168 (1989) 237–243

Fortunato, V., St. D. Bocciarelli, G. Auriti: Contributo allo studio della morfologia ossea dell'area cribrosa dell'etmoide. Clin. otorinolaring. 22 (1970) 3–15

Hadziselimovic, H., M. Cus, V. Tomic: Appearance of the sigmoid groove and jugular foramen in relation to the configuration of the human skull. Acta Anat. (Basel) 77 (1970) 501–507

Honee, G. L. J. M.: The Musculus pterygoideus lateralis. Thesis, Amsterdam 1970 (S. 1–152)

Ingervall, B., B. Thilander: The human sphenooccipital synchondrosis. 1. The time of closure appraised macroscopically. Acta Odont. Scand. 30 (1972) 349–356

Isley, C. L., J. V. Basmajian: Electromyography of human cheeks and lips. Anat. Rec. 176 (1973) 143–148

Lang, J.: Structure and postnatal organization of heretofore uninvestigated and infrequent ossifications of the sella turcica region. Acta Anat. (Basel) 99 (1977) 121–139

Lang, J., S. Niederfeilner: Über Flächenwerte der Kiefergelenkspalte. Anat. Anz. 141 (1977) 398–400

Lang, J., K. Tisch-Rottensteiner: Lage und Form der Foramina der Fossa cranii media. Verh. Anat. Ges. (Jena) 70 (1976) 557–565

Melsen, B.: Time and mode of closure of the sphenooccipital synchondrosis determined on human autopsy material. Acta Anat. (Basel) 83 (1972) 112–118

Oberg, T., G. E. Carlsson, C. M. Fajers: The temporomandibular joint. A morphologic study on human autopsy material. Acta Odont. Scand. 29 (1971) 349–384

Platzer, W.: Zur Anatomie der „Sellabrücke" und ihrer Beziehung zur A. carotis interna. Fortschr. Röntgenstr. 87 (1957) 613–616

Pomaroli, A.: Ramus mandibulae. Bedeutung in Anatomie und Klinik. Hüthig, Heidelberg 1987

Porter, M. R.: The attachment of the lateral pterygoid muscle to the meniscus. J. Prosth. Dent. 24 (1970) 555–562

Proctor, A. D., J. P. de Vincenzo: Masseter muscle position relative to dentofacial form. Angle Orthodont. 40 (1970) 37–44

Putz, R.: Schädelform und Pyramiden. Anat. Anz. 135 (1974) 252–266

Shapiro, R., F. Robinson: The foramina of the middle fossa. A phylogenetic, anatomic and pathologic study. Amer. J. Roentgenol. 101 (1967) 779–794

Schelling, F.: Die Emissarien des menschlichen Schädels. Anat. Anz. 143 (1978) 340–382

Stofft, E.: Zur Morphometrie der Gelenkflächen des oberen Kopfgelenkes (Beitrag zur Statik der zerviko-okzipitalen Übergangsregion. Verh. Anat. Ges. (Jena) 70 (1976) 575–584

Vitti, M., M. Fujiwara, J. V. Basmajian, M. Lida: The integrated roles of longus colli and sternocleidomastoid muscles: an electromyographic study. Anat. Rec. 177 (1973) 471–484

Weisengreen, H. H.: Observation of the articular disc. Oral. Surg. 40 (1975) 113–121

Wentges, R. T.: Surgical anatomy of the pterygopalatine fossa. J. Laryngol. 89 (1975) 35–45

Wright, D. M., B. C. Moffett jr.: The postnatal development of the human temporomandibular joint. Amer. J. Anat. 141 (1974) 235–249

Zenker, W.: Das retroartikuläre plastische Polster des Kiefergelenkes und seine mechanische Bedeutung. Z. Anat. Entwickl.-Gesch. 119 (1956) 375–388

Peripheral Nerves and Vessels

Beaton, L. E., B. J. Anson: The relation of the sciatic nerve and of its subdivisions to the piriformis muscle. Anat. Rec. 70 (1937) 1

Fasol, P., P. Munk, M. Strickner: Blutgfäßversorgung des Handkahnbeins. Acta Anat. (Basel) 100 (1978) 27–33

Hilty, H.: Die makroskopische Gefäßvariabilität im Mündungsgebiet der V. saphena magna des Menschen. Schwabe, Basel 1955

Lahlaidi, A.: Vascularisation arterielle des ligaments intra-articulaires du genou chez l'homme. Folia Angiol. (Pisa) 23 (1975) 178–181

Lauritzen, J.: The arterial supply to the femoral head in children. Acta Orthop. Scand. 45 (1974) 724–736

Lippert, H.: Arterienvarietäten, Klinische Tabellen. Beilage in Med. Klin. 1967–1969, 18–32

May, R.: Chirurgie der Bein- und Beckenvenen. Thieme, Stuttgart 1974

May, R., R. Nißl: Die Phlebographie der unteren Extremität, 2. Aufl. Thieme, Stuttgart 1973

Mercier, R., Ph. Fouques, N. Portal, G. Vanneuville: Anatomie chirurgicale de la veine saphene externe. J. Chir. 93 (1967) 59

Miller, M. R., H. J. Ralston, M. Kasahara: The pattern of innervation of the human hand. Amer. J. Anat. 102 (1958) 183–218

Moosmann, A., W. Hartwell jr.: The surgical significance of the subfascial course of the lesser saphenous vein. Surg. Gynec. Obstet. 118 (1964) 761

Ogden jr., A.: Changing patterns of proximal femoral vascularity. J. Bone Jt. Surg. 56-A (1974) 941–950

Poisel, S., D. Golth: Zur Variabilität der großen Arterien im Trigonum caroticum. Wien. Med. Wschr. 124 (1974) 229–232

Schmidt, H.-M.: Topographisch-klinische Anatomie der Guyon'schen Loge an der menschlichen Hand. Acta Anat. 120 (1984) 66

Sirang, H.: Ursprung, Verlauf und Äste des N. saphenus. Anat. Anz. 130 (1972) 158–169

Tillmann, B., K. Gretenkord: Verlauf des N. medianus im Canalis carpi. Morphol. Med. 1 (1981) 61–69

Wallace, W. A., R. E. Coupland: Variations in the nerves of the thumb and index finger. J. Bone Jt Surg. 57-B (1975) 491–494

Weber, J., R. May: Funktionelle Phlebologie. Thieme, Stuttgart 1989

Wladimirov, B.: Über die Blutversorgung des Kniegelenkknorpels beim Menschen. Anat. Anz. 140 (1976) 469–476

Index

Boldface page numbers indicate extensive coverage of the subject.